$$\frac{7 \text{ bgs mortar}}{1000 \text{ brick}}$$

$$\frac{0.2 T \text{ sand}}{\text{bg mortar}}$$

$$\frac{40 \text{ block}}{\text{bg mortar}}$$

$7 \frac{3}{4}" \times 8 \frac{1}{2}"$

ESTIMATING CONSTRUCTION COSTS

McGraw-Hill Series in Construction Engineering and Project Management

Consulting Editor: *Raymond E. Levitt (Stanford University)*

Barrie and Paulson: *Professional Construction Management*
Jervis and Levin: *Construction Law: Principles and Practice*
Koerner: *Construction and Geotechnical Methods in Foundation Engineering*
Levitt and Samelson: *Construction Safety Management*
Oglesby, Parker, and Howell: *Productivity Improvement in Construction*
Peurifoy and Ledbetter: *Construction Planning, Equipment, and Methods*
Peurifoy and Oberlender: *Estimating Construction Costs*
Shuttleworth: *Mechanical and Electrical Systems for Construction*

Also Available from McGraw-Hill

Schaum's Outline Series in Civil Engineering

Each outline includes basic theory, definitions, and hundreds of solved problems and supplementary problems with answers.

Current List Includes:

Advanced Structural Analysis
Basic Equations of Engineering Science
Descriptive Geometry
Dynamic Structural Analysis
Engineering Mechanics, 4th edition
Fluid Dynamics
Fluid Mechanics & Hydraulics
Introduction to Engineering Calculations
Introductory Surveying
Reinforced Concrete Design, 2d edition
Space Structural Analysis
Statics and Strength of Materials
Strength of Materials, 2d edition
Structural Analysis
Theoretical Mechanics

Available at Your College Bookstore

ESTIMATING CONSTRUCTION COSTS

Fourth Edition

R. L. Peurifoy, P.E.

Formerly Professor of Construction Engineering
Texas A & M University
and
Oklahoma State University

G. D. Oberlender, P.E.

Professor of Civil Engineering
Oklahoma State University

McGraw-Hill Book Company

New York   St. Louis   San Francisco   Auckland   Bogotá   Caracas   Colorado Springs
Hamburg   Lisbon   London   Madrid   Mexico   Milan   Montreal   New Delhi   Oklahoma City
Panama   Paris   San Juan   São Paulo   Singapore   Sydney   Tokyo   Toronto

This book was set in Times Roman.
The editors were Anne T. Brown, Lyn Beamesderfer, and Steven Tenney;
the production supervisor was Denise L. Puryear.
The cover was designed by Rafael Hernandez.
Arcata Graphics/Halliday was printer and binder.

ESTIMATING CONSTRUCTION COSTS

1234567890 HAL HAL 89432109

ISBN 0-07-049740-0

Library of Congress Cataloging-in-Publication Data

Peurifoy, R. L. (Robert Leroy), (date).
    Estimating construction costs/R. L. Peurifoy, G. D. Oberlender.—
    4th ed.
    p.   cm.—(McGraw Hill series in construction engineering
    and project management)
    Includes index.
    ISBN 0-07-049740-0
    1. Building—Estimates—United States.   I. Oberlender, Garold D.
    II. Title.   III. Series.
    TH435.P47   1989
    692'.5—dc19   88-30398

ABOUT THE AUTHORS

Robert L. Peurifoy is an author and consulting engineer. He received his M.S. from the University of Texas. He has taught civil engineering at the University of Texas and Texas A&I College and construction engineering at Texas A&M University and Oklahoma State University. Mr. Peurifoy has served as a highway engineer for the U.S. Bureau of Public Roads and was a contributing editor to *Roads and Streets Magazine*. He is the author of two other books, *Construction Planning, Equipment, and Methods* and *Formwork for Concrete Structures*, and has written over 50 magazine articles dealing with construction. Mr. Peurifoy is a registered professional engineer and a member of the American Society of Civil Engineers and the American Society for Engineering Education.

Garold D. Oberlender is professor and coordinator of the graduate program in Construction Engineering and Management in the School of Civil Engineering at Oklahoma State University. He received his Ph.D. in civil engineering from the University of Texas at Arlington. Dr. Oberlender has conducted research and presented seminars on a variety of topics related to construction engineering and project management. A civil engineer with more than twenty-five years of experience, Dr. Oberlender has been a consultant to numerous companies in the application of computers in the design and construction of projects. He is a registered professional engineer and a member of the American Society of Civil Engineers, the National Society of Professional Engineers, and the American Society for Engineering Education.

CONTENTS

prefabricated forms/Forms for concrete columns/Materials required for forms for concrete columns/Cost of lumber for forms/Quantity of nails required for forms/Cost of adjustable steel column clamps/Labor making and erecting forms for concrete columns/Economy of reducing the size of concrete columns/Shores/Wood shores/Adjustable shores/Quantity of lumber required for forms for concrete beams using wood shores/ Quantity of lumber required for forms for concrete beams using adjustable shores/Labor required to build forms for concrete beams/ Forms for flat-slab-type concrete floors/Design of forms for flat-slab-type concrete floors/Quantity of lumber required for forms for flat-slab-type concrete floors/Labor required to build forms for flat-slab-type concrete floors/Forms for slabs for beam-and-slab type of concrete floors/Quantity of lumber required for forms for beam-and-slab type of concrete floors/ Labor required to build forms for beam-and-slab type of concrete floors/ Forms for metal-pan and concrete-joist type of concrete floors/Lumber required for metal-pan and concrete-joist construction/Labor required to build forms and install metal pans for concrete floors/Concrete stairs/ Lumber required for forms for concrete stairs/Labor required to build forms for concrete stairs/Reinforcing Steel/Types and sources of reinforcing steel/Properties of reinforcing bars/Estimating the quantity of reinforcing steel/Cost of reinforcing steel/Size extras for reinforcing steel/ Quantity extras for reinforcing steel/Cost for detailing and listing reinforcing steel/Cost for fabricating reinforcing steel/Cost for reinforcing steel delivered to a project/Labor placing reinforcing steel bars/Welded-wire fabric/Labor placing welded wire-fabric/Concrete/Cost of concrete/ Quantities of materials for concrete/Output of concrete mixers/Labor mixing and placing concrete/Ready-mixed concrete/Labor placing ready-mixed concrete/Lightweight concrete/Perlite concrete aggregate/Cost of Perlite concrete/Tilt-up concrete walls/General description/Concrete bridge piers/Forms for piers/Steel forms/Wood forms.

Combined corrugated-steel forms and reinforcement for floor systems/
Description/Installing corrugated sheets/Labor installing corrugated
sheets.

structural steel/Labor erecting structural steel/Labor bolting structural steel/Welded structures/Advantages of welded connections/Erection equipment required for welded-steel structures/Erecting steel structures with welded connections/Types of welds Arc-welding terminology/Methods of producing the most economical welds/Electrodes/Cost of welding/Quantity of details and welds required for a structural-steel frame building/Analysis of the cost of welded connections compared with bolted connections/Painting structural steel.

PREFACE

In preparing the fourth edition of this book, we have revised and rewritten the material in the third edition to include more reasonable prices for the cost of labor, equipment, and materials. Also we added some more material that should enable estimators to improve their estimating methods.

Although the costs used in the book are representative of costs in 1989, clearly costs vary with the location of the project and with time. However, the methods and procedures described here for estimating construction costs will continue to apply regardless of changes in the costs of labor, equipment, and materials.

Professional estimators who have applied the methods and procedures presented in earlier editions have advised us that using the book has helped them to prepare more accurate estimates. We hope this benefit will continue with the fourth edition.

McGraw-Hill and the authors would like to thank the following reviewers for their many useful comments and suggestions: R. James Diegel; Daniel Halpin, Purdue University, Donald Hancher, Texas A&M University; Raymond Levitt, Stanford University; Dohn Mehlenbacker, Illinois Institute of Technology; James O'Connor, University of Texas—Austin; David Rogge, Oregon State University; and Jerald Rounds, Arizona State University.

Comments from readers will be welcomed.

R. L. Peurifoy, P.E.
G. D. Oberlender. P.E.

LIST OF SYMBOLS AND ABBREVIATIONS

ASA = American Standards Association

AWWA = American Water Works Association

bhp = brake horsepower

bm = bank measure, volume of earth before loosening

ft³/min = cubic feet per minute

ft³ = cubic feet

yd³ = cubic yard

cwt = 100 lb, or hundredweight

d = pennyweight for nails

D and M = dressed and matched lumber

dbhp = drawbar horsepower

deg = degree

diam. = diameter

ed. = edition

°F = degrees Fahrenheit

fbm = feet board measure of lumber

f.o.b = free on board

ft/min = feet per minute

ft = foot

ft · lb = foot-pound

fwhp = flywheel horsepower

gal = gallon

gal/min = gallons per minute

hp = horsepower

hp · h = horsepower-hour

h = hour

in = inch

$$kW = kilowatt$$
$$kWh = kilowatthour$$
$$lb = pound$$
$$lin\ ft = linear\ feet$$
$$M = 1,000$$
$$min = minute$$
$$mi/h = miles\ per\ hour$$
$$no = number$$
$$pc = piece$$
$$lb/ft^2 = pounds\ per\ square\ foot$$
$$lb/in^2 = pounds\ per\ square\ inch$$
$$pt = pint$$
$$S4S = surfaced\ on\ four\ sides$$
$$in^2 = square\ inch$$
$$ton/h = tons\ per\ hour$$

ESTIMATING CONSTRUCTION COSTS

CHAPTER
1

INTRODUCTION

Purpose of This Book

The primary purpose of this book is to enable the user to gain fundamental knowledge of estimating the cost of projects to be constructed. Experienced estimators agree that the procedures used for estimating vary from company to company and even among individuals within a company. Although there are these variations, fundamental concepts that are universally applicable do exist. The information contained in this book presents the fundamental concepts to assist the user in understanding the estimating procedures developed by others. This book can also serve as a guide for assisting the user in the development of her or his own estimating procedures.

There are so many variations in the costs of materials, labor, and equipment from one location to another, and over time, that no book can dependably give costs that may be applied for bidding purposes. However, the estimator who learns to determine the quantities of materials, labor, and equipment for a given project and who applies proper unit costs to these items should be able to estimate the direct costs accurately.

Purpose of Estimating

The purpose of estimating is to determine the forecast costs required to complete a project in accordance with the contract plans and specifications. For any given project, the estimator can determine with reasonable accuracy the direct costs for materials, labor, and equipment. The bid price can then be determined by adding to the direct costs the costs for overhead (indirect costs required to build the project), contingencies (costs for any potential unforeseen work), and profit (costs for compensation for performing the work). The bid price of a project should be high enough to allow the contractor to complete the project with a reasonable profit, yet low enough to be within the owner's budget.

There are two distinct tasks in estimating: to determine the probable real cost and to determine the probable real time to build a project. With an increased emphasis on project planning and scheduling, the estimator is often requested to provide production rates, crew sizes, equipment spreads, and the estimated time required to perform various individual work items. This information, combined with costs, allows an integration of the estimating and scheduling functions of construction project management.

Because construction estimates are prepared before a project is constructed, an estimate is, at best, a close approximation of the actual costs. The true value of the project will not be known until the project has been completed and all costs have been recorded. Thus, the estimator does not establish the cost of a project; he or she simply establishes the amount which the contractor will receive for constructing the project.

Types of Estimates

Cost estimates may be divided into at least two different types, depending on the purposes for which they are prepared and the amount of information known when the estimates are prepared. There are approximate estimates (sometimes called *preliminary, conceptual,* or *budget estimates*) and detailed estimates (sometimes called *final* or *definitive estimates*). Each of these may be subdivided.

Although each project is unique, generally three parties are involved: the owner, the designer, and the contractor. Each has responsibility for estimating costs during various phases of the project. Early in a project, prior to the design, the prospective owner may wish to know the approximate cost of a project before making a decision to construct it. As the design of the project progresses, the designer must determine the costs of various design alternatives in order to finalize the design to satisfy the owner's budget and desired use of the project. The contractor must know the costs required to perform all work in accordance with the final contract documents.

Approximate Estimates

The prospective owner of a project establishes the budget for a project. For example, a government agency will need to know the approximate cost before holding a bond election, or it may need to know the approximate cost to ensure that the cost of a project does not exceed the funds appropriated during a fiscal year. The prospective owner of a private construction project will generally conduct a feasibility study during the developmental phase of a project. As a part of the study, an economic analysis is undertaken to compare the cost of construction with potential earnings that can be obtained upon completion of the project. Privately owned utility companies, and other types of multibuilder owners, prepare annual construction budgets for all projects proposed during a fiscal year. An approximate estimate is sufficiently accurate for these purposes.

The designer of a project must determine the costs of various design alternatives in order to obtain an economical design that meets the owner's budget. An architect will reduce a building to square feet of area, or cubic feet of volume, and then multiply the number of units by the estimated cost per unit. An engineer will multiply the number of cubic yards of concrete in a structure by the estimated cost per cubic yard to determine the probable cost of the project. Considerable experience and judgment are required to obtain a dependable approximate estimate of the cost, for the estimator must adjust the unit costs resulting from the qualities of materials, workmanship, location, and construction difficulties. Approximate estimates are sufficiently accurate for the evaluation of design alternatives or the presentation of preliminary construction estimates to the owner, but are not sufficiently accurate for bid purposes.

When the time for completion of a project is important, the owner may select a construction contractor before the design has been completed. A contractor may be asked to provide an approximate cost estimate based on the limited information known about the project. The contractor determines the approximate costs of the various work items, such as the cost per cubic yard for foundations, cost per pound of structural steel, and cost per square foot of finished room. Based on preliminary quantities of work, the contractor calculates a preliminary cost for construction of the project. The owner may then negotiate a construction contract with the contractor based on the approximate cost estimate. Construction projects of this type require owners who are knowledgeable in project management and contractors who have developed good project record keeping.

Detailed Estimates

A detailed estimate of the cost of a project is prepared by determining the costs of the materials, labor, equipment, subcontract work, overhead, and profit. Detailed estimates are generally prepared by contractors from a complete set of contract documents prior to submission of the bid or formal proposal to the owner. The detailed estimate is important to both the owner and the contractor because it represents the bid price—the amount of money the owner must pay for completion of the project and the amount of money the contractor will receive for building the project.

The preparation of the detailed estimate generally follows a systematic procedure that has been developed by the contractor for her or his unique construction operations. The process begins with a thorough review of the complete set of contract documents—the bidding and contract requirements, drawings, and technical specifications. It is also desirable to visit the proposed project site to observe factors that can influence the cost of construction, such as available space for storage of materials, control of traffic, security, and existing underground utilities.

The compilation of costs begins with a well-organized checklist of all work items necessary to construct the project. Appendix A provides an illustration of an organized checklist of work items for a building construction project. The estimator prepares a *material quantity takeoff* of all materials from the drawings. This involves tabulating the quantity and unit of measure of all work required during construction. Upon completion of the quantity takeoff, an extension of prices is performed. The quantity of material multiplied by the unit cost of the material yields the material cost. The quantity of work required of equipment is divided by the equipment production rate and then multiplied by the unit cost of equipment to obtain the total cost of equipment. Similarly, the cost of labor can be obtained by dividing the quantity of work required of labor by the labor production rate and then multiplying by the unit cost of labor.

For many projects a significant amount of work is performed by subcontractors who specialize in a particular area. Examples are clearing, dry-wall, painting, and roofing contractors. The estimator provides a set of drawings and specifications to potential subcontractors and requests a bid from them for their particular work. The cost of their work is normally quoted by telephone with a subsequent confirmation by letter. The suppliers and subcontractors usually present their costs just prior to final submission of the bid.

The direct costs of a project include materials, labor, equipment, and subcontractor costs. Upon completion of the estimate of direct costs, the estimator must determine the indirect costs of taxes, bonds, insurance, and overhead required to complete the project. It is also desirable to include some reasonable amount in the final estimate for contingencies. Regardless of the effort and amount of detailed estimating of a project, there is almost always some unforeseen work that develops during construction. Caution must be used in assigning contingency to an estimate. A contingency that is too low might reduce the profits in a project while a contingency that is too high may prevent the bid from being competitive.

Upon calculation of the direct and indirect costs for a project, a reasonable amount of profit is added to the estimate to establish the bid price. The amount of profit can vary considerably, depending on numerous factors, such as the size and complexity of the project, amount of work in progress by the contractor, accuracy and completeness of bid documents, competition for work, availability of money, and volume of construction activity in the project area. The profit may be as low as 5 percent for large projects or as high as 30 percent for small projects or projects that are high risks or remodels of existing projects. Table 1-1 lists the steps required to compile a detailed estimate for a project.

Organization of Estimates

A comprehensive and well-defined organization of work items is essential to the preparation of an estimate for any project. Each contractor develops her or his own

TABLE 1-1
Steps for preparing a detailed estimate

1. *Review the scope of the project.* Consider the effect of location, security, traffic, available storage space, etc., on costs.

2. *Determine quantities.* Perform a material quantity takeoff for all work items in the project, and record the quantity and the unit of measure for each item.

3. *Price material.* Extend material costs:

$$\text{Material cost} = \text{Quantity} \times \text{Unit price}$$

4. *Price labor.* Based on probable labor production rates and crew sizes, determine labor costs:

$$\text{Labor cost} = \frac{\text{Quantity}}{\text{Labor production rate}} \times \text{Labor rate}$$

5. *Price equipment.* Based on probable equipment production rates and equipment spreads, determine equipment costs:

$$\text{Equipment cost} = \frac{\text{Quantity}}{\text{Equipment production rate}} \times \text{Equipment rate}$$

6. *Obtain specialty contractors' bid.* Receive and tabulate the costs for each specialty contractor on the project.

7. *Obtain suppliers' bids.* Receive and tabulate the costs for each supplier on the project.

8. *Calculate taxes, bonds, insurance, and overhead.* Tabulate the costs of material and labor taxes, bonds, insurance, and job overhead.

9. *Contingency and markup.* Add costs for potential unforeseen work based on the amount of risk.

10. *Profit.* Add costs for compensation for performing the work in accordance with the bid documents.

---

procedures to compile the cost of construction for the type of work the company performs. The contractor's system of estimating and use of forms develop over years of experience in that type of work. Two basic approaches have evolved to organize work items for estimating. One approach is to identify work by the categories contained in the project's written specifications, such as those of the Construction Specification Institute (CSI) for building construction projects. The other approach uses a *work breakdown structure* (WBS) to identify work items by their location on the project.

Building Construction Projects

Building construction contractors organize their estimates in a format that closely follows the CSI division of work. The organization of work items is defined by 16 major divisions that are recognized as the industry standard for building construction. A typical summary of an estimate for a building construction project is shown in Table 1-2.

Each major division is subdivided into smaller items of work. For example, the work required for division 2 sitework is subdivided into clearing, excavation,

TABLE 1-2

Example of building construction project bid summary using the CSI organization of work

Item	Division	Material	Labor	Subcontract	Total
1	General requirement	$ 16,435.00	$ 36,355.00	$ 4,882.00	$ 57,672.00
2	Sitework	15,070.00	20,123.00	146,186.00	181,389.00
3	Concrete	97,176.00	51,524.00	0.00	148,700.00
4	Masonry	0.00	0.00	212,724.00	212,724.00
5	Metals	212,724.00	59,321.00	0.00	272,045.00
6	Wood and plastics	38,753.00	10,496.00	4,908.00	54,157.00
7	Thermal and moisture	0.00	0.00	138,072.00	138,072.00
8	Doors and windows	36,821.00	32,115.00	0.00	68,936.00
9	Finishes	172,587.00	187,922.00	0.00	360,509.00
10	Specialties	15,748.00	11,104.00	9,525.00	36,377.00
11	Equipment	0.00	0.00	45,729.00	45,729.00
12	Furnishings	0.00	0.00	0.00	0.00
13	Special construction	0.00	0.00	0.00	0.00
14	Conveying systems	0.00	0.00	0.00	0.00
15	Mechanical	0.00	0.00	641,673.00	641,673.00
16	Electrical	0.00	0.00	354,661.00	354.661.00
	Total direct costs	$605,314.00	$408,960.00	$1,558,360.00	$2,572,644.00
	Material tax (5%)	30,266.00			2,602,910.00
	Labor tax (18%)		73,613.00		2,676,523.00
	Contingency (2%)			53,530.00	2,730,053.00
	Bonds/Insurance			34,091.00	2,764,144.00
	Profit (10%)			276,414.00	3,040,558.00
				Bid price =	$3,040,558.00

compaction, handwork, termite control, drilled piers, foundation drains, landscape, and paving. Table 1-3 gives the direct costs of material and labor for these items.

A complete list of the CSI master data format is contained in App. A. This organization of work for building construction provides a systematic checklist and serves as a guide for the quantity takeoff, price extensions, and summary of costs for the final estimate. An estimate will usually have from 10 to 20 items in each division, so a comprehensive estimate will have from 160 to 320 line items of costs.

Heavy Engineering Construction Projects

Heavy engineering construction contractors generally organize their estimates in a WBS unique to the project to be constructed. An example of the WBS organization of an estimate for an electric power construction project is illustrated in Tables 1-4 to 1-6. Major areas of the project are defined by groups: switch station, transmission lines, substation, etc., as shown in Table 1-4. Each group is subdivided into divisions of work required to construct the group. For example, Table 1-5 provides

TABLE 1-3
Division 2 estimate for sitework

Cost code	Description	Quantity	Material	Labor	Subcontract	Total
2110	Clearing	L.S.	$ 0.00	$ 0.0	$ 3,694.00	$ 3,694.00
2220	Excavation	8,800 yd^3	0.00	11,880.00	9,416.00	21,296.00
2250	Compaction	950 yd^3	0.00	2,223.00	722.00	2,945.00
2294	Handwork	500 yd^2	0.00	1,750.00	0.00	1,750.00
2281	Termite control	L.S.	0.00	0.00	3,475.00	3,475.00
2372	Drilled piers	1,632 lin ft	14,580.00	2,800.00	14,524.00	31,904.00
2411	Foundation drains	14 ea.	490.00	1,470.00	0.00	1,960.00
2480	Landscape	L.S.	0.00	0.00	8,722.00	8,722.00
2515	Paving	4,850 yd^2	0.00	0.00	105,633.00	105,633.00
			$15,070.00	$20,123.00	$146,186.00	$181,389.00

TABLE 1-4
Example of electric power construction bid summary using the WBS organization of work†
Group-level report

No.	Group	Material	Labor and equipment	Subcontract	Total
1100	Switch station	$1,257,295.00	$ 323,521.00	$3,548,343.00	$ 5,128,167.00
2100	Transmission line A	3,381,625.00	1,260,837.00	0.00	4,641.462.00
2300	Transmission line B	1,744,395.00	0.00	614,740.00	2,358,135.00
3100	Substation at spring creek	572,874.00	116,403.00	1,860,355.00	2,549,632.00
4200	Distribution line A	403,297.00	54,273.00	215,040.00	672,610.00
4400	Distribution line B	227,599.00	98,675.00	102,387.00	427,661.00
4500	Distribution line C	398,463.00	21,498.00	113,547.00	532,508.00
		$7,985,548.00	$1,872,215.00	$6,453,412.00	$16,311,175.00

† For large projects the costs are sometimes rounded to the nearest $100 or $1,000. Tables 1-4 to 1-6 show full dollars to illustrate the transfer of costs among the component, division, and group levels of an estimate.

TABLE 1-5

Example of electric power construction estimate using the WBS organization of work

Division-level report for transmission line A

Cost item	Description	Material	Labor	Equipment	Total
2100	TRANSMISSION LINE A				
2210	Fabrication of steel towers	$ 692,775.00	$ 0.0	$ 0.0	$ 692,775.00
2370	Tower foundations	83,262.00	62,126.00	71,210.00	216,598.00
2570	Erection of steel towers	0.00	144,141.00	382,998.00	527,139.00
2620	Insulators and conductors	2,605,588.00	183,163.00	274,744.00	3,063,495.00
2650	Shield wire installation	0.00	78,164.00	63,291.00	141,455.00
	Total for 2100	$3,381,625.00	$467,594.00	$792,243.00	$4,641,462.00

a work breakdown for all the divisions of work required to construct group 2100, transmission line A: steel fabrication, tower foundations, steel erection, etc. Each division is further broken down into components of work required to construct each division. For example, Table 1-6 provides a work breakdown for all the components of work required to construct division 2370, tower foundations: drilling, reinforcing steel, foundation concrete, and stub angles. The WBS provides a systematic organization of all the information necessary to derive an estimate for the project.

TABLE 1-6

Example of electric power construction estimate using the WBS organization of work

Component-level report for tower foundations

Cost item	Description	Quantity	Material	Labor	Equipment	Total
2370	TOWER FOUNDATIONS					
2372	Drilling foundations	4,196 lin ft	$ 0.00	$25,428.00	$44,897.00	$ 70,325.00
2374	Reinforcing steel	37.5 tons	28,951.00	22,050.00	15,376.00	66,377.00
2376	Foundation concrete	870 yd^3	53,306.00	13,831.00	10,143.00	77,280.00
2378	Stub angles	3,142 lb	1,005.00	817.00	794.00	2,616.00
	Total for 2370		$83,262.00	$62,126.00	$71,210.00	$216,598.00

Other types of heavy engineering projects, such as for highways, utilities, and petrochemical and industrial plants, are organized in a WBS that is unique to their particular types of work. The total estimate is a compilation of costs in a WBS that matches the project to be constructed.

Regardless of the system of estimating selected, either CSI or WBS, to each work item in the estimate a code number should be assigned which is reserved exclusively for that work item for all estimates within the contractor's organization. This same number should also be used in the accounting, job costing, and scheduling functions, to enable one to track the work item during construction.

The computations and organization of information in the estimate lend themselves well to computer applications by use of electronic spread sheets. Electronic spread sheets have rapidly replaced the traditional estimating forms for quantity takeoffs, price extensions, recapitalization sheets, and bid summary sheets that were used in earlier years.

Material Takeoff

The estimator prepares the cost estimate from the plans and specifications for the project. The first step is a quantity takeoff. This involves all materials in the project plus earth excavation and fill.

The quantity of material in a project can be accurately determined from the drawings. The estimator must review each sheet of the drawings, calculate the quantity of material, and record the amount and unit of measure on the appropriate line item in the estimate. As previously discussed, before starting the quantity takeoff, the estimator must prepare a well-organized checklist of all items required to construct the project. The unit costs of different materials should be obtained from material suppliers and used as the basis of estimating the costs of materials for the project. If the prices quoted for materials do not include delivery, the estimator must include appropriate costs for transporting materials to the project. The cost for taxes on materials should be added to the total cost of all materials at the end of the estimate.

Each estimator must develop a system of quantity takeoff that ensures that a quantity is not omitted or calculated twice. A common error in estimating is to completely omit an item or to count an item twice. A well-organized checklist of work will help reduce the chances of omitting an item. A careful recheck of the quantity calculations will detect those items that might be counted twice. The estimator must also add an appropriate percentage for waste for those items where waste is likely to occur during construction. For example, a 5 percent waste might be added to the volume of mortar that is calculated for bricklaying.

The material quantity takeoff is extremely important for cost estimating because it often establishes the quantity and unit of measure for the costs of labor and the contractor's equipment. For example, the quantity of concrete material for piers might be calculated as 20,000 cubic yards (yd^3). The labor-hours and the cost of labor required to place the concrete would also be based on 20,000 yd^3 of material. Also the number of days and the cost of the contractor's equipment that would be required to install the concrete would be based upon 20,000 yd³ of material. Therefore, the estimator must carefully and accurately calculate the quantity and unit of measure of all materials in the project.

Labor and Equipment Costs

Labor and equipment costs are discussed in Chap. 2.

Forms for Preparing Estimates

Experienced estimators will readily agree that it is very important to use a good form in preparing an estimate. As previously stated, the form should treat each operation to be performed in constructing a project. For each operation there should be a systematic listing of materials, equipment, labor, and any other items, with space for all calculations, number of units, unit costs, and total cost.

Each operation should be assigned a code number, and this number should be reserved exclusively for that operation on this estimate as well as on estimates for other projects within a given construction organization. For example, in Table 1-3, item 2250 refers to compaction, whereas item 2372 refers to drilled piers. The accounting department should use the same item number in preparing cost records.

> **Example 1-1.** This example illustrates a form which might be used in preparing an estimate.†
>
> 4210 Common brick
>
> | | | |
> |---|---|---|
> | Brick, 8,000 ft² × 6 bricks/ft² | = | 48.00 M |
> | Add for waste 1 % × 48 M | = | 0.48 M |
> | Total | = | 48.48 M |
>
> | | |
> |---|---|
> | Cost of bricks, 48.48 M × $105.00 per M | = $5,090.40 |
> | Bricklayers, 48 M × 10 h per M = 480 h @ $16.50 | = $7,920.00 |
> | Helpers, 48 M × 6 h per M = 288 h @ $8.75 | = $2,520.00 |

Checklist of Operations

To prepare an estimate, an estimator should use a checklist which includes all the operations necessary to construct the project. Before completing an estimate, one should check this list to be sure that no operations have been omitted. The CSI master format, shown in App. A, provides a uniform approach for organizing project information, and it can be used for preparing an estimate for a building project. It is desirable for the operations to appear, as nearly as possible, in the same order in which they will be performed during construction of the project.

---

† Here M means 1,000. Also these units are abbreviated: square feet, ft²; hour, h.

Other checklists, serving the same purpose, should be prepared for projects involving highways, water systems, sewerage systems, etc.

The checklist may be used to summarize the costs of a project by providing a space for entering the cost of each operation, as illustrated in Tables 1-2 to 1-6. A suitable symbol should be used to show no cost for those operations which are not required. The total cost should include the costs for material, equipment, and labor for the particular operation, as determined in the detailed estimate.

Tables 1-2 and 1-3 illustrate the bid summary for an estimate that was prepared from the checklist in App. A.

Lump-Sum Estimates

For projects with a complete set of plans and specifications that have been prepared prior to construction, the estimate is normally prepared for the purpose of submitting a lump-sum bid on the project. When the cost of a project is estimated on this basis, only one final total-cost figure is quoted. Unless there are revisions in the plans or specifications, this figure represents the amount which the owner will pay to the contractor for the completed project.

It is common practice for government-funded projects to have one or more "alternates" attached to the bid documents of lump-sum contracts. The alternate may be to add or deduct a work item from the base lump-sum bid. This allows the owner the option of selecting the number of alternates so that the total bid cost will be within the amount specified in the owner's budget.

A lump-sum estimate must include the cost of all materials, labor, equipment, overhead, taxes, bonds, insurance, and profit. It is desirable to estimate the costs of materials, labor, and equipment separately for each operation; to obtain a subtotal of these costs for the entire project; and then to estimate the cost of overhead, taxes, bonds, insurance, and profit.

Unit-Price Estimates

Many projects are bid on a unit-price basis. Such projects involve pavements, curbs and gutters, earthwork, various kinds of pipelines, clearing and grubbing land, etc. The cost per unit, submitted in a bid, includes the furnishing of materials, labor, equipment, supervision, insurance, taxes, profit, and bonds, as required, for completely installing a unit. The units designated include square yards (yd^2), cubic yards (yd^3), lineal feet (lin ft) or feet (ft), tons, acres, etc. A separate estimate should be prepared for each type or size unit.

The costs of materials, equipment, and labor are determined for each unit. These are called *direct costs*. To these costs must be added a proportionate part of the indirect costs, such as moving in, temporary construction, overhead, insurance, taxes, profit, and bonds, since indirect costs are not bid separately.

For a unit-price contract, the cost that the owner will pay to the contractor is not determined until the project has been completed. The final cost is determined by multiplying the bid cost per unit by the actual quantity of work that is placed by the contractor. Note that the contractor is paid only for work items that are in the list of pay quantities in the bid documents. If an item is required that is not on the bid list, such as traffic control or laboratory testing, then the estimator must include the cost for this work in one of the unit-cost bid items in the bid documents.

A unit-cost bid might have the following form:

48 acres clearing and grubbing @ $420 per acre
6,240 lin ft of 6-in class B cast-iron pipe in place @ $11.40 per lin ft
8,564 lin ft of 8-in class B cast-iron pipe in place @ $16.40 per lin ft

Negotiated Work

Sometimes the owner will negotiate the work for a project with a construction firm prior to completion of a set of plans and specifications. This is usually done when the owner wants to start construction at the earliest possible date in order to benefit by an early completion and use of the project. The estimates for these types of projects are usually prepared by the contractor as approximate estimates. A representative of the owner works with the contractor to evaluate alternatives in order to obtain a project configuration that meets the needs of the owner, yet with a cost within the owner's allowable budget. The final contract agreement usually is a cost plus a fixed fee with a guaranteed maximum amount. If the actual cost is above or below the guaranteed maximum amount, then a splitting of the difference is agreed on by the owner and the contractor.

Contract Documents

The end result of the design process is the production of a set of contract documents for the project to be constructed. These documents contain all the drawings and written specifications required for preparation of a complete estimate and are used throughout the construction process. Written specifications may be divided into two general parts; one part addresses the legal aspects between the owner and contractor while the other part addresses the technical requirements of the project. The legal part of the written specifications contains at least four items that are important to the estimator: procedures for receipt and opening of bids, qualifications required of bidders, owner's bid forms, and bonds and insurance required for the project.

Owner's Bid Forms

Owners develop bid forms unique to each project for their use in evaluating and comparing contractors' bids. When the contractor has completed the cost estimate, using her or his system of estimating, the contractor must reorganize the estimate to conform to the owner's bid forms.

Bid forms for lump-sum contracts are the simplest since only one final total-cost figure is requested. However, there may be one or more alternates in the bid forms for a lump-sum contract. An alternate is an addition to or a deduction from the base bid. For example, the bid form may request a base cost to construct an office building with two alternates. Alternate 1 might be to eliminate the landscape, and alternate 2 might be to add the parking lot. Upon receipt of the contractor's base bid and alternates, the owner can select the base bid with the combination of alternates that is within the owner's allowable budget for the project.

The owner's bid form for unit-price contracts, such as for highways, may have from one to as many as a hundred line items. Each line item lists a description, quantity of work, and unit of measure for a bid item, with a blank space provided for the estimator to record the unit-price bid. A contractor may incur many costs that do not appear in the unit-price bid forms, such as for control of traffic, temporary construction, etc. Because these costs are not bid separately, the estimator must add them to one or more line items of the owner's unit-price bid forms.

Overhead

The overhead costs chargeable to a project involve many items which cannot be classified as materials, construction equipment, or labor. Some firms divide overhead into two categories: job overhead and general overhead.

Job overhead includes costs which can be charged specifically to a project. These costs are the salaries of the project superintendent and other staff personnel and the costs of utilities, supplies, engineering, tests, drawings, rents, permits, insurance, etc., which can be charged directly to the project.

General overhead is a share of the costs incurred at the general office of the company. These costs include salaries, office rent, utilities, insurance, taxes, shops and yards, and other company expenses not chargeable to a specific project.

Some estimators follow the practice of multiplying the direct costs of a project, materials, equipment, and labor by an assumed percentage to determine the probable cost of overhead. Although this method gives quick results, it may not be sufficiently accurate for most estimates.

While it is possible to estimate the cost of job overhead for a given project, it is usually not possible to estimate accurately the cost of general overhead chargeable to a project. Since the cost of general overhead is incurred in operating all the projects constructed by a contractor, it is reasonable to charge to each project a portion of this cost. The actual amount charged may be based on the duration of the project, the amount of the contract, or a combination of the two.

Example 1-2. This example illustrates a method of determining the amount of general overhead chargeable to a given project.

Average annual value of construction = $6,000,000
Average annual cost of general overhead = 240,000
Amount of general overhead chargeable

to a project $\dfrac{\$240,000 \times 100}{\$6,000,000}$ = 4% of total project cost

Materials Taxes

After the direct costs for materials and labor have been determined, the estimator must include the applicable taxes for each. The tax rate for materials will vary depending on the location. Generally a 3 percent state tax and a 2 to 3 percent city or county tax are assessed on materials. Therefore the tax on materials will range from 3 to 6 percent. It is the responsibility of the estimator to include the appropriate amount of tax in the summary of the estimate.

Labor Taxes

There are two basic types of taxes on labor. The federal government requires a 7.51 percent tax on all wages up to $45,000 per year. In addition, an unemployment tax of approximately 3 percent may be required. Therefore the total tax on labor is approximately 12 percent. The estimator must determine the appropriate tax on labor and include that amount in the summary of the estimate. A discussion of labor costs is presented in Chap. 2.

Basic Builder's Risk Insurance

This insurance affords a contractor protection against loss resulting from fire and lightning damage during the period of construction. Under the terms of this coverage, the insurance is based on the estimated completed value of the project. However, because the actual value varies from zero at the beginning of construction to the full value when the project has been completed, the premium rate usually is set at 60 percent of the completed value. In the event of a loss, the recovery is limited to the actual value at the time of the loss.

   Although the cost of protection will vary with the type of structure and its location, the following rates are representative costs for basic builder's risk:

Type of construction	Cost of premiums per $100.00 of value for basic builder's risk
Framed construction	$1.154
Joisted masonry	0.757
Noncombustible	0.272

Basic Builder's Risk, Extended Coverage

This insurance is a supplement to the basic builder's risk insurance, and it provides protection against wind, smoke, explosion, and vandalism. The premium rate for this insurance is about $0.84 per $100.00 of completed value of the project.

Comprehensive General Liability Insurance

As the result of construction operations, it is possible that persons not employed by the contractor may be injured or killed. Also property not belonging to the contractor may be damaged. Comprehensive general liability insurance should be carried as a protection against loss resulting from such injuries or damage. This insurance provides the protection ordinarily obtainable through public liability and property damage insurance. The coverage should be large enough to provide the necessary protection for the given project. The premium rate varies with the limits of liability specified in the policy.

Contractor's Protective Liability Insurance

This is a contingent insurance that protects a contractor against claims resulting from accidents caused by subcontractors or their employees, for which the contractor may be held liable.

Installation Floater Policy

This insurance provides protection to the contractor against loss resulting from the collapse of a structure during erection.

Contractor's Equipment Floater

This insurance provides protection to the contractor against loss or damage to equipment because of fire, lightning, tornado, flood, collapse of bridges, perils of transportation, collision, theft, landslide, overturning, riot, strike, and civil commotion.

The cost of this insurance, which will vary with the location, should be about $0.62 per $100.00 of equipment value per year.

Bid Bond

It is common practice to require each bidder on a project to furnish with the bid a bid bond, a cashier's check, or a certified check in an amount equal to 5 to 20 percent of the amount of the bid. In the event that the contract to construct the project is tendered to a bidder and the bidder refuses or fails to sign the contract, the owner may retain the bond or check as liquidated damages. Bid bonds covering construction usually cost $25.00.

For some projects, cashier's checks are specified instead of bid bonds. These checks, which are issued to the owner of the project by a bank, are purchased by the bidder. They can be cashed easily, whereas it is necessary for the owner to secure payment on a bid bond through the surety, and the surety may challenge the payment. The use of cashier's checks requires bidders to tie up considerable sums of money for periods which may vary from a few days to several weeks in some instances.

There is no uniform charge for cashier's checks. Some banks charge $0.25 per $100.00 for small checks, with reduced rates for large checks, while others make no charge for checks furnished for regular customers. However, the interest cost for a cashier's check for $100,000 for 2 weeks at 6 percent interest will amount to approximately $240.00.

Contractor's Performance Bond

All government agencies and many private owners require a contractor to furnish a performance bond to last for the period of construction of a project. The bond is furnished by an acceptable surety to assure the owner that the contract will be completed at the specified cost and that all wages and bills for materials will be paid. In the event a contractor fails to complete a project, it is the responsibility of the surety to secure completion. Although the penalty under a performance bond is specified as 25, 50, or 100 percent of the amount of the contract, the cost of the bond usually is based on the amount of the contract.

Representative costs of performance bonds per $1,000.00 are as follows:

For buildings and similar projects:

First	$ 500,000 =	$14.40
Next	2,000,000 =	8.70
Next	2,500,000 =	6.90
Next	2,500,000 =	6.30
All over 7,500,000 =		5.75

For highways and engineering construction:

First	$ 500,000 =	$12.00
Next	2,000,000 =	7.50
Next	2,500,000 =	5.75
Next	2,500,000 =	5.25
All over 7,500,000 =		4.80

Profit

Profit is defined as the amount of money, if any, which a contractor retains after he or she has completed a project and has paid all costs for materials, equipment, labor, overhead, taxes, insurance, etc. The amount included in a bid for profit is subject to considerable variation, depending on the size of the project, extent of risk involved, desire of the contractor to get the job, extent of competition, and other factors. A contractor might include 3 to 6 percent profit on a $1,000,000 highway paving project when the risk is low and competition is high, whereas the contractor might include 20 to 30 percent profit or more on a foundation or river project when the risk is high and there is little competition.

Representative Estimates

Numerous examples of estimates are presented in this book to illustrate the steps to follow in determining the probable cost of the project. Nominal amounts are included for overhead and profit in some instances to give examples of complete estimates for bid purposes. In other instances only the costs of materials, construction equipment, and labor are included. The latter three costs are referred to as *direct costs*. They represent the most difficult costs to estimate, and they are our primary concern.

In preparing the sample estimates, unit prices for materials, equipment, and labor are used primarily to show how an estimate is prepared. Note that these unit costs will vary with the time and location of a project. An estimator must obtain and use unit prices which are correct for the particular project. Estimators do not establish prices; they simply use them.

Remember that estimating is not an exact science. Experience, judgment, and care should enable an estimator to prepare an estimate which will reasonably approximate the ultimate cost of the project.

Figure 1-1 illustrates a form which might be used in preparing a detailed estimate. When a project includes several operations, the direct costs for material, equipment, and labor should be estimated separately for each operation, then the indirect costs, for insurance, taxes, overhead, profit, performance bond, etc., should be estimated for the entire project.

Instructions to the Reader

In the examples in this book, a uniform method is used of calculating and expressing the time units for equipment and labor and the total cost. For equipment the time is expressed in equipment-hours, and for labor it is expressed in labor-hours. A labor-hour is one person working 1 h or two people working $\frac{1}{2}$ h.

If a job which requires the use of four trucks lasts 16 h, the time units for the truck are the product of the number of trucks and the length of the job, expressed in hours. The unit of cost is for 1 truck-hour. The calculations are as follows:

Trucks, 4 × 16 h = 64 truck-hours @ $14.50 = $928.00

In a similar manner, the time and cost for the truck drivers are

Truck drivers, 4 × 16 h = 64 labor-hours @ $9.50 = $608.00

The terms "64 truck-hours" and "64 labor-hours" may be shortened to read 64 h without producing ambiguity.

Sample estimate

Item no.		Description	Calculations
2350	0	Furnish and drive 200 creosote-treated piles. Drive piles to full penetration into normal soil. Length of piles, 50 ft. Size of piles, 14-in butt and 6-in-tip diameters.	
	10	Materials	
		Piles; add 5 for possible breakage.	205 piles × 50 ft
	20	Equipment	
		Moving to and away from the job	
		Crane, 12-ton crawler type	200 piles ÷ $2\frac{1}{2}$ per h
		Hammer, single-acting, 15,000 foot-pounds (ft·lb)	
		Boiler, water, fuel, etc.	
		Leads and sundry equipment	
	30	Labor, add 16 h to set up and take down equipment	
		Foreman	80 + 16
		Fireman	
		Crane operator	
		Crane oiler	
		Workers on hammer	2 × 96
		Helpers	2 × 96
		Subtotal, direct cost	
	50	Overhead	10% × $60,198.00
	60	Social security tax	7.51% × $9,504.00
	62	Workers' compensation insurance	8.68% × $9,504.00
	64	Unemployment tax	3% × $9,504.00
	70	Subtotal cost	
	80	Profit	10% × $68,037.62
	85	Subtotal cost	
	90	Performance bond	1.2% × $74,841.38
	94	Total cost, amount of bid	
	95	Cost per lin ft	$75,739.48 ÷ 10,000 lin ft

FIGURE 1-1
Form used to estimate construction costs.

No. units	Unit	Unit cost		Material cost		Equipment cost		Labor cost		Total cost	
10,250	lin ft	4	20	43,050	00					43,050	00
		Lump sum				3,500	00			3,500	00
80	h	37	25			2,980	00			2,980	00
80	h	7	50			600	00			600	00
80	h	4	25			340	00			340	00
80	h	2	80			224	00			224	00
96	h	21	20					2,035	20	2,035	20
96	h	14	30					1,372	80	1,372	80
96	h	17	25					1,656	00	1,656	00
96	h	11	25					1,080	00	1,080	00
192	h	9	40					1,804	80	1,804	80
192	h	8	10					1,555	20	1,555	20
				43,050	00	7,644	00	9,504	00	60,198	00
										6,019	80
										713	75
										824	95
										281	12
										68,037	62
										6,803	76
										74,841	38
										898	10
										75,739	48
										7	57

Production Rates

To determine the time required to perform a given quantity of work, it is necessary to estimate the probable rates of production of the equipment or labor. These rates are subject to considerable variation, depending on the difficulty of the work, job and management conditions, and the condition of the equipment.

A production rate is the number of units of work produced by a unit of equipment or a person in a specified unit of time. The time is usually 1 h. The rate may be determined during an interval when production is progressing at the maximum possible speed. It is obvious that such a rate cannot be maintained for a long time. There will always be interruptions and delays which reduce the average production rates to less than the ideal rates. If a machine works at full speed only 45 min/h, the average production rate will be 0.75 of the ideal rate. The figure 0.75 is defined as an operating factor.

A shovel with a 1-yd^3 dipper may be capable of handling 3 dippers per minute under ideal conditions. However, on a given job the average volume per dipper may be only 0.8 yd^3, with the shovel actually operating only 45 minutes per hour (min/h):

The ideal output is $180 \times 1 = 180$ yd^3/h.
The dipper factor is 0.8.
The time factor is 0.75.
The combined operating factor is $0.8 \times 0.75 = 0.6$.
The average output is $0.6 \times 180 = 108$ yd^3/h.

The average output should be used in computing the time required to complete a job.

Tables of Production Rates

In this book numerous tables give production rates for equipment and workers. In all tables the rates are adjusted to include an operating factor, usually based on a 45- to 50-min working hour. If this factor is too high for a given job, the rates should be reduced to more appropriate values.

An estimator who has access to production rates obtained on projects constructed under conditions similar to the conditions that will exist on a project for which he or she is preparing an estimate should use them instead of the rates appearing in tables prepared by someone else.

Computer Applications

An estimator must assemble a large amount of information in an organized manner and perform numerous calculations to prepare a cost estimate. The computer can be used by the estimator to organize, store, and retrieve information and to perform the many calculations necessary to prepare an estimate. It can be an effective tool for decreasing the preparation time and increasing the accuracy of cost estimating.

The computer is used for estimating in at least five different applications: quantity takeoff, price extensions and bid summary, historical cost data base, labor and equipment productivity data base, and supplier data base. Each application may be subdivided and should be linked together in an overall, integrated system.

An electronic digitizer can be used to obtain the quantities of materials from the construction drawings of a project. The estimator can use a digitizer pen to trace the lines on the drawings to obtain information such as the square yards of paving, square feet of brick, linear feet of pipe, or number of windows. Using a digitizer to calculate the quantity takeoff automates the process and provides the information to the estimator in an organized form.

The most common computer application for cost estimating is the electronic spread sheet. An electronic spread-sheet program can be used to perform the numerous price extension calculations and the bid summary of an estimate. Spread-sheet programs can be easily developed by the estimator using her or his system of estimating for the particular type of construction project.

Numerous computer data bases can be developed by the estimator to automate and standardize the estimating function. The estimator can develop a historical cost data base from the cost records of projects that have been completed by the company. This information can be stored as unit costs in the data base and organized in a CSI or WBS system with a cost code for each item. The estimator can retrieve information from the historical cost data base for the preparation of estimates for future projects. As new information is obtained from current projects, the estimator can update the historical cost data base.

Labor and equipment productivity data bases can be developed from records of previously completed projects. For example, the labor-hours per square foot of formwork, the number of cubic yards of earth per equipment-day, etc., can be organized and stored for specific job conditions. The estimator can retrieve the labor or equipment productivity figures from the data base for the preparation of a cost estimate for a prospective project. Adjustments to the stored productivity can be made by the estimator to reflect unique job conditions.

The estimator can prepare a data base of information for the material suppliers and subcontractors who perform work for the company. The data base can be organized by type, size, and location of the supplier. During the preparation of an estimate for a project, the estimator can retrieve supplier and subcontractor information pertinent to the project. For example, the estimator can sort and list all dry-wall contractors capable of performing $80,000 of work at a particular job location. The name, address, and phone number of the contact person for each potential supplier or subcontractor can be retrieved from the computer.

Much of the information used in the operation of a construction company is accessed by different individuals at different times. For example, a project is planned and scheduled based on the time and cost information prepared by the estimator. Likewise, the project budget control system is developed from the cost estimate. The computer can be used to link the information from the estimating function to the planning function and to the budget control function of a contractor's operation. Common cost codes can be used for each operation to integrate, automate, and standardize the operations of a construction firm.

CHAPTER
2

COST OF CONSTRUCTION LABOR AND EQUIPMENT

CONSTRUCTION LABOR

Construction laborers influence every part of a project. They operate equipment, fabricate and install materials, and make decisions that have a major effect on the project. Most individuals involved in construction will readily agree that people are the most important resource on a project. The cost to hire a laborer includes the straight-time wage plus any overtime pay, workers' compensation insurance, social security, unemployment compensation tax, public liability and property damage, and any fringe benefits.

Sources of Labor Rates

Wage rates vary considerably with the locations of projects and with the various types of crafts. The hourly rate of construction laborers is determined by one of three means: union wage, open-shop wage, or prevailing wage. Construction workers who are members of a labor union are paid a wage rate established by a labor contract between their local union and the construction contractor's management. Construction workers who are not members of a union are paid an open-shop wage agreed to by each individual employee and the employer. For construction employees who work on projects funded with state or federal money, their wage rate is established by the prevailing wage at the project location. The federal government, and many states, establishes a prevailing wage for each construction craft. The prevailing wage rate is determined for each craft by a wage survey for each geographic location.

Cost of Labor

The rates listed in Table 2-1 are representative base rates. In addition to paying the base rate, an employer must pay or contribute amounts for such items as social security tax, unemployment tax, workers' compensation insurance, public liability and property damage insurance, and any fringe benefits. Fringe benefits include such items as apprenticeship plans, pension plans, and health and welfare insurance. Base rates normally apply to work done during the 40-h workweek, 8 h/day and 5 days/week. For work in excess of 8 h/day or 40 h/week, the base rate is

TABLE 2-1
Representative base wage rates in the United States for 1988

Craft	Wage rate, $/h
Bricklayers	16.50
Carpenters	16.70
Cement masons	16.20
Electricians	18.75
Electrician's helpers	11.25
Glaziers	18.75
Glazier helpers	11.25
Plumbers	18.75
Plumber's helpers	11.25
Ironworkers, reinforcing steel	15.80
Ironworkers, structural steel	15.80
Painters	16.10
Stone masons	17.60
Sheet-metal workers	15.56
Welders	14.66
Power shovel operators	17.25
Power shovel oilers	11.25
Hoisting engineers	17.00
Air compressor operators	12.35
Air tool operators	10.80
Tractor operators	15.86
Truck drivers, single-axle	8.10
Truck drivers, tandem-axle	9.45

generally increased to $1\frac{1}{2}$ or 2 times the base rate. The base rates in Table 2-1 are used to determine the costs in the examples in this book. The effect of fringe benefits, taxes, and insurance is not included in the rates.

Social Security Tax

The federal government requires an employer to pay a tax for the purpose of providing retirement benefits to persons who become eligible. At present, the employer must pay 7.51 percent of the gross earnings of an employee, up to $45,000 per year. The employee contributes an equal amount through the employer. This rate is subject to change by Congress.

Unemployment Compensation Tax

This tax, which is collected by the states, is for the purpose of providing funds with which to compensate workers during periods of unemployment. The base cost of this tax is usually 3 percent of the wages paid to the employees, all of which is paid by the contractor. This rate may be reduced by establishing a high degree of employment stability, with few layoffs, during a specified period.

Workers' Compensation and Employer's Liability Insurance

Most states require contractors to carry workers' compensation and employer's liability insurance as a protection to the workers on a project. In the event of an injury or death of an employee working on the project, the insurance carrier will provide financial assistance to the injured person or to his or her family. Although the extent of financial benefits varies within several states, in general they cover reasonable medical expenses plus the payment of reduced wages during the period of injury. Each state which requires this coverage has jurisdiction, through a designated agency, over the insurance to the extent of specifying the minimum amounts to be carried, the extent of the benefits, and the premium rates paid by the employer.

The base or manual rates for workers' compensation insurance vary considerably among states, and within a state they vary according to the classification of work performed by an employee. A higher premium rate is charged for work that subjects workers to a greater risk of injury. A contractor who establishes a low record of accidents on jobs for a specified period will be granted a credit, which will reduce the cost of the insurance. A contractor who establishes a high record of accidents over a period will be required to pay a rate higher than the base rate, thus increasing the cost of the insurance.

The premium rate for this insurance is specified to be a designated amount for each $100.00 of wages paid under each classification of work. The rate normally varies from about $5.00 per $100.00 of wages paid for lowrisk crafts to approximately $12.00 per $100.00 of wages paid for high-risk crafts. To determine the cost of this insurance for a given project, it is necessary to estimate the amount of wages that will be paid under each classification of work and then to apply the appropriate rate to each wage classification. Since the base rates are subject to changes, an estimator should verify them before preparing an estimate.

Public Liability and Property Damage Insurance

This insurance protects the contractor against injuries to the general public or damage to public property due to actions of the employee while performing work during construction. The cost for this insurance is specified as a rate for each $100.00 of base wages. The rate can vary from $2.00 to $5.00 per $100.00 of base wages, depending on the craft and the safety record of the contractor. Due to the large variations in the premium rates for this insurance, it is necessary for the estimator to obtain the rate from an insurance company before estimating the cost of this insurance.

Fringe Benefits

As a part of the agreement of employment, the contractor often agrees to pay benefits for the employee. Examples are health insurance, pension plan, training programs, etc. The cost of these fringe benefits will depend on the number of different coverages and the amount of coverage. Generally the costs range from $0.75 to $2.10 per hour.

ILLUSTRATIVE EXAMPLE. The example which follows illustrates the costs that a contractor must incur to hire an employee. For estimating and bidding purposes, the estimator normally determines the total cost of all labor, using the base wages. The final cost of labor is then determined by multiplying the base-wage costs by a percentage to account for taxes and insurance.

> **Example 2-1.** An ironworker works 10 h/day, 6 days/week. A base wage of $15.80 per hour is paid for all straight-time work, 8 h/day, 5 days/week. An overtime rate of time and one-half is paid for all hours over 8 h/day, Monday through Friday, and double time is paid for all Saturday work. The social security tax is 7.51 percent, and the unemployment tax is 3 percent of actual wages. The rate for workers' compensation insurance is $8.50 per $100.00 of base wages, and the public liability and property damage insurance rate is $3.25 per $100.00 of base wages. Fringe benefits are $1.27 per hour. Calculate the average hourly cost to hire the ironworker.

$$\text{Average hourly pay} = (\text{Pay hours} \div \text{Actual hours}) \times \text{Base wage}$$

Straight-time pay, 8 h/day × 5 days @ 1.0	= 40 h
Weekday overtime pay, 2 h/day × 5 days @ 1.5	= 15 h
Saturday overtime pay, 10 h/day × 1 day @ 2.0	= <u>20 h</u>
Total pay hours	= 75 h

Average hourly pay, $75 \div 60 \times \$15.80$	=	$19.7500
Social security tax, $7.51\% \times \$19.75$	=	1.4832
Unemployment tax, $3.0\% \times \$19.75$	=	0.5925
Workers' compensation, $\$8.50 \div \$100.00 \times \$15.80$	=	1.3430
Public liability/property damage, $\$3.25 \div \$100.00 \times \$15.80$ =		0.5135
Fringe benefits	=	1.2700
Average hourly cost	=	$24.9522

Daily cost, 10 h @ $24.9522 = $249.52 per day
Weekly cost, 60 h @ $24.9522 = $1,497.13 per week
Monthly cost, $60 \times (52 \div 12)$ h @ $24.9522 = $6,487.57 per month

If there are five ironworkers in a crew and the crew can place 6,500 pounds (lb) per day (lb/day) of reinforcing steel, then the cost of labor per pound of steel in place could be determined as follows:

Cost per unit installed = Unit crew cost ÷ Crew production rate
Crew cost per day, 5 workers @ $249.52 = $1,247.60 per day
Labor cost per pound, $1,247.60 per day ÷ 6,500 lb/day = $0.19

Production Rates for Labor

A *production rate* is defined as the number of units of work produced by a person in a specified time, usually an hour or a day. Production rates may also specify the time in labor-hours or labor-days required to produce a specified number of units of work, such as 12 labor-hours to lay 1,000 bricks. This book uses h as the unit of time. Production rates should be realistic to the extent of including an allowance for the fact that a person usually will not work 60 min/h.

The time which a laborer will consume in performing a unit of work will vary between laborers and between projects and with climatic conditions, job supervision, complexities of the operation, and other factors. It requires more time to fabricate and erect lumber forms for concrete stairs than for concrete foundation walls. An estimator must analyze each operation to determine the probable time required for the operation.

Information on the rates at which work has been performed on similar projects is very helpful. Such data may be obtained by keeping accurate records of the production of labor on projects as construction progresses. For the information to be most valuable to an estimator, there should be submitted with each production report an accurate record showing the number of units of work completed, the number of laborers employed, by classification, the time required to complete the work, and a description of job conditions, climatic conditions, and any other conditions or factors which might affect the production of labor. The reports should be for relatively short periods, such as a day or week, so that the conditions described will accurately represent the true conditions for the given period. Reports covering a complete project, lasting for several months, will give average production rates but will fail to indicate varying rates resulting from changes in working conditions. It is not sufficiently accurate for an estimator to know that a bricklayer laid an average of 800 bricks per day on a project. The estimator should know the rate at which each type of brick was laid under different working conditions, considering the climatic and any other factors which might have affected production rates. All experienced construction workers know that the production of labor is usually low during the early stages of construction. As the organization becomes more efficient, the production rates will improve; then as the construction enters the final stages, there will usually be a reduction in the production rates. This is important to an estimator. For a small job it is possible that labor will never reach its most efficient rate of production because there will not be sufficient time. If a job is of such a type that laborers must frequently be transferred from one operation to another or if there are frequent interruptions, then the production rates will be lower than when the laborers remain on one operation for long periods without interruptions.

In this book numerous tables give the rates at which laborers should perform various operations. These rates include an adjustment for nonproductive time, by assuming that a person will actually work about 45 to 50 min/h. Conditions on some projects may justify a further adjustment in the rates. The frequent use of a range in rates instead of a single rate will permit the estimator to select the rate which she or he believes is most appropriate for the project.

ILLUSTRATIVE EXAMPLES. The examples which follow illustrate methods of estimating the labor required to perform units of work for different types of projects under varying conditions.

Example 2-2. This example illustrates the probable rates of excavating earth by hand under different conditions.

For the first project, the soil is a sandy loam which requires light loosening with a pick before shoveling. The maximum depth of the trench or pit will be 4 ft. Climatic conditions are good, with a temperature of about 70°F. A laborer should easily loosen $\frac{1}{2}$ yd^3 of earth per hour. If a long-handled round-pointed shovel is used, it should require about 150 loads to remove 1 yd^3 of earth, bank measure. The laborer who can handle $2\frac{1}{2}$ shovel loads per minute will remove 1 yd^3/h. These rates of production will require 2 labor-hours to loosen and 1 labor-hour to remove 1 yd^3 of earth from the trench, or a total of 3 labor-hours/yd^3, bank measure.

For a second project, the soil is a tough clay, which is difficult to dig and which lumps badly. The maximum depth of the trench or pit will be 5 ft. The temperature will be about 100°F, which will reduce the operating efficiency of the laborers. Under these conditions a laborer may not loosen more than $\frac{1}{4}$ yd^3/h. Because of the physical condition of the loosened earth, it will require about 180 shovel loads to remove 1 yd^3 of earth. If a laborer can handle 2 shovel loads per minute, he or she can remove 120 shovel loads per hour. This is equivalent to $\frac{2}{3}$ yd^3 per labor-hour or $1\frac{1}{2}$ labor-hours/yd^3, bank measure. For these rates of production it will require 4 labor-hours to loosen and $1\frac{1}{2}$ labor-hours to remove 1 yd^3, or a total of $5\frac{1}{2}$ labor-hours/yd^3.

Example 2-3. This example illustrates a method of determining the probable rate of placing reinforcing steel for a given project.

Steel bars are to be used to reinforce a concrete slab 57 ft wide and 70 ft long. The bars will be $\frac{1}{2}$ in in diameter, with no bends, maximum length limited to 20 ft, and spaced 12 in apart both ways. All laps will be 18 in. Precast concrete blocks, spaced not over 6 ft apart each way, will be used to support the reinforcing. The bars will be tied at each intersection, by bar ties. The steel will be stored in orderly stock piles, according to length, about 80-ft average distance from the center of the slab. The slab will be constructed on the ground.

The length of the bars parallel to the 57-ft side will be

Length of the side	=	57 ft
Length of laps, 2 × 18 in	=	3 ft
Total length of bars per row	=	60 ft
Use 3 bars 20 ft long	=	60 ft
Total number of bars required, 3 × 70	=	210
Total length of the bars, 210 × 20 ft	=	4,200 ft

The length of the bars parallel to the 70-ft side will be

Length of side	= 70 ft
Length of laps, 3 × 18 in	= 4 ft 6 in
Total length of bars per row	= 74 ft 6 in
Use 4 bars 18 ft 7 in long	= 74 ft 6 in
Total number of bars required, 4 × 57	= 228
Total length of the bars, 228 × 18 ft 7 in	= 4,246 ft

The weight of the reinforcing will be

20-ft bars, 4,200 ft @ 0.668 lb/ft	= 2,806 lb
18-ft 7-in bars, 4,246 ft @ 0.668 lb/ft	= 2,836 lb
Total weight	= 5,642 lb

The number of intersections of bars will be 57 × 70 = 3,990

The time required to place the reinforcing, using 2 steel setters, should be about as follows:

Carrying bars to slab site:

Time to walk 160 ft @ 100 ft/min	= 1.6 min
Add time to pick up and put down reinforcing	= 1.0 min
Time for round trip	= 2.6 min

Assume that the 2 persons can carry 6 bars, weighing approximately 80 lb each trip

Number of trips required, 438 bars ÷ 6 bars per trip = 73

Total time to carry reinforcing, 73 × 2.6 ÷ 60   = 3.17 h

Placing the bars on blocks and spacing them:

Assume that 2 persons working together can place 2 bars/min, or 120 bars/h

Time to place, 438 bars ÷ 120 bars/h   = 3.67

Tying the bars at intersections:
    Assume that a person can make 5 ties per min
    2 people will make $2 \times 5 \times 60 = 600$ ties per h
    Time to tie reinforcing, $3,990$ ties $\div 600$ per h $= 6.65$ h

The total working time will be
Carrying the reinforcing	$= 3.17$ h
Placing the reinforcing	$= 3.67$ h
Tying the reinforcing	$= 6.65$ h
Total working time	$= 13.49$ h

On a project a worker will seldom work more than 45 to 50 min/h, because of necessary delays. Based on a 45-min hour, the total clock time to handle and place the reinforcing will be

$$\frac{13.49 \times 60}{45} = 18.0 \text{ h}$$

Total labor-hours for the job, 2×18 $= 36$
No. of tons placed, $5,642 \div 2,000$ $= 2.821$
Labor-hours per ton placed, $36 \div 2,821 = 12.75$

If the reinforcing steel in Example 2-3 is bent to furnish negative reinforcing over the beams or if it has hooks on the ends and if, in addition, it must be hoisted to the second floor of a building, then it will require extra time to hoist and place it. An estimator should adjust the production rate accordingly instead of using flat rates for all projects.

CONSTRUCTION EQUIPMENT

Most projects involve the use of construction equipment. The purchase of equipment represents a capital investment by the owner for the purpose of accomplishing the work which it will do and at the same time making a profit on the investment. If a profit is to be realized from the use of equipment, it is first necessary for the owner to recover from the use of the equipment during its useful life sufficient money to pay the entire cost of the equipment, plus the cost of maintenance and repairs, interest, insurance, taxes, storage, fuel, lubrication, etc., plus an additional amount for profit. Any estimate must provide for the cost of equipment used on the project.

Sources of Equipment

The use of equipment may be secured through purchase or rental. For each method there are several plans.

When equipment is purchased, either one of the following plans may be used:

1. Cash purchase
2. Purchase on a deferred-payment plan

Equipment may be rented under one of the following plans:

1. The lessee will pay a specified price per month, week, day, or hour for the use of each unit.
   a. The lessee pays for the operator, fuel, lubrication, and all repairs.
   b. The lessor pays for the operator, fuel, lubrication, and all repairs.
   c. Some other combination of a and b.
2. The lessee will pay a specified price for each unit of work performed by the equipment.
3. The lessee will pay a specified rental rate for the use of the equipment, with an option to purchase the equipment at a later date, with the provision that all or a part of the money paid for rent shall apply toward the purchase of the equipment.

Equipment Costs

When equipment is to be rented, the estimator should include the cost in the estimate. Reference 1 provides a comprehensive list of the rental and operating costs for all types of construction equipment. The list is periodically updated and has become the industry standard for equipment rental rates.

When equipment is purchased, it is necessary to determine the cost of owning and operating each unit, which will include all or several of the following items:

1. Depreciation
2. Maintenance and repairs
3. Investment
4. Fuel and lubrication or another type of energy, such as electricity

In this book the general practice is to charge on an hourly basis for equipment that is owned. Discussions and examples which follow will illustrate methods of estimating the hourly cost of owning and operating equipment. These examples are intended to show the estimator how to determine the probable hourly cost for any type of construction equipment. The costs which are determined in these examples apply for the given conditions only, but by following the same procedure and using appropriate prices for the particular equipment, the estimator may determine hourly costs which are suitable for use on any project.

Depreciation Costs

Depreciation is the loss in value of equipment resulting from use and age. The owner of equipment must recover the original cost of the equipment during its useful life or sustain an equipment loss on those projects where the equipment is used. The cost of a unit of equipment should include the purchase price and the cost of transporting it to the purchaser, plus the cost of unloading and assembling it at its destination.

Although any reasonable method may be used for determining the cost of depreciation, the following three are most commonly used:

1. Straight-line method
2. Declining-balance method
3. Sum-of-the-year's-digits method

A construction company may select the depreciation method that best suits its operations and financial situation. The straight-line method is most frequently chosen for assigning costs to an estimate owing to its simplicity and uniform distribution of costs throughout the useful life of the equipment.

The cost of depreciation for construction equipment is commonly determined for two purposes. An estimator determines the depreciation costs that must be applied to estimates of projects on which the equipment is to be used. An accountant must determine the depreciation costs by methods that are approved by the U.S. Internal Revenue Service (IRS) for tax purposes. Therefore it is common to utilize two different depreciation methods on a particular unit of equipment, reporting one value to the estimator for use in estimating construction projects and another value to the IRS to obtain the most favorable tax benefits.

Straight-Line Depreciation

When the cost of depreciation is determined by this method, it is assumed that a unit of equipment will decrease in value from its original total cost at a uniform rate. The depreciation rate may be expressed as a cost per unit of time, or it may be expressed as a cost per unit of work produced. The depreciation cost per unit of time is obtained by dividing the original cost, less the estimated salvage value to be realized at the time it will be disposed of, by the estimated useful life, expressed in the desired units of time, which may be years, months, weeks, days, or hours. For example, a given unit of equipment, whose original cost is $12,000 may have a useful life of 2,000 h/yr for 5 yr and a salvage value of $2,000. The cost of depreciation is determined as follows:

Total depreciation, $12,000 − $2,000   = $10,000
Annual cost of depreciation, $10,000 ÷ 5   =   2,000
Hourly cost of depreciation, $2,000 ÷ 2,000 =   1.00

Another method of estimating the straight-line cost of depreciation is to divide the original cost, less the estimated salvage value, by the probable number of units of work which it will produce during its useful life. This method is satisfactory for equipment whose life is determined by the rate at which it is used instead of by time. Examples of such equipment include the pump and discharge pipe on a hydraulic dredge, rock crushers, rock-drilling equipment, rubber tires, and conveyer belts.

Declining-Balance Method

Under this method of determining the cost of depreciation, the estimated life of the equipment in years will give the average percentage of depreciation per year. This percentage is doubled for the 200 percent declining-balance method. The value of the depreciation during any given year is determined by multiplying the resulting percentage by the value of the equipment at the beginning of that year. While the estimated salvage value is not considered in determining depreciation, the depreciated value is not permitted to drop below a reasonable salvage value.

When the cumulative sum of all costs of depreciation is deducted from the original total cost, the remaining value is designated the book value. Thus, if a unit of equipment whose original cost was $10,000 has been depreciated a total of $6,000, the book value will be $4,000.

Example 2-4. This example illustrates how the declining-balance method of determining the cost of depreciation may be applied to a unit of equipment.

Total cost, $10,000
Estimated salvage value, $1,000
Estimated life, 5 yr
Average rate of depreciation, 20% per yr
Double this rate of depreciation, 2×20 = 40%
Cost of depreciation, first year, $0.40 \times \$10,000$ = $4,000
Book value at the start of the second year = 6,000
Cost of depreciation, second year, $0.40 \times \$6,000$ = 2,400

Table 2-2 gives the schedule of depreciation costs for this equipment.

This method may be used for any reasonable useful life. The depreciation per year may be continued until the book value of the equipment is reduced to a reasonable salvage value.

Sum-of-the-Year's-Digits Method

Under this method of determining the cost of depreciation, all the digits representing each year of the estimated life of the equipment are totaled. For an estimated life of 5 yr, the sum of the digits will be $1 + 2 + 3 + 4 + 5 = 15$. Deduct the estimated salvage value from the total cost of the equipment. During the first year, the cost of depreciation will be $\frac{5}{15}$ of the cost less salvage value. During the second

TABLE 2-2
Annual cost of depreciation using the declining-balance method

End of year	Percentage of depreciation	Depreciation for the year	Book value
0	0	$ 0	$10,000.00
1	40	4,000.00	6,000.00
2	40	2,400.00	3,600.00
3	40	1,440.00	2,160.00
4	40	864.00	1,296.00
5	40	518.40	777.60
5†	...	296.00	1,000.00

† The value of the equipment may not be depreciated below a reasonable minimum salvage value. If this value is $1,000, the lower figures will apply; otherwise, the upper figures will apply.

TABLE 2-3
Annual cost of depreciation using the sum-of-the-year's-digits method

End of year	Depreciation ratio	Total depreciation	Depreciation for the year	Book value
0	0	$9,000	$ 0	$10,000
1	$\frac{5}{15}$	9,000	3,000	7,000
2	$\frac{4}{15}$	9,000	2,400	4,600
3	$\frac{3}{15}$	9,000	1,800	2,800
4	$\frac{2}{15}$	9,000	1,200	1,600
5	$\frac{1}{15}$	9,000	600	1,000

year, the cost of depreciation will be $\frac{4}{15}$ of the cost less salvage value. Continue this process for each year through the fifth year. Table 2-3 gives the schedule of depreciation costs for a unit of equipment under the stated conditions.

Total cost = $10,000
Estimated salvage value = 1,000
Total cost of depreciation = $ 9,000
Estimated useful life, 5 yr
Sum of the year's digits, 15

Costs of Maintenance and Repairs

The costs for maintenance and repairs include the expenditures for replacement parts and the labor required to keep the equipment in good working condition. These costs vary considerably with the type of equipment and the service for which it is used. If a power shovel is used to excavate soft earth, the replacement of parts will be considerably less often than when the same shovel is used to excavate hard clay or rock. The extent of variation in costs for repairs is illustrated by the following examples, which are based on the cost records kept by the owner of the equipment.

Example 2-5. This example covers information for two draglines of identical make and size, each operated for 6 yr.

| Unit | Time operated, h | Cost of repairs | |
		Total	Per h
1	9,768	$39,472	$4.03
2	12,448	16,316	1.31

Example 2-6. This example covers information for two motor graders of identical make and size, each operated for 5 yr.

| Unit | Time operated, h | Cost of repairs | |
		Total	Per h
1	11,680	$8,239	$0.71
2	12,168	3,696	$0.31

The manufacturers of tractors have released information showing that the average costs for maintenance and repairs for crawler tractors are approximately 100 percent of the cost of depreciation during a 5-yr period of use. The costs for individual units will run substantially higher or lower than the average depending on the service provided and the types of jobs on which the equipment is used.

The Power Crane and Shovel Association (Ref. 2) suggests that the rates shown in the table on page 35 be applied in determining the average costs for maintenance, repairs, and supplies for the indicated equipment.

The costs for maintenance and repairs given in this article should be used as a guide only. Estimators should increase or decrease the suggested rates when they believe that operating conditions will be more or less severe, respectively, than average conditions.

Cost of Rubber Tires

Many types of construction equipment use rubber tires, whose life usually will not be the same as that for the equipment on which they are used. The cost of depreciation and repairs for tires should be estimated separately from that for the equipment.

	Useful life		Percentage of total cost	
	Yr	H	Per yr	Per h
For shovels and hoes				
Size, yd^3:				
$\frac{3}{8}-\frac{3}{4}$	5	10,000	20.00	0.0100
$1-1\frac{1}{2}$	6	12,000	16.67	0.0083
$2-2\frac{1}{2}$	8	16,000	12.50	0.00625
For draglines and clamshells				
Size, yd^3:				
$\frac{3}{8}-\frac{3}{4}$	5	10,000	16.00	0.00800
$1-1\frac{1}{2}$	9	18,000	8.89	0.00445
$2-2\frac{1}{2}$	12	24,000	6.66	0.00333
For lifting cranes				
Capacity, tons:				
$2\frac{1}{2}-5$	5	10,000	12.00	0.00600
10–15	9	18,000	6.67	0.00333
20 and over	12	24,000	5.00	0.00250

A set of tires for a unit of equipment, whose cost is $1,200.00, may have an estimated life of 5,000 h, with the repairs during the life of the tires costing 15 percent of the initial cost of the tires. The cost is determined as follows:

Depreciation, $1,200 ÷ 5,000 h = $0.24 per h
Repairs, 0.15 × $0.24   =   0.04 per h
    Total cost   = $0.28 per h

Investment Costs

It costs money to own equipment, regardless of the extent to which it is used. These costs, which are frequently classified as investment costs, include interest on the money invested in the equipment and taxes of all types which are assessed against the equipment, insurance, and storage. See Fig. 2-1. The rates for these items will vary somewhat among different owners, with location, and for other reasons.

There are several methods of determining the cost of interest paid on the money invested in the equipment. Even though the owners pay cash for equipment, they should charge interest on the investment, because the money spent for the equipment could be invested in some other asset which would produce interest for the owner if it were not invested in equipment.

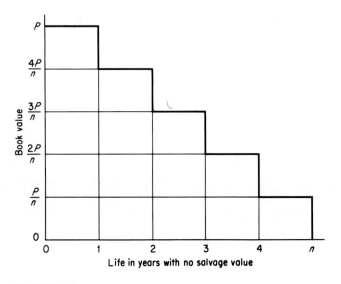

FIGURE 2-1
Value of equipment without salvage value by year.

The average annual cost of interest should be based on the average value of the equipment during its useful life. This value may be obtained by establishing a schedule of values for the beginning of each year that the equipment will be used. The calculations given below illustrate a method of determining the average value of equipment:

Original cost of equipment, $25,000
Estimated useful life, 5 yr
Average annual cost of depreciation, $25,000 ÷ 5 = $5,000

Beginning of year	Cumulative depreciation	Value of equipment
1	0	$25,000
2	$ 5,000	20,000
3	10,000	15,000
4	15,000	10,000
5	20,000	5,000
6	25,000	0

Total of values in column 3   = $75,000
Average value, $75,000 ÷ 5   = $15,000

$$\text{Average value as \% of original cost, } \frac{\$15,000 \times 100}{\$25,000} = 60\%$$

The average value of equipment may be determined from the following equation:

$$\bar{P} = \frac{P(n + 1)}{2n} \tag{2-1}$$

where P = total initial cost
   $\bar{P}$ = average value
   n = useful life, yr

Equation (2-1) assumes that a unit of equipment will have no salvage value at the end of its useful life. If a unit of equipment will have salvage value when it is disposed of, the average value during its life can be obtained from Eq. (2-2):

$$\bar{P} = \frac{P(n + 1) + S(n - 1)}{2n} \tag{2-2}$$

where P = total initial cost
   $\bar{P}$ = average value
   S = salvage value
   n = useful life, yr

Example 2-7. Consider a unit of equipment costing $25,000 with an estimated salvage value of $5,000 after 5 yr. Using Eq. (2-2), we get

$$\bar{P} = \frac{25,000(5 + 1) + 5,000(5 - 1)}{2 \times 5}$$

$$= \frac{150,000 + 20,000}{10}$$

$$= \$17,000$$

Because insurance and taxes are usually paid on the depreciated value of equipment, it is proper to use the average value in determining the average annual cost of insurance and taxes. See Table 2-4.

It is common practice to combine the costs of insurance, interest, taxes, and storage and to estimate them as a fixed percentage of the average value of the equipment. The present national average rate is about 14 percent, which includes interest at 9 percent, insurance, taxes, and storage at 5 percent of the average value per year.

Operating Costs

Construction equipment which is driven by internal combustion engines requires fuel and lubricating oil, which should be considered as an operating cost. Whereas the amounts consumed and the unit cost of each will vary with the type and size of equipment, the conditions under which it is operated, and the location, it is possible to estimate the cost reasonably accurately for a given condition.

An estimator should be reasonably familiar with the conditions under which a unit of equipment will be operated. Whereas a tractor may be equipped with a

TABLE 2-4
Average value of equipment with no
Salvage value after useful life

Estimated life, yr	Average value as percentage of original cost
2	75.00
3	66.67
4	62.50
5	60.00
6	58.33
7	57.14
8	56.25
9	55.55
10	55.00
11	54.54
12	54.17

200-horsepower (hp) engine, this tractor will not demand the full power of the engine at all times, possibly only when it is used to load a scraper or to negotiate a steep hill. Also, equipment is seldom, if ever, used 60 min/h. Thus, the fuel consumed should be based on the actual operating conditions. Perhaps the average demand on an engine might be 50 percent of its maximum power for an average of 45 min/h.

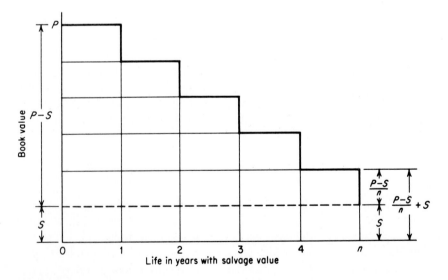

FIGURE 2-2
Value of equipment with salvage value by year.

Fuel Consumed

When operating under standard conditions, namely at a barometric pressure of 29.9 in of mercury and at a temperature of 60°F, a gasoline engine will consume approximately 0.06 gallon of fuel for each actual horsepower-hour developed [0.06 gal/(hp·h)]. A diesel engine will consume approximately 0.04 gal of fuel for each actual horsepower-hour developed.

Consider a power shovel with a diesel engine rated at 160 hp. During a cycle of 20 seconds (s), the engine may be operated at full power while filling the dipper in tough ground, requiring 5 s. During the balance of the cycle, the engine may be operated at not more than 50 percent of its rated power. Also, the shovel may not operate more than 45 min/h on average. For this condition the approximate amount of fuel consumed during 1 h is determined as follows:

Rated power, 160 hp
Engine factor:
    Filling the dipper, $\frac{5}{20}$ = 0.250
    Rest of cycle, $\frac{15}{20} \times 0.5$ = 0.375
    Total factor   = 0.625
Time factor, $\frac{45}{60}$ = 0.75
Operating factor, 0.625×0.75 = 0.47
Fuel consumed per h, $0.47 \times 160 \times 0.04 = 3.1$ gal

For other operating conditions the quantity of fuel consumed can be estimated in a similar manner.

Lubricating Oil Consumed

The quantity of lubricating oil consumed by an engine will vary with the size of the engine, the capacity of the crankcase, the condition of the pistons, and the number of hours between oil changes. It is common practice to change oil every 100 to 200 h, unless extreme dust makes more frequent changes desirable. The quantity of oil consumed by an engine during a change cycle includes the amount added at the time of change plus the makeup oil added between changes.

Equation (2-3) may be used to estimate the quantity of oil consumed:

$$Q = \frac{\text{hp} \times 0.6 \times 0.006 \text{ lb/(hp·h)}}{7.4 \text{ lb/gal}} + \frac{c}{t} \qquad (2\text{-}3)$$

where Q = quantity consumed, gal/h
    hp = rated horsepower of engine
    c = capacity of crankcase, gal
    t = hours between oil changes

Equation (2-3) is based on an operating factor of 0.60, or 60 percent. It assumes that the quantity of oil consumed between oil changes will be 0.006 gal per rated horsepower-hour. Using this equation for a 100-hp engine with a crankcase capacity of 4 gal requiring a change every 100 h, we find that the quantity consumed per hour is

$$Q = \frac{100 \times 0.6 \times 0.006}{7.4} + \frac{4}{100} = 0.089 \text{ gal}$$

Lubricants other than crankcase oil are required for motor-driven equipment. Although the costs of such lubricants will vary, an average cost equal to 50 percent of the cost of the crankcase oil is satisfactory.

EXAMPLES ILLUSTRATING THE COST OF OWNING AND OPERATING EQUIPMENT.

Example 2-8. Determine the probable cost per hour of owning and operating a 160-hp diesel-engine-powered crawler-type tractor. The following information will apply:

Engine, 160 hp
Crankcase capacity, 6 gal
Hours between oil changes, 100 h
Operating factor, 0.60
Fuel consumed per h, $0.6 \times 0.04 \times 160 = 3.9$ gal
Lubricating oil consumed,

$$\frac{160 \times 0.6 \times 0.006}{7.4} + \frac{6}{100} = 0.138 \text{ gal/h}$$

Shipping weight, 31,000 lb
Useful life, 6 yr
Hours used per year, 1,800 h

Cost to owner:

Price f.o.b. factory	=	$145,860
Freight, 31,000 × $2.40 per cwt†	=	744
Unloading and assembling	=	358
Taxes, 6% × $145,860	=	8,752
Total cost to owner	=	$155,714
Average investment, 0.5833 × $155,714 =		90,828

Annual cost:

Depreciation, $155,714 ÷ 6 yr	=	$25,952
Maintenance and repairs, 100% of depreciation	=	25,952
Investment, 0.14 × $90,828	=	12,716
Total annual fixed costs	=	$64,620

Hourly cost:

Fixed cost, $64,620 ÷ 1800 h	=	$35.90
Fuel, 3.9 gal @ $0.95	=	3.71
Lubricating oil, 0.138 gal @ $1.30	=	0.18
Other lubricants, 0.5 × $0.18	=	0.09
Total cost per h, excluding labor	=	$39.88

---

† One hundredweight (abbreviated cwt) = 100 lb.

Example 2-9. Determine the probable cost per hour for owning and operating a 25-yd³ heaped-capacity bottom-dump wagon with six rubber tires. Because the tires will have a different life from the wagon, they are treated separately. The following information will apply:

Engine, 250 hp, diesel
Crankcase capacity, 14 gal
Hours between oil changes, 80 h
Operating factor, 0.60
Fuel consumed per h, $0.6 \times 0.04 \times 250 = 6.0$ gal
Lubricating oil consumed,

$$\frac{250 \times 0.60 \times 0.006}{7.4} + \frac{14}{80} = 0.30 \text{ gal/h}$$

Useful life, 5 yr
Hours used per year, 2,000 h
Maintenance and repairs, 50% of depreciation
Life of tires, 5,000 h
Repairs to tires, 15% of depreciation of tires

Cost to owner:

Cost delivered to owner	= $82,700
Less cost of tires	= 15,620
Net cost less tires	= $67,080
Average investment, 0.6 × $82,700	= 49,620

Annual cost:

Depreciation, $67,080 ÷ 5 yr	= $13,416
Maintenance and repairs, 0.5 × $13,416 =	6,708
Investment, 0.14 × $49,620	= 6,947
Total annual fixed cost	= $27,071

Hourly cost:

Fixed cost, $27,071 ÷ 2,000	= $13.54
Tire depreciation, $15,620 ÷ 5,000	= 3.12
Tire repairs, 0.15 × $3.12	= 0.47
Fuel, 6.0 gal @ $0.40	= 2.40
Lubricating oil, 0.30 gal @ $1.20	= 0.36
Other lubricants, 0.5 × $0.36	= 0.18
Total cost per h, excluding labor	= $20.07

As noted, the costs calculated in the previous examples do not include allowances for the salvage value of the equipment at the end of its useful life. Any piece of construction equipment will have some value at trade-in time or some resale value at the conclusion of its use. However, many owners prefer to depreciate their equipment to zero and to use the residual value, or salvage, as a trade-in for acquisition of replacement equipment.

The hourly cost of owning and operating construction equipment, as illustrated in previous examples, will vary with the conditions under which the equipment is operated. The job planner should analyze each job to determine the probable conditions that will affect the cost and should appropriately adjust the costs.

If a crawler tractor is used on rock surfaces, the life of the undercarriage will be significantly reduced. Undercarriage costs can represent a major portion of the operating costs of track-type machines. Undercarriage costs can be determined similarly to tire costs, based on the cost of undercarriage replacement and its useful life. The costs should be based on local costs of parts and labor.

If trucks are operated over smooth and level haul roads, the cost of repairs will be significantly lower than when the same trucks are operated over rough haul roads. Therefore, the estimator must know the job site conditions before appropriate costs can be assigned to construction equipment.

PROBLEMS

2-1. The original cost of a small backhoe is $64,820. The estimated useful life is assumed to be 5 yr, with no salvage value. Determine its book value at the end of 2 yr determining depreciation by (*a*) straight-line, (*b*) declining-balance, and (*c*) sum-of-the-year's-digits methods.

2-2. If the backhoe of Prob. 2-1 is assumed to have a salvage value of $6,500 at the end of 5 yr, find the book value at the end of 4 yr, using the straight-line, declining-balance, and sum-of-the-year's-digits methods.

2-3. A $1\frac{1}{2}$-yd backhoe, whose total cost is $109,750, is assumed to have a useful life of 12,000 h. Using the method suggested by the Power Crane and Shovel Association, determine the probable cost per year and per hour for maintenance, repairs, and supplies.

2-4. What is the average annual investment cost for a tractor whose original cost is $142,500 and whose useful life is estimated to be 5 yr, with no salvage value? What is the average annual investment cost of this tractor if it is assumed to have a salvage value of $15,200 at the end of the 5 yr? Use an investment cost equal to 14 percent of the average value of the tractor.

2-5. Determine the probable cost per hour for owning and operating a crawler tractor for the given conditions.

> Engine, 325-flywheel-horsepower (fwhp) diesel
> Shipping weight, 69,900 lb
> Freight rate, $2.40 per cwt
> Capacity of crankcase, 9 gal
> Estimated life, 6 yr
> Estimated operating time, 2,000 h/yr
> Time between oil changes, 120 h
> Assume a 50% operating factor and a 45-min hour
> Cost, f.o.b. factory, $289,000
> Probable salvage value, $32,000
> Investment cost, 14% of average value per yr
> Cost of diesel fuel, $0.95 per gal
> Cost of lubricating oil, $3.85 per gal
> Assume that repairs cost 75% of depreciation

2-6. Determine the probable cost per hour for owning and operating a wheel-type tractor and scraper for the given conditions.

> Engine, 330-fwhp diesel
> Shipping weight, 64,000 lb
> Freight rate, $2.10 per cwt
> Capacity of crankcase, 9 gal
> Time between oil changes, 100 h
> Estimated life, 5 yr
> Operating time, 2,000 h/yr
> Assume an operating factor of 55% and a 45-min hour
> Cost, f.o.b. factory, $275,000
> Cost of tires, $13,600
> Life of tires, 5,000 h
> Cost of repairing tires, 15% of tire cost per h
> Probable salvage value, $41,250
> Investment cost, 14% of average value per yr
> Cost of diesel fuel, $0.95 per gal
> Cost of lubricating oil, $3.85 per gal
> Assume that repairs cost 80% of depreciation

2-7. Determine the probable cost per hour for owning and operating the tractor-scraper of Prob. 2-6 when the operating time is 1,200 h/yr with a useful life of 6 yr. The other conditions are the same as those given in Prob. 2-6.

2-8. Determine the probable cost per hour for owning and operating the tractor of Prob. 2-5 when the unit will be operated 1,200, 1,600, 2,000, and 2,400 h/yr. All other conditions are the same as those given in Prob. 2-5.

REFERENCES

1. *Rental Rate Blue Book*, Dataquest Incorporated, Equipment Guide-Book Company, Palo Alto, Calif.
2. "Operating Cost Guide," *Technical Bulletin* 2, Power Crane and Shovel Association, Milwaukee, Wis.

CHAPTER

3

HANDLING AND TRANSPORTING MATERIAL

In many instances, construction materials are delivered by the seller or producer directly to the project in trucks. However, in other instances the materials must be obtained by the contractor at the storage yard of the supplier.

Some projects require the use of aggregates, sand and gravel, or crushed stone, which are produced from natural deposits or quarries and hauled to the project in trucks or tractor-pulled wagons. The handling and hauling may be done by a contractor, using his or her laborers and equipment, or it may be accomplished through a subcontractor. Regardless of the method used, it will involve a cost which must be included in the estimate for a project.

When estimating the time required by a truck for a round trip, the estimator should divide the round-trip time into four elements:

1. Loading
2. Hauling, loaded
3. Unloading
4. Returning, empty

44

The time required for each element should be estimated. If elements 2 and 4 require the same time, they may be combined. Since the time required for hauling and returning will depend on the distance and the effective speed, it is necessary to determine the probable speed at which a vehicle can travel along the given haul road for the conditions that will exist. Speeds are dependent on the vehicle, traffic congestion, condition of the road, and other factors. An appropriate operating factor should be used in determining production rates. For example, if a truck will operate only 45 min/h, this time should be used in determining the number of round trips the truck will make in 1 h.

While the discussions and examples given in this chapter do not include all types of handling and transporting, they should serve as guides to illustrate methods which may be applied to any job.

Hauling Lumber to the Job

Lumber is usually loaded by laborers directly onto flatbed trucks, hauled to the job, and stacked according to size.

A laborer should be able to handle lumber at a rate of 2,000 to 4,000 feet board measure per hour (fbm/h), using the lower rate for small pieces and the higher rate for large pieces. A fair average rate should be about 3,000 fbm/h.

Trucks of the type generally used will haul 2 to 6 tons, corresponding to 1,000 to 3,000 fbm per load. The average speed will vary with the distance, type of road, traffic congestion, and weather.

> **Example 3-1.** Estimate the cost of transporting 40,000 fbm of lumber from a lumberyard to a job site. The lumber will be transported by trucks which can carry 2,000 fbm per load. The job site is 2 miles (mi) from the lumberyard.
>
> An examination of the haul road indicates an average speed of 20 mi/h for the trucks, including necessary delays to check oil, gasoline, water, etc. Assume that a worker will handle 3,000 fbm/h of lumber.
>
> Based on using two workers, a truck driver and a laborer, and one truck, the length of the job can be determined as follows:

$$
\begin{aligned}
\text{Rate of loading a truck, } 2 \times 3,000 &= 6,000 \text{ fbm/h} \\
\text{Time to load a truck, } 2,000 \div 6,000 &= 0.33 \text{ h} \\
\text{Time to unload a truck, } 2,000 \div 6,000 &= 0.33 \text{ h} \\
\text{Travel time, round trip, } 4 \text{ mi} \div 20 \text{ mi/h} &= \underline{0.20 \text{ h}} \\
\text{Total time per load} &= 0.86 \text{ h} \\
\text{No. trips per h, } 1 \div 0.86 &= 1.16 \\
\text{Quantity hauled, } 1.16 \times 2,000 &= 2,320 \text{ fbm/h} \\
\text{Total time for the job, } 40,000 \div 2,320 &= 17.2 \text{ h}
\end{aligned}
$$

> An alternative method of determining the length of the job is as follows:

$$
\begin{aligned}
\text{No. truckloads required, } 40,000 \div 2,000 &= 20 \\
\text{Round-trip time per load, } &0.86 \text{ h} \\
\text{Total time for the job, } 20 \times 0.86 &= 17.2 \text{ h}
\end{aligned}
$$

The cost will be

Truck, 17.2 h @ $10.58	= $181.98
Truck driver, 17.2 h @ $8.10	= 139.32
Laborer, 17.2 h @ $7.50	= 129.00
Total cost	= $450.30

Cost per 1,000 fbm, $450.30 ÷ 40 = $ 11.26

Based on the use of two laborers who remain at the yard, two laborers who unload the trucks at the job, and a driver for each truck, with the drivers to assist in loading and unloading the trucks, the costs can be determined as follows:

Rate of loading truck, 3 × 3,000 = 9,000 fbm/h	
Time to load a truck, 2,000 ÷ 9,000	= 0.222 h
Time to unload a truck, 2,000 ÷ 9,000	= 0.222 h
Travel time, round trip, 4 mi ÷ 20 mi/h	= 0.200 h
Total time per load	= 0.644 h

No. trips per h per truck, 1 ÷ 0.644 = 1.55

$$\text{No. trucks required,} \frac{\text{Round-trip time}}{\text{Time to load}} = \frac{0.644}{0.222} = 2.9$$

An alternate method of determining the number of trucks required is as follows:

No. trips per h per truck	= 1.55
Quantity of lumber hauled per h per truck, 1.55 trips × 2,000 fbm = 3,100 fbm	
No. trucks required, 9,000 ÷ 3,100	= 2.9
Determine the cost, using 3 trucks	

Total time for the job, 40,000 fbm ÷ 9,000 fbm/h = 4.44 h

The cost will be

Trucks, 3 × 4.44 = 13.32 h @ $10.58	= $140.93
Truck drivers, 13.32 h @ $8.10	= 107.89
Laborers, 4 × 4.44 = 17.76 h @ $7.50	= 133.20
Total cost	= $382.02

Determine the cost, using 2 trucks. Since there are not enough trucks to keep the laborers busy, the rate of hauling will determine the length of the job.

Quantity hauled per truck, 1.55 × 2,000	= 3,100 fbm/h
Quantity hauled by two trucks	= 6,200 fbm/h
Total time required for job, 40,000 ÷ 6,200 = 6.45 h	

The cost will be

Trucks, 2 × 6.45 = 12.9 h @ $10.58	= $136.48
Truck drivers, 12.9 h @ $8.10	= 104.49
Laborers, 4 × 6.45 = 25.8 h @ $7.50	= 193.50
Total cost	= $434.47

Note that the total costs are based on labor and truck costs for the actual time at work, with no allowance for laborers and trucks while they are waiting their turn to start working. If there will be costs during such nonproductive times, the total cost of the job will be higher than the values calculated.

Transporting Sand and Gravel

Sand and gravel are strip-mined by companies and stockpiled for use on construction projects. Sand is generally excavated from riverbeds by draglines, loaded into trucks, and transported to a central location for later distribution to prospective buyers. Similarly, gravel is strip-mined from a rock quarry, crushed in a rock-crushing machine, screened, and transported by trucks to a central gravel yard. Sand and gravel can be handled by several types of equipment, such as clamshells, front-end loaders, or portable conveyers. Small quantities may be handled by common laborers.

Portable belt conveyers are used frequently for handling and transporting material such as earth, sand, gravel, crushed stone, concrete, etc. Because of the continuous flow of material at relatively high speeds, belt conveyers have high capacities for handling material.

The amount of material that can be handled by a conveyer depends on the width and speed of the belt and on the angle of repose for the material. Portable belt conveyers are available in lengths of 33 to 60 ft, with belt widths of 18 to 30 in. The maximum speeds of conveyer belts range from 250 to 450 ft/min. A 300-ft/min belt speed is representative for many job sites. Table 3-1 gives the areas of cross section of materials with various angles of repose.

> **Example 3-2.** A portable belt conveyer is used to load sand from a stockpile into trucks. The conveyer has a 24-in-wide belt which has a travel speed of 300 ft/min. The conveyer will load 12-yd^3 dump trucks that will haul the sand 4 mi at an average travel speed of 30 mi/h. Assume a dump time of 2 min and an angle of repose of the sand of 20°. Determine the number of trucks required to balance the production rate of the belt conveyer.
>
> Production rate of conveyer, 300 ft/min × 0.331 ft^2 = 99.3 ft^3/min
> The hourly production rate will be
> (99.3 ft^3/min ÷ 27 ft^3/yd^3) × 60 min/h = 220.7 yd^3/h

TABLE 3-1
Areas of cross section of materials for loaded belts, ft^2

Width of belt, in	Angle of repose		
	10°	20°	30°
18	0.134	0.174	0.214
24	0.257	0.331	0.410
30	0.421	0.541	0.668

The cycle time for a truck is

Time to load, 12 yd^3 ÷ 220.7 yd^3/h = 0.05 h
Time to haul, 4 mi ÷ 30 mi/h   = 0.13 h
Time to dump, 2 min ÷ 60 min/h   = 0.03 h
Time to return, 4 mi ÷ 30 mi/h   = 0.13 h
   Round-trip time per truck   = 0.34 h

Quantity hauled per hour, 12 yd^3 ÷ 0.34 h = 35.3 yd^3/h
Number of trucks required, 220.7 yd^3/h ÷ 35.3 yd^3/h = 6.2
Therefore 6 or 7 trucks are required

The following example illustrates a method of determining the cost of unloading and hauling the material to a job.

Example 3-3. Estimate the probable cost of loading 160 yd^3 of gravel from a stockpile into trucks and hauling it 3 mi to a project. Use a $\frac{1}{2}$-yd^3 truck-mounted clamshell to load 6-yd^3 dump trucks. The trucks can maintain an average speed of 30 mi/h hauling and returning to the cars. The truck time at the dump will average 8 min, including the time required to check, service, and refuel a truck.

The clamshell should handle an average of 32 yd^3/h,† allowing time for moving the clamshell and for other minor delays. The estimated cost can be determined as follows:

Time to load a truck, 6 yd^3 ÷ 32 yd^3/h = 0.188 h
Travel time, round trip, 6 mi ÷ 30 mi/h = 0.200 h
Time at dump, 8 min ÷ 60 min/h   = 0.133 h
   Total round-trip time   = 0.521 h
No. loads per h per truck, 1.00 ÷ 0.521 = 1.92
Volume hauled per truck per h, 6 × 1.92 = 11.52 yd^3
No. trucks required, 32 yd^3 ÷ 11.52 = 2.78

If 3 trucks are used, the time required to complete the job will be 160 yd^3 ÷ 32 yd^3/h = 5 h.

The cost will be

Moving clamshell to and from job   = $ 420.00
Cost of clamshell, 5 h @ $57.50   =   287.50
Trucks, 3 × 5 = 15 h @ $16.80   =   252.00
Clamshell operator, 5 h @ $17.25   =   86.25
Clamshell oiler, 5 h @ $11.25   =   56.25
Truck drivers, 3 × 5 = 15 h @ $8.10 =   121.50
Laborers, 2 × 5 = 10 h @ $7.50   =   75.00
   Total cost   = $1,298.50

---

† See Table 4-12 for the output of a clamshell.

If 2 trucks are used, the length of the job will be determined by the rate at which the trucks haul the gravel. As previously determined, a truck should haul 11.58 yd³/h. Two trucks should haul $2 \times 11.58 = 23.16$ yd³/h.

The length of the job will be $160 \div 23.16 = 6.9$ h.
The cost will be

Moving clamshell to and from job	=	$ 420.00
Cost of clamshell, 6.9 h @ $57.50	=	396.75
Trucks, $2 \times 6.9 = 13.8$ h @ $16.80	=	231.84
Clamshell operator, 6.9 h @ $17.25	=	119.03
Clamshell oiler, 6.9 h @ $11.25	=	77.63
Truck drivers, $2 \times 6.9 = 13.8$ h @ $8.10	=	111.78
Laborers, $2 \times 6.9 = 13.8$ h @ $7.50	=	103.50
Total cost	=	$1,460.53

Thus it is cheaper to use 3 trucks.

If a foreman is used to supervise the job, her or his salary should be added to the cost of the project.

The costs determined in previous examples are the minimum amounts, based on the assumption that the cost durations can be limited to 5 or 6.9 h. It may be impossible or impractical to limit the costs to these respective durations. If such is the case, the appropriate durations should be used.

Handling and Transporting Bricks

Bricks are generally loaded at the brick supplier by forklifts or small cranes that are mounted on the bed of flatbed trucks and hauled to a project. The capacity of the trucks is normally 2,000 to 3,000 bricks per load. Popular sizes of building bricks weigh about 4 lb each.

Upon arrival at the project, the bricks are unloaded by a small crane that is mounted on the truck, or by a forklift, onto small four-wheel tractors that transport the bricks around the perimeter of the structure where the bricks are to be laid.

Workers, using brick tongs, can carry 6 to 10 bricks per load to the brick mason for laying. A worker should be able to pick up a load, walk to the location of the brick mason and deposit the load, and return for another load in $\frac{1}{2}$ to 1 min per trip. If it is assumed that the average time for a trip is $\frac{3}{4}$ min and the work carries 8 bricks per trip, in 1 h he or she will handle 640 bricks. The actual number of bricks that a worker can handle will depend on the job site conditions at a particular project. Table 3-2 provides rates for various job conditions.

Example 3-4. The cost for 60,000 bricks from a supplier includes delivering, unloading, and stacking the bricks at a central location at the job site. Estimate the cost of transporting the bricks from the central location where they are stacked by the supplier to the structure where they will be installed by a brick mason. The bricks will be hauled by a small four-wheel tractor at a rate of 1800 bricks per hour.

TABLE 3-2
Rates of handling bricks

Bricks carried per trip	Trip time, min	Bricks handled per hour	Hours per 1,000 bricks
6	0.5	720	1.39
8	0.5	960	1.04
10	0.5	1,200	0.83
6	0.75	480	2.08
8	0.75	640	1.56
10	0.75	800	1.25
6	1.00	360	2.78
8	1.00	480	2.08
10	1.00	600	1.67

Laborers will be stationed around the perimeter of the structure to carry the bricks to the brick masons. Assume each laborer will carry 8 bricks per load and will average $\frac{3}{4}$ min per trip each way.

No. bricks handled per labor-hour will be $8 \times 60 \div \frac{3}{4} = 640$
Time to haul bricks by tractor $= 60,000 \div 1,800$ per h $= 33.33$ h
Time to haul bricks by laborers $= 60,000 \div 640$ per h $= 93.75$ h
The costs will be
Tractor, 33.33 h @ \$19.00 $= \$ 633.27$
Operator, 33.33 h @ \$9.50 $= 316.64$
Laborers, 93.75 h @ \$7.50 $= \underline{703.13}$
Total cost $ = \$1,653.04$

The cost for handling bricks is $\$1,653.04 \div 60,000 = \$0.028$ per brick.

Example 3-5. Estimate the cost of hauling 60,000 bricks from the brickyard of a supplier to a job 4 mi distant. Each truck will haul 3,000 bricks per load and can travel at an average speed of 20 mi/h loaded and at 30 mi/h empty, allowing for lost time. Three laborers will be stationed at the yard to load the trucks, and another three will be at the job to unload the trucks. The truck drivers will assist in loading and unloading the trucks. Each laborer will carry 8 bricks at the yard and at the job. Assume that a laborer will average $\frac{3}{4}$ min per trip at the yard and at the job.

No. bricks handled per labor-hour, $8 \times 60 \div \frac{3}{4} = 640$
Rate of loading a truck, $4 \times 640 = 2,560$ per h
Time to load a truck, $3,000 \div 2,560 = 1.170$ h
Time to unload a truck, same $ = 1.170$ h
Time to drive from yard to job, $4 \div 20 = 0.200$ h
Time to drive from job to yard, $4 \div 30 = \underline{0.133}$ h
Time for round trip $ = 2.673$ h
No. bricks hauled per h per truck, $3,000 \div 2.673 = 1,122$
No. trucks required to haul 2,560 bricks per h, $2,560 \div 1,122 = 2.28$

It will be necessary to use either 2 or 3 trucks. If 2 trucks are used, the cost will be determined as follows.

Rate of hauling bricks, 2 × 1,122 = 2,244 per h
Time required to finish the job, 60,000 ÷ 2,244 = 26.7 h
The cost will be
    Trucks, 2 × 26.7 = 53.4 h @ $22.80   = $1,217.52
    Truck drivers, 53.4 h @ $8.10   =   432.54
    Laborers, 6 × 26.7 = 160.2 h @ $7.50 = 1,201.50
        Total cost   = $2,851.56

If 3 trucks are used, the rate of loading the trucks, as previously determined, will be 2,560 bricks per hour.

Time required to finish the job, 60,000 ÷ 2,560 = 23.5 h
The cost will be
    Trucks, 3 × 23.5 = 70.5 h @ $22.80   = $1,607.40
    Truck drivers, 70.5 h @ $8.10   =   571.05
    Laborers, 6 × 23.5 = 141.0 h @ $7.50 = 1,057.50
        Total cost   = $3,235.95

It will be more economical to use 2 trucks.

Unloading and Hauling Cast-Iron Pipe

This project involves loading cast-iron pipe onto trucks and hauling it to a project, where it will be laid on the ground along city streets. The average haul distance will be 5 mi.

The pipe will be 12 in. in diameter and 18 ft long and weigh 1,140 lb per joint of 18 ft. A 4-ton truck-mounted crane will be used to load the pipe onto 8-ton trucks, which will haul 14 joints per load. The pipe will be unloaded from the trucks by an 80-hp crawler tractor, with a side boom, which will move along with the trucks.

The trucks can average 25 mi/h loaded and 30 mi/h empty.
The total labor crew will consist of the following persons:

At the loading site
  1 crane operator
  1 crane oiler
  2 laborers with the crane
  2 laborers on the truck
At the unloading site
  1 tractor operator
  2 laborers helping unload pipe

Determine the number of trucks required and the direct cost per foot for transporting the pipe.

The time required by a truck for a round trip will be

Loading truck, 14 joints × 3 min/joint = 42 min ÷ 60   = 0.70 h
Hauling to job, 5 mi ÷ 25 mi/h   = 0.20 h
Unloading truck, 14 joints × 3 min/joint = 42 min ÷ 60 = 0.70 h
Returning to car, 5 mi ÷ 30 mi/h   = 0.17 h
                                                         ─────────
  Total time   = 1.77 h

No. trips per h, 1 ÷ 1.77 = 0.565
No. joints hauled per h per truck, 0.565 × 14 = 7.9
No. joints unloaded per h, 60 ÷ 3 = 20
No. trucks required, 20 ÷ 7.9 = 2.54
Use 3 trucks

Assume that the crane and truck will operate 50 min/h. The average number of joints hauled per hour will be 20 × 50 ÷ 60 = 16.67.

The cost per h will be

Crane, 1 h @ $62.75   = $ 62.75
Trucks, 3 h @ $22.50   =   67.50
Tractor with boom, 1 h @ $18.20   =   18.20
Crane operator, 1 h @ $17.25   =   17.25
Crane oiler, 1 h @ $11.25   =   11.25
Tractor operator, 1 h @ $15.86   =   15.86
Truck drivers, 3 h @ $8.10   =   24.30
Laborers, 6 h @ $7.50   =   45.00
Foreman, 1 h @ $17.30   =   17.30
                                             ────────
  Total cost   = $279.41
Cost per lin ft, $279.41 ÷ (18 × 16.67) =   0.931

Any cost of moving equipment to the job and back to the storage yard should be prorated to the total length of pipe handled, and added to the unit cost determined above, to obtain the total cost per unit length.

PROBLEMS

3-1. Estimate the total direct cost and the cost per 1,000 fbm of lumber for transporting 50,000 fbm of lumber from a lumberyard to a job which is 3 mi from the yard. A 5-ton stake-body truck will haul 3,000 fbm per load. The trucks can average 25 mi/h loaded and 40 mi/h empty.

One laborer plus the truck driver will load lumber onto the truck at the lumber yard, and another laborer plus the truck driver will unload the truck and stack the lumber at the job. Each worker will handle 1,200 fbm/h when working.

Assume that the laborers and the trucks operate 50 min/h. Determine the most economical number of trucks to use.

Will placing an additional laborer at the yard and at the job reduce the cost of handling and hauling the lumber?

Labor and truck costs per hour will be

Trucks	$16.40
Truck drivers	9.25
Laborers	7.30

3-2. Using the information given in Prob. 3-1, is it less expensive to keep two laborers at the yard and two other laborers at the job site or to use only two laborers, who will ride with the truck driver to load and unload a truck?

3-3. The owner of a sand and gravel pit is considering the purchase of a fleet of trucks to deliver aggregate to customers. Two sizes are being considered, namely, 10- and 15-yd^3 diesel-engine-powered dump trucks. The haul distances will vary from 6 to 20 mi, with an average distance of about 12 mi.

It is estimated that the 10-yd^3 trucks can travel at an average speed of 40 mi/h loaded and at 50 mi/h empty, while the 15-yd^3 trucks can travel at an average speed of 35 mi/h loaded and at 45 mi/h empty. The trucks will be loaded from stockpiles of aggregate, using a 2-yd^3 clamshell with an angle of swing averaging 90°. See Table 4-12 for the output for the clamshell. The average truck time at the dump will be 4 min for the 10-yd^3 truck and 5 min for the 15-yd^3 truck. Assume that the trucks and the clamshell will operate 45 min/h.

Determine which size truck is more economical.

The cost of a 10-yd^3 truck is $27.10 per hour, and that of a 15-yd^3 truck is $43.55 per hour. A truck driver will be paid $9.50 per hour. Because the cost of the clamshell will be the same for each size truck, this cost need not be considered.

3-4. Is it more economical to use a 1- or a 2-yd^3 clamshell to load sand and gravel from stockpiles into 10-yd^3 diesel-engine-powered dump trucks? Assume an average angle of swing of 90° from the clamshell.

Use Table 4-12 to determine the production rate for each clamshell.

In addition to the time required by a clamshell to load a truck, there will be an average delay of 2 min waiting for another truck to move into position for loading. Assume a 50-min hour for both the clamshell and the trucks. The costs per hour will be

1-yd^3 clamshell	$38.85
2-yd^3 clamshell	72.15
Truck	29.50
Truck driver	9.50
Clamshell operator	18.20
Clamshell oiler	12.40

3-5. A total of 140,000 bricks is to be hauled from a brickyard by 5-ton gasoline-engine-powered stake-body trucks to a project 4 mi from the yard. A truck, which can haul 3,000 bricks per load, will average 30 mi/h loaded and 40 mi/h empty. Assume that the trucks operate 45 min/h.

Determine the total cost and the cost per 1,000 bricks for each of the stated conditions.

(a) Three laborers plus a truck driver will each load 450 bricks per hour onto a truck, and three other laborers plus a truck driver will unload the bricks from a truck at the job at the same rate.

(b) Three laborers plus a truck driver will each load 450 bricks per hour onto a truck; then the truck driver and the three laborers will ride on the truck to the job, where they and the truck driver will unload the bricks at the same rate as at the yard. The costs per hour will be

Truck, each	$15.75
Truck driver	9.50
Laborers, each	7.25

3-6. The operator of a gravel pit is invited to bid on furnishing 2,500 yd^3 of bank-run gravel for a job. The gravel is to be delivered to a job, which is 12 mi from the pit. The gravel will be loaded into trucks by a hydraulic excavator which will load at a rate of 80 yd^3/h. The trucks will haul 10 yd^3 per load and average 35 mi/h loaded and 45 mi/h empty. Assume that the hydraulic excavator and the trucks will operate 45 min/h.

The following costs will apply:

Royalty paid for gravel, $0.42 per yd^3
General overhead, $0.30 per yd^3
Profit, $0.55 per yd^3
Hydraulic excavator, $63.00 per h
Trucks, each, $26.00 per h
Excavator operator, $18.10 per h
Excavator oiler, $12.50 per h
Truck drivers, $9.10 per h
Foreman, $18.50 per h
Labor taxes, 15% of wages paid, including foreman

What price per cubic yard should be bid?

CHAPTER

4

EARTHWORK AND
EXCAVATION

Most projects constructed today involve excavation to some extent. The extent of excavation varies from a few cubic yards for footings and trenches for pipes to millions of cubic yards for large earth-filled dams. It is usually a relatively simple operation to determine the quantity of material to be excavated. It is much more difficult to estimate the rate at which it will be handled by earth-excavating equipment. A great many factors affect the rates of production. These factors may be divided into two groups, job and management.

Job Factors

Job factors involve the type of material, extent of water present, weather, freedom of workers and equipment to operate on the job, size of the job, length of haul for disposal, condition of the haul road, etc. It is difficult for the constructor to change job conditions.

Management Factors

Management factors involve organization for the job, maintaining good morale among the workers, selecting and using suitable equipment and methods, care in servicing equipment, maintaining production records, and others. These factors are under the control of the constructor.

Methods of Excavating

Methods of excavating vary from hand digging and shoveling for small jobs to that done by trenching machines, power shovels, draglines, clamshells, tractor-pulled scrapers, bulldozers, elevating loaders, drilling machines, and dredges. Some material, such as rock, is so hard that it is necessary to drill holes and loosen it with explosives prior to excavating it.

Hauling Materials

When it is necessary to move the excavated material some distance for disposal or to build an earth structure, transporting equipment such as wheelbarrows, trucks, and tractor-pulled wagons should be used. Many types and sizes are available to the constructor. The size of the hauling unit should be balanced with the output of the excavating equipment if possible.

Physical Properties of Earth

To estimate the cost of excavating and hauling earth intelligently, it is necessary to have a knowledge of the physical properties of earth.

In its undisturbed condition, prior to excavating, earth will weigh about 100 lb/ft^3, or about 2,700 lb/yd^3. The weight of solid rock is approximately 150 lb/ft^3, or about 4,000 lb/yd^3. However, the weight varies with the type of earth or rock, and if a more exact weight is required, it must be determined for the particular project.

When earth and rock are loosened during excavation, they assume a larger volume and a corresponding reduction in weight per unit of volume. This increase in volume is described as *swell* and is usually expressed as a percentage gain compared with the original volume. If earth is placed in a fill and compacted with modern equipment, it usually occupies a smaller volume than in its natural state in the cut or borrow pit. This decrease in volume is described as *shrinkage* and is expressed as a percentage of the original volume.

The amount of shrinkage depends on the type and moisture content of the soil and on the type and number of passes of the compaction equipment. Shrinkage can vary from 5 to 15 percent, depending on these factors. The amount of swell depends on the type of soil and the amount that the soil is loosened during excavation. Table 4-1 indicates the percentage swell for various soils. The swell and shrinkage values may be obtained from Eqs. (4-1) and (4-2):

By weight

$$S_w = \left(\frac{B}{L} - 1\right) \times 100 \qquad S_h = \left(1 - \frac{B}{C}\right) \times 100 \qquad \text{(4-1)}$$

where S_w = percentage of swell
S_h = percentage of shrinkage
B = unit weight of undisturbed soil
L = unit weight of loose soil
C = unit weight of compacted soil

TABLE 4-1
Swell for different classes of earth

Material	Swell, %
Sand or gravel	14–16
Loam	16–25
Ordinary earth	20–30
Dense clay	25–40
Solid rock	50–75

By volume

$$L = \left(1 + \frac{S_w}{100}\right) \times B \qquad C = \left(1 - \frac{S_h}{100}\right) \times B \qquad (4\text{-}2)$$

where S_w = percentage of swell
S_h = percentage of shrinkage
B = volume of undisturbed soil
L = volume of loose soil
C = volume of compacted soil

Excavating by Hand

Numerous types and sizes of excavating equipment are available. The selection of excavating equipment for a particular project depends on many factors: the type and quantity of material to be excavated, depth of excavation, amount of groundwater in the construction area, required haul distance, and the space available for the operation of the equipment.

Generally it is desirable to use excavating equipment instead of excavation by laborers; however, at some job sites the space is not sufficient for equipment to operate. For example, excavation for a motor/pump foundation for a unit in a refinery may be located in a confined space that prevents access by an excavator. An excavator may be able to access an excavation area, but may not be able to operate because of an overhead pipe rack that blocks the clearance necessary for the equipment to operate. Excavation may be required in an area where there are numerous underground electric or telephone cables. For each of the above situations, excavating by hand methods may be necessary.

The rate at which a laborer can excavate earth varies with the type of earth, extent of digging required, height to which it must be lifted, and climatic conditions. If loosening is necessary, a pick is most commonly used. Lifting is generally done with a round-pointed, long-handled shovel. It requires 150 to 200 shovels of earth to excavate 1 yd^3 in its natural state. Representative rates of excavating and hauling earth are given in Table 4-2.

TABLE 4-2
Rates of handling earth by hand

Operation	yd³/h	h/yd³
Excavating sandy loam	1–2	0.5–1.0
Shoveling loose earth into a truck	$\frac{1}{2}$–1	1.0–2.0
Loosening with a pick	$\frac{1}{4}$–$\frac{1}{2}$	2.0–4.0
Shoveling from trenches to 6 ft 0 in deep	$\frac{1}{2}$–1	1.0–2.0
Shoveling from pits to 6 ft 0 in deep	$\frac{1}{2}$–1	1.0–2.0
Backfilling	$1\frac{1}{2}$–$2\frac{1}{2}$	0.4–0.7
Spreading loose earth	4–7	0.15–0.25

For hours per cubic yard, the lower values should be used for sandy loam
and the higher values for heavy soils and clays.

If the excavated earth is to be hauled distances up to 100 ft, wheelbarrows are
frequently used. A wheelbarrow will hold about a 3 ft³ loose volume. A worker
should be able to haul the earth 100 ft and return in $2\frac{1}{2}$ min if the haul path is
reasonably firm and smooth. Filling the wheelbarrow with loose earth will require
about $2\frac{1}{2}$ min. Thus it will require about 5 min to load and haul 3 ft³ of earth up to
100 ft. This corresponds to about 1 h/yd³ bank measure.

Example 4-1. Excavation for a trench in a confined area of a refinery requires
excavation by laborers. Estimate the cost of excavating and backfilling the trench
which is 3 ft wide, 4 ft deep, and 150 ft long in sandy loam.

Volume of earth, $(3 \times 4 \times 150) \div 27 = 66.7$ yd³
The labor-hours will be
  Loosening earth,

$$66.7 \text{ yd}^3 \times 2 \text{ h/yd}^3 \quad = 133.4 \text{ h}$$

Shoveling earth from trench,

$$66.7 \text{ yd}^3 \times 1 \text{ h yd}^3 \quad = 66.7 \text{ h}$$

Shoveling earth back from trench,

$$66.7 \text{ yd}^3 \times 0.5 \text{ h/yd}^3 = 33.4 \text{ h}$$

Backfilling trench,

$$66.7 \text{ yd}^3 \times 0.5 \text{ h/yd}^3 = \underline{33.4 \text{ h}}$$
$$\text{Total labor-hours} = 266.9 \text{ h}$$

The cost will be
  Total cost, 266.9 h @ $7.50 per h   = $2,001.75
  Cost per yd³, $2,001.75 ÷ 66.7 yd³   =   30.01
  Cost per lin ft, $2.001.75 ÷ 150 lin ft =   13.35

Excavating with Trenching Machines

Even though it may be economical to excavate short sections of trenches with hand labor, a trenching machine is more economical for larger jobs. Once the machine is transported to the job and put into operation, the cost of excavating is considerably less than the cost by hand. For a given job, the savings in excavating costs resulting from the use of the machine as compared with hand excavating must be sufficient to offset the cost of transporting the machine to the job and back to storage after the job is completed. Otherwise, hand labor is more economical.

Trenching machines may be purchased or rented. Several types are available.

Self-Transporting Trenching Machines

For shallow trenches such as those required for grade beams, underground electric cables, telephone lines, or television cables, a trenching machine with a dozer blade attached to the front, as illustrated in Fig. 4-1, is frequently used. This type of equipment is highly maneuverable because of its rubber tires and small size. Equipment of this type can be used to dig narrow trenches 8 to 12 in wide to depths of 8 ft, as shown in Table 4-3.

FIGURE 4-1
Self-transporting trenching machine. (*The Charles Machine Works.*)

TABLE 4-3
Representative sizes of trenches for self-transporting trenching machines

Depth of trench, in	Maximum width of trench, in	
48	24	
60	20	
72	16	
84	14	
96	12	for soft soil
96	10	for firm soil
96	8	for hard soil

Source: The Charles Machines Works, Inc.

The rate of trenching will depend on the type of soil, width and depth of trench, and horsepower of the trencher. Equation (4-3) can be used to approximate the rate of trenching for this type of equipment:

$$S = \frac{C \times \text{hp}}{D \times W} \qquad (4\text{-}3)$$

where S = digging speed, ft/min
$\quad C$ = soil factor
$\quad D$ = depth of trench, in
$\quad W$ = width of trench, in
$\quad \text{hp}$ = flywheel horsepower of engine

The soil factor C can be approximated as shown in Table 4-4.

TABLE 4-4
Values of soil factor C for self-transporting trenchers

Type of soil	C
Sandstone	20
Hard clay	40
Firm clay	60
Soft clay	90

Source: The Charles Machines Works, Inc.

Example 4-2. Estimate the total cost and the cost per linear foot for excavating a trench for a 1,600-ft-long underground electric cable in hard clay. The depth of the trench will average 5 ft 6 in, and it will be 9 in wide. A rubber-tire self-transporting trenching machine with 65 hp will be used. There will be no obstructions to retard the progress of the machine.

From Eq. (4-3) the probable digging speed will be

$$S = \frac{40(65)}{66(9)} = 4.4 \text{ ft/min}$$

The time required for trenching will be

$$(1,600 \text{ ft} \div 4.4 \text{ ft/min}) \div 60 \text{ min/h} = 6.1 \text{ h}$$

The cost will be

Moving equipment to and from job	=	$350.00
Trenching machine, 6.1 h @ $32.50	=	198.25
Utility truck, 6.1 h @ $12.50	=	76.25
Machine operator, 6.1 h @ $16.50	=	100.65
Truck driver, 6.1 h @ $8.10	=	49.41
Laborer, 6.1 h @ $6.90	=	42.09
Foreman, 6.1 h @ $18.30	=	111.63
Total cost	=	$928.28
Cost per lin ft, $928.28 ÷ 1,600	=	0.58

Wheel-Type Trenching Machines

For shallow trenches such as those required for water mains and gas and oil pipe lines, a wheel-type machine (Fig. 4-2) is frequently used. The wheel rotates at the

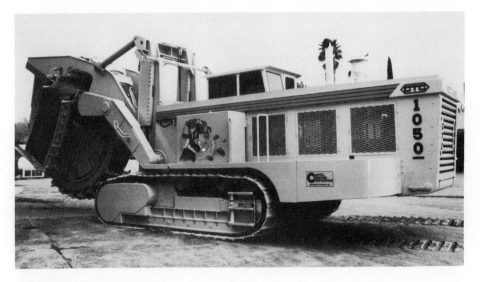

FIGURE 4-2
Wheel-type trenching machine. (*Caterpillar, Inc.*)

rear of the machine, which is mounted on crawler tracks. A combination of teeth and buckets attached to the wheel loosens and removes the earth from the trench as the machine advances. The earth is cast into a windrow along the trench. This machine may be used to excavate trenches whose depths do not exceed approximately 8 ft.

Ladder-Type Trenching Machines

For deep trenches (Fig. 4-3), such as those required for sewer pipes and other utilities, the ladder-type machine should be used. Machines with inclined or vertical booms are available. The boom is mounted at the rear of the machine. Cutter teeth and buckets are attached to endless chains which travel along the boom. As the machine advances, the earth is excavated and cast along the trench. The depth of cut is adjusted by raising or lowering the boom. By adding side cutters, the width of the trench may be increased considerably. This type of machine is available in a number of sizes, making it possible to excavate trenches 20 ft or more deep and 6 ft or more wide.

Table 4-5 gives information on various types of trenching machines. Use the higher speeds for soft earth such as sandy loam and the lower speeds for heavy tight earth and clay.

FIGURE 4-3
Ladder-type trenching machine.

TABLE 4-5
Data on trenching machines

Depth of trench, ft	Width of trench, in	Digging speed, ft
Wheel type:		
2–4	16, 18, 20	150–600
	24, 24, 26	90–300
	28, 30	60–180
4–6	16, 18, 20	40–120
	22, 24, 26	25–90
	28, 30	15–40
Ladder type:		
4–6	16, 20, 24	100–300
	22, 26, 30	75–200
	28, 32, 36	40–125
6–8	16, 20, 24	40–125
	22, 26, 30	30–60
	28, 32, 36	25–50
8–12	18, 24, 30	30–75
	30, 33, 36	15–40

Example 4-3. Estimate the total cost and the cost per linear foot for excavating a trench 3 ft 0 in wide, 6 ft 6 in average depth, and 2,940 ft long in ordinary earth. A ladder-type trenching machine will be used.

There will be no obstructions to retard the progress of the machine. Table 4-5 indicates a speed between 30 and 50 ft/h. Assume an average speed of 40 ft/h.

Time required to complete job, 2,940 ÷ 40 = 73.5 h.
The cost will be

Transporting machine to and from job	=	$ 1,260.00
Trenching machine, 73.5 h @ $86.25	=	6,339.38
Utility truck, 73.5 h @ $12.50	=	918.75
Machine operator, 73.5 h @ $16.50	=	1,212.75
Truck driver, 73.5 h @ $8.10	=	595.35
Laborers, 2 × 73.5 = 147 h @ $6.90	=	1,014.30
Foreman, 73.5 h @ $17.30	=	1,271.55
Total cost	=	$12,612.08
Cost per lin ft, $12,612.08 ÷ 2,940	=	4.29

Excavating with Power Shovels

Power shovels (Fig. 4-4) are excavating machines. They will handle all classes of earth without prior loosening; but in excavating solid rock it is necessary to loosen the rock first, usually by drilling holes and discharging explosives in them. The excavated material is loaded into trucks, tractor-pulled wagons, or cars, which haul

FIGURE 4-4
Power shovel on a typical job.

it to its final destination. For the shovel to maintain its maximum output, sufficient hauling units must be provided.

Output of Power Shovels

In estimating the output of a power shovel, it is necessary to know the class of earth to be excavated, the depth of dig, the ease with which hauling equipment can approach the shovel, the angle of swing from digging to emptying the dipper, and the size of the dipper. The size of a power shovel is designated by the struck capacity of the dipper, expressed in cubic yards, loose measure.

When hard earth is dug, the output will be less than for soft earth. If the face against which the shovel is digging is too shallow, it will not be possible for the shovel to fill the dipper in a single cut, which will reduce the output. If the face is too deep, the dipper will be filled before it reaches the top of the face, which will necessitate emptying the dipper and returning to a part face operation. As the angle of swing for the dipper from digging to dumping is increased, the time required for a cycle will be increased, which will reduce the output of the shovel.

The optimum depth of cut for a power shovel is that depth at which the dipper comes to the surface of the ground with a full dipper without overcrowding or undercrowding the dipper. The optimum depth varies with the class of soil and the size of the dipper. Values of optimum depths for various classes of soils and sizes of dippers are given in Table 4-6.

TABLE 4-6
Ideal outputs of power shovels per 60-min hour, bank measure

Class of material	Size of shovel, yd³								
	$\frac{3}{8}$	$\frac{1}{2}$	$\frac{3}{4}$	1	$1\frac{1}{4}$	$1\frac{1}{2}$	$1\frac{3}{4}$	2	$2\frac{1}{2}$
Moist loam or light	3.8	4.6	5.3	6.0	6.5	7.0	7.4	7.8	8.4
sandy clay*	85	115	165	205	250	285	320	355	405
Sand and gravel*	3.8	4.6	5.3	6.0	6.5	7.0	7.4	7.8	8.4
	80	110	155	200	230	270	300	330	390
Good common earth*	4.5	5.7	6.8	7.8	8.5	9.2	9.7	10.2	11.2
	70	95	135	175	210	240	270	300	350
Hard, tough clay*	6.0	7.0	8.0	9.0	9.8	10.7	11.5	12.2	13.3
	50	75	110	145	180	210	235	265	310
Well-blasted rock	40	60	95	125	155	180	205	230	275
Wet, sticky clay*	6.0	7.0	8.0	9.0	9.8	10.7	11.5	12.2	13.3
	25	40	70	95	120	145	165	185	230
Poorly blasted rock	15	25	50	75	95	115	140	160	195

Source: Power Crane and Shovel Association.
* The upper number indicates the optimum depth in ft. The lower number indicates the ideal output.

The outputs for power shovels given in Table 4-6 are based on operating a 60-min hour, at an angle of swing of 90°, and at optimum depth. Table 4-7 gives factors that correct outputs for other depths and angles of swing.

Example 4-4. This example illustrates a method of determining the probable output of a power shovel for the stated conditions.

> Size of shovel, 1 yd³
> Class of soil, good common earth
> Depth of cut, 9.5 ft
> Angle of swing, 120°
> Operating factor, 50-min h
> From Table 4-6 the ideal output is 175 yd³/h
> Optimum depth is 7.8 ft
> Percentage of optimum depth, (9.5/7.8) × 100 = 122%
> From Table 4-7 the conversion factor for depth and swing will be 0.86
> Probable output, 175 × 0.86 = 150.5 yd³/h
> Probable output corrected for a 50-min h will be 150.5 × $\frac{50}{60}$ = 125.5 yd³, bank measure

Cost of Owning and Operating a Power Shovel

The cost of owning and operating a power shovel, as with other construction equipment, includes such items as depreciation, interest, taxes, insurance, maintenance and repairs, fuel, lubrication, greasing, supplies, and labor. Some of these

TABLE 4-7
Conversion factors for depth of cut and angle of swing for a power shovel

Percentage of optimum depth	Angle of swing						
	45°	60°	75°	90°	120°	150°	180°
40	0.93	0.89	0.85	0.80	0.72	0.65	0.59
60	1.10	1.03	0.96	0.91	0.81	0.73	0.66
80	1.22	1.12	1.04	0.98	0.86	0.77	0.69
100	1.26	1.16	1.07	1.00	0.88	0.79	0.71
120	1.20	1.11	1.03	0.97	0.86	0.77	0.70
140	1.12	1.04	0.97	0.91	0.81	0.73	0.66
160	1.03	0.96	0.90	0.85	0.75	0.67	0.62

Source: Power Crane and Shovel Association.

items vary with use, while others do not. Methods of determining the costs of owning and operating power shovels are discussed in Chap. 2.

To estimate the total cost of excavating a given job, it is necessary to determine the cost of transporting the shovel to and from the job, the labor cost of setting up the shovel for operation at the job, the equipment and labor costs of excavating the material, and the labor cost of removing the shovel at the end of the job. Depending on the size of the shovel and the job conditions, it is generally necessary to use one or two helpers in addition to the shovel operator. Also a foreman usually supervises the excavating and hauling. A portion of her or his salary should be charged to excavation.

Hauling Excavated Materials

Trucks and tractor-pulled wagons or trailers are used to haul the materials excavated by power shovels. The capacity of a hauling unit may be expressed in tons or cubic yards. The latter capacity may be expressed as struck or heaped. The *struck capacity* is the volume which a unit will hold when it is filled even with, but not above, the sides. This volume depends on the length, width, and depth of the unit. The *heaped capacity* is the volume which a unit will hold when the earth is piled above the sides. While the struck capacity of a given unit is fixed, the heaped capacity will depend on the depth of the earth above the sides and on the area of the bed. Some manufacturers specify the heaped capacity based on a 1:1 slope of the earth above the sides. The actual capacity of a unit should be determined by measuring the volume of earth in several representative loads, then using the average of these values. Units are available with capacities varying from 2 to 30 yd^3 or more.

If it is desirable to express the volume in bank measure, a shrinkage factor must be applied to the loose volume. For ordinary earth, whose swell is 25 percent, the loose volume will be 1.25 times the bank volume. The shrinkage factor will be $1 \div 1.25 = 0.8$. Thus a truck hauling 10 yd^3 of loose measure will have a bank-measure load of $10 \times 0.8 = 8$ yd^3, or this may be expressed as $10 \div 1.25 = 8$ yd^3.

The size of the hauling units should be balanced against the dipper capacity of the shovel. For best results, considering output and economy, the capacity of a hauling unit should be 4 to 6 times the dipper capacity.

The volume which a truck can haul in a given time depends on the volume per load and the number of trips it can make in that time. The number of trips depends on the distance, speed, time at loading, time at the dump, and time required for servicing. Higher travel speeds are possible on good open highways than on streets with heavy traffic. For example, top speeds in excess of 50 mi/h may be possible on some paved highways, whereas speeds on crowded city streets may be not more than 10 to 15 mi/h. Since delays resulting from lost time at loading, dumping and servicing trucks may reduce the actual operating time to 45 or 50 min/h, an appropriate operating factor should be applied in determining the number of trips per unit of time.

Example 4-5. Estimate the probable total cost and the cost per cubic yard for excavating and hauling 58,640 yd^3, bank measure (bm), of good common earth for the stated conditions.

> Pit conditions, good
> Shovel, 1-yd^3 diesel-engine-operated crawler type
> Trucks, 6 yd^3, loose measure, gasoline engine
> Truck time at dump, 4 min
> Truck time waiting at shovel to move into loading position, 3 min average
> Average truck speed, 30 mi/h
> Average depth of cut, 8 ft
> Average angle of swing, 120°
> Operating factor for shovel and trucks, 45-min h
> Haul distance, one way, 4 mi
> Probable ideal output of shovel, 175 yd^3/h
> Percentage of optimum depth, $(8.0/7.8) \times 100 = 102.5\%$
> Conversion factor for depth of cut and angle of swing, 0.88
> Corrected output, $0.88 \times 175 = 154$ yd^3 for 60-min h†
> Output for 45-min h, $(45/60) \times 154 = 116$ yd^3, bm
> Volume of truck, $6 \div 1.25 = 4.8$ yd^3, bm
> Truck cycle time:
>   Loading truck, $4.8 \div 154$ $= 0.031$ h
>   Traveling, $8 \div 30$ $= 0.267$ h

† Note that the time required to load a truck is based on using the corrected output of the shovel for a 60-min hour. The reason for using this rate is that it will be the rate of production for the shovel when it is loading a truck.

At dump and waiting, $7 \div 60 = \underline{0.117\ h}$

Round-trip time $= 0.415\ h$

No. trips per h, $1 \div 0.415 = 2.41$

No. trips per 45-min h, $2.41 \times (45/60) = 1.8$

Volume hauled per truck per h, $1.8 \times 4.8 = 8.64\ yd^3$, bm

No. trucks required, $116 \div 8.64 = 13.4$

Use 14 trucks, possibly another for a standby

Time to complete the job, $58,640 \div 116 = 505\ h$

The cost will be

Transporting shovel to and from job	= \$ 1,850.00
Shovel, 505 h @ \$71.50	= 36,107.50
Trucks, $14 \times 505 = 7,070$ h @ \$18.20	= 128,674.00
Labor setting up and dismantling shovel:	
Shovel operator, 16 h @ \$17.25	= 276.00
Shovel oiler, 16 h @ \$11.25	= 180.00
Laborer, 16 h @ \$7.50	= 120.00
Foreman, 16 h @ \$18.50	= 296.00
Labor excavating and hauling earth:	
Shovel operator, 505 h @ \$17.25	= 8,711.25
Shovel oiler, 505 h @ \$11.25	= 5,681.25
Truck drivers, 7,070 h @ \$8.10	= 57,267.00
Laborer, 505 h @ \$7.50	= 3,787.50
Foreman, 505 h @ \$18.50	= 9,342.50
Total cost	= \$252,293.00
Cost per yd^3, \$252,293.00 $\div$ 58,640 =	4.30

Excavating with Hydraulic Excavators

The term *hydraulic excavator* applies to an excavating machine of the power shovel group. There are two basic types of hydraulic excavators, depending on their type of digging action.

Hydraulic excavators that have their digging action toward the front of the excavator are commonly referred to as *front shovels* (Fig. 4-5). They are used mostly for pit excavation where the bucket load is obtained from the vertical face of the excavation pit above and in front of the excavator. The excavated material is loaded into trucks and hauled to another location.

Hydraulic excavators that have their digging action toward the back of the excavator are referred to by several names, such as *hoe, backhoe, back shovel,* or *pull shovel* (Fig. 4-6). They are used primarily to excavate below the natural surface, such as trenches, pits for basements, and general excavation which requires precise control of depths.

Hydraulic excavators have become one of the most widely used types of excavating equipment. Because of their rigidity they are superior to draglines for loading into dump trucks. Due to their direct pull of the bucket, they may exert greater tooth pressure than power shovels. In some respects they are superior to trenching machines, especially for digging utility trenches whose banks are permitted to establish natural slopes and for which trench shoring will not be used.

FIGURE 4-5
Hydraulic excavator with front shovel. (*Caterpillar, Inc.*)

FIGURE 4-6
Hydraulic excavator with back shovel. (*Caterpillar, Inc.*)

Hydraulic excavators can remove the earth as it caves in to establish natural slopes, whereas trenching machines cannot do this easily.

When a hydraulic excavator is used to dig at moderate depths, the production rate is comparable to that of the power shovel. However, the production rate will decrease considerably as the depth of excavation increases. The most effective production rate is obtained when the dipper stick is at right angles (90°) to the boom.

The production rate depends on the bucket payload, average cycle time, and job efficiency. If an estimator can predict the excavator cycle time and the bucket payload, the probable production rate for excavation can be determined.

The excavator cycle time will depend on the particular job conditions, such as the difficulty in loosening the soil, the angle of swing, the size of the truck that the excavator must load, and the skill of the operator. The estimator must use judgment and knowledge of actual job conditions to predict the cycle time for a particular job.

The *average bucket payload* is equal to the heaped bucket capacity payload multiplied by the bucket fill factor. The bucket fill factor will depend on the type of soil to be excavated. Table 4-8 provides representative values of the bucket fill factor.

> **Example 4-6.** Estimate the probable production rate of a backhoe with a $1\frac{1}{4}$-yd^3 heaped-capacity bucket operating in hard, tough clay. The estimated cycle time is 20 s per load, or 3 loads per minute. Assume the job conditions are represented by a 50-min working hour.
>
> The average bucket payload will be the heaped bucket capacity multiplied by the bucket fill factor:
>
> $$\text{Average bucket payload } 1.25 \times 85\% = 1.06 \text{ yd}^3$$
>
> The probable production rate will be
>
> $$\frac{1.06 \text{ yd}^3}{1 \text{ cycle}} \times \frac{1 \text{ cycle}}{20 \text{ s}} \times \frac{60 \text{ s}}{1 \text{ min}} \times \frac{50 \text{ min}}{1 \text{ h}} = 159 \text{ yd}^3/\text{h}$$

Table 4-9 provides representative ranges of production rates for excavating with backhoes. The information contained in this table is based on a 60-min hour, or 100 percent efficiency. The estimator should apply a job efficiency factor to the figures shown in the table based on personal judgment or knowledge of actual job conditions. The upper limit corresponds to the fastest practical cycle time: easy digging earth, low angles of swing, and no obstructions. The lower limit corresponds to the toughest digging, deep depths, large angles of swing, loading into small trucks, and obstructions in the work area.

Excavating with Draglines

For excavating on some projects, the dragline (Fig. 4-7) is more suitable than the power shovel. When one is digging ditches or building levees, it can excavate and

TABLE 4-8
Representative range of bucket fill factors

Type of material	Bucket-fill factor range, % of heaped bucket capacity
Moist loam or sandy clay	100–110
Sand and gravel	95–100
Hard, tough clay	80–90
Rock, well blasted	60–75
Rock, poorly blasted	40–50

Source: Caterpillar, Inc.

TABLE 4-9
Probable output of backhoes

Bucket payload, yd^3 loose	Output of backhoe, yd^3/h loose measure
$\frac{1}{2}$	75–135
$\frac{3}{4}$	90–202
1	120–270
$1\frac{1}{4}$	150–300
$1\frac{1}{2}$	154–360
$1\frac{3}{4}$	180–420
2	205–420
$2\frac{1}{4}$	231–472

Source: Caterpillar, Inc.

transport the earth within casting limits, thus eliminating hauling equipment. The dragline can operate on wet ground and can dig earth out of pits containing water. It cannot excavate rock as well as a power shovel. It has a lower output than a power shovel of the same size. Many units can be converted from power shovels to draglines by changing the boom and substituting a bucket for the shovel dipper. The cost of a dragline is somewhat less than that for a shovel of the same size. The

FIGURE 4-7
Dragline on a typical job.

TABLE 4-10
Ideal output of short-boom draglines, yd^3/h, bank measure

Class of material	Size of bucket, yd^3								
	$\frac{3}{8}$	$\frac{1}{2}$	$\frac{3}{4}$	1	$1\frac{1}{4}$	$1\frac{1}{2}$	$1\frac{3}{4}$	2	$2\frac{1}{2}$
Moist loam or light	5.0	5.5	6.0	6.6	7.0	7.4	7.7	8.0	8.5
sandy clay*	70	95	130	160	195	220	245	265	305
Sand and gravel*	5.0	5.5	6.0	6.6	7.0	7.4	7.7	8.0	8.5
	65	90	125	155	185	210	235	255	295
Good common earth*	6.0	6.7	7.4	8.0	8.5	9.0	9.5	9.9	10.5
	55	75	105	135	165	190	210	230	265
Hard, tough clay*	7.3	8.0	8.7	9.3	10.0	10.7	11.3	11.8	12.3
	35	55	90	110	135	160	180	195	230
Wet, sticky clay*	7.3	8.0	8.7	9.3	10.0	10.7	11.3	11.8	12.3
	20	30	55	75	95	110	130	145	175

Source: Power Crane and Shovel Association.
* The upper number indicates the optimum depth in ft. The lower number indicates the ideal output.

hourly operating costs are 5 to 10 percent less than for a shovel. The size of a dragline is indicated by the size of the bucket, expressed in cubic yards.

Table 4-10 gives ideal outputs for short-boom draglines when one is excavating at optimum depth with an angle of swing of 90°, based on a 60-min hour. Table 4-11 gives factors that may be used to determine outputs for other depths and angles of swing. The production determined from Tables 4-10 and 4-11 must be corrected for an operation less than 60 min per hour.

Example 4-6A. Estimate the total cost and the cost per cubic yard for excavating a drainage ditch 12 ft wide at the bottom, 36 ft wide at the top, 12 ft deep, and 12,110 ft

TABLE 4-11
Effect of depth of cut and angle of swing on output of draglines

Percentage of optimum depth	Angle of swing							
	30°	45°	60°	75°	90°	120°	150°	180°
20	1.06	0.99	0.94	0.90	0.87	0.81	0.75	0.70
40	1.17	1.08	1.02	0.97	0.93	0.85	0.78	0.72
60	1.24	1.13	1.06	1.01	0.97	0.88	0.80	0.74
80	1.29	1.17	1.09	1.04	0.99	0.90	0.82	0.76
100	1.32	1.19	1.11	1.05	1.00	0.91	0.83	0.77
120	1.29	1.17	1.09	1.03	0.98	0.90	0.82	0.76
140	1.25	1.14	1.06	1.00	0.96	0.88	0.81	0.75
160	1.20	1.10	1.02	0.97	0.93	0.85	0.79	0.73
180	1.15	1.05	0.98	0.94	0.90	0.82	0.76	0.71
200	1.10	1.00	0.94	0.90	0.87	0.79	0.73	0.69

Source: Power Crane and Shovel Association.

long in good common earth. The excavated earth can be cast on one or both sides of the ditch. Use a $1\frac{1}{2}$-yd^3 diesel-engine-powered crawler-mounted dragline. The average angle of swing will be 120°.

The volume of earth excavated will be $\dfrac{12 + 36}{2} \times 12 \times 12{,}110 \div 27 = 129{,}173$ yd^3 bm

Reference to Table 4-10 indicates a percent of optimum depth of $\dfrac{12.0}{9.0} \times 100 =$ 1.33

From Table 4-10 the ideal output will be 190 yd^3 per hr bm

From Table 4-11 the depth-swing factor will be 0.89

Applying these quantities gives an output of $0.89 \times 190 = 169.1$ yd^3 per h for a 60-min h

If the dragline operates an average of 50 min per h, the corrected output will be $50/60 \times 169.1 = 141$ yd^3 per h bm

The time required to excavate the earth will be $129{,}173 \div 141 = 916$ h

The cost will be

Transporting dragline to and from job =		$2,300.00
Dragline, 916 h @ $57.69	=	$52,844.04
Labor setting up and removing dragline:		
Operator, 16 h @ $17.25	=	276.00
Oiler, 16 h @ $11.25	=	180.00
Helper, 16 h @ $7.50	=	120.00
Foreman, 16 h @ $17.30	=	276.80
Labor excavating:		
Operator, 916 h @ $17.25	=	15,801.00
Oiler, 916 h @ $11.25	=	10,305.00
Helper, 916 h @ $7.50	=	6,870.00
Foreman, 916 h @ $17.30	=	15,846.80
Total direct cost	=	$104,819.64
Cost per yd^3, $104,819.64 ÷ 129,173	=	0.81

Handling Materials with a Clamshell

Most draglines in the popular sizes can be converted to clamshells by replacing the dragline bucket with a clamshell bucket. The size of a clamshell is indicated by the size of the bucket, expressed in cubic yards.

The clamshell is a satisfactory machine for handling sand, gravel, crushed stone, sandy loam, and other loose materials. Because of the difficulty which it encounters in loosening solid earth, it is not very satisfactory for handling compacted earth, clay, and other solid materials.

The bucket is lowered into the material to be handled with the jaws open. Its weight will cause the bucket to sink into the material as the jaws are closed. Then it is lifted vertically, swung to the emptying position, over a truck, to a spoil pile or elsewhere, and the jaws are opened to permit the load to flow out. All these operations are controlled by the clamshell operator. The output of the clamshell is affected by the looseness of the materials being handled, type of material, height of lift, angle of swing, method of disposing of the materials, and skill of the operator.

TABLE 4-12
Representative output of clamshells

Size of bucket, yd³	Angle of swing, deg	Material, yd³		
		Light loam	Sand gravel	Crushed stone
½	45	48	43	38
	90	40	36	31
	180	31	28	24
¾	45	63	56	49
	90	53	48	42
	180	41	37	32
1	45	81	73	63
	90	68	61	53
	180	54	48	42
2	45	134	120	104
	90	113	102	88
	180	87	78	68

Table 4-12 gives the probable output of clamshells of different sizes for various angles of swing and materials. These rates should apply for jobs which do not require frequent interruptions.

The hourly cost of owning and operating a clamshell is about the same as for a dragline of the same size.

Excavating and Hauling Earth with Tractors And Scrapers

Tractor-pulled scrapers are used to excavate and haul earth for such projects as dams, levees, highways, airports, and canals. Since these units perform both excavating and hauling operations, they are independent of the operations of other equipment. Thus, if one of several units breaks down, the rest of the units can continue to operate, whereas if a power shovel breaks down, the entire project must stop until the shovel is repaired.

These units are rugged and can operate under adverse conditions. They are available in a wide range of sizes and capacities. The capacity of a scraper is designated by the volume of earth that it will carry, either struck or heaped, expressed in cubic yards, loose measure. For example, the capacity might be designated as 12 yd³ struck, 15 yd³ heaped.

Crawler-Tractor-Pulled Scrapers

For short-haul distances, the crawler-type tractor pulling a rubber-tired self-loading scraper is sometimes used (Fig. 4-8). The crawler tractor has a high drawbar pull for loading the scraper, has good traction with the ground, and can

FIGURE 4-8
Crawler and scraper.

operate over muddy haul roads, but it has a low travel speed, which is a disadvantage for long hauls. The top speed is approximately 6 to 7 mi/h depending on the unit.

Although it is usually desirable to provide an auxiliary tractor to help during the loading operation, a crawler tractor can load a scraper without additional help, but at a reduced rate.

The size of a crawler tractor usually is expressed in terms of the drawbar horsepower, which is the power available at the drawbar, when it is operated at sea level on a level haul road having a rolling resistance of 110 lb/ton of gross load.

Wheel-Type Tractor-Pulled Scrapers

For longer haul distances in excess of approximately 600 ft, the wheel-type tractor pulling a rubber-tired self-loading scraper is more economical than the crawler-tractor-pulled unit. While the wheel-type tractor cannot deliver as great a tractive effort in loading a scraper, the higher travel speed, which may exceed 30 mi/h for some units, will offset the disadvantage in loading when the haul distance is sufficiently long. Both two- and four-wheel tractors are available. A helper tractor, such as a bulldozer, should be used to help load the scrapers.

The size of a rubber-tired tractor usually is designated by the brake horsepower of the engine.

FIGURE 4-9
Two-wheeled tractor scraper loading with the assistance of a bulldozer. (*Caterpillar, Inc.*)

Cost of Owning and Operating a Wheel-Type Tractor and Scraper

This example illustrates a method of determining the cost of owning and operating a wheel-type tractor and scraper. The unit is similar to the one illustrated in Fig. 4-9. The specified cost includes freight charges from the factory to the owner.

FIGURE 4-10
Two-wheeled tractor scraper hauling its load. (*Caterpillar, Inc.*)

Because the life of the tires will usually be different from that of the basic unit, the cost of the tires should be estimated separately. The following conditions will apply.

Tractor, diesel engine, 275 fwhp
Scraper, 21 yd³, struck capacity
Life of basic unit, 5 yr at 2,000 h per yr
Salvage value after 5 yr, none
Life of tires, 5,000 h

Cost of unit fob owner	= $268,520
Cost of tires	= 12,345
Cost less tires	= $256,175
Average investment, 0.6 × $268,520 =	161,112

Annual cost:

Depreciation, $256,175 ÷ 5	=	51,235
Maintenance and repairs, 75% of $51,235 =		38,426
Investment, 14% × $161,112	=	22,556
Total annual cost	=	$112,217

Hourly cost:

Fixed cost, $112,217 ÷ 2,000	=	$56.11
Fuel, 9 gal @ $0.95	=	8.55
Lubricating oil, 0.4 gal @ $1.30	=	0.52
Other lubrication, 50% of $0.52	=	0.26
Depreciation of tires, $12,345 ÷ 5,000 h =		2.47
Tire repairs, 15% of $2.47	=	0.37
Total cost per h	=	$68.28

Production Rates for Tractor-Pulled Scrapers

The production rate for a tractor-pulled scraper will equal the number of trips per hour times the net volume per trip. An appropriate operating factor, such as a 45- or 50-min hour, should be used in determining the number of trips per hour, to allow for nonproductive time.

Frequently it is desirable to express the speed of a tractor in feet per minute as well as miles per hour. A speed of 1 mi/h is equal to 5,280 ft in 60 min, which equals 88 ft/min. Table 4-13 gives speeds in miles per hour and feet per minute.

TABLE 4-13
Speeds in miles per hour and feet per minute

mi/h	ft/min	mi/h	ft/min
1	88	7	616
2	176	8	704
3	264	9	792
4	352	10	880
5	440	15	1,320
6	528	20	1,760

Example 4-7. Determine the probable production rate for a crawler tractor, with 125 drawbar hp, and a wheel-type scraper, whose struck capacity is 10 yd^3. Assume that the scraper load will be heaped to give a volume of 11 yd^3. The material will be ordinary earth.

Haul distance, 500 ft at average speed of 4 mi/h
Return distance, 600 ft at average speed of 4.5 mi/h
Assume a 50-min hour
Net load, 11 ÷ 1.25 = 8.8 yd^3, bm

The fixed time per trip will be
Loading, 100 ft ÷ 88 ft/min	= 1.14 min
Dumping	= 0.80 min
Turning and delays	= 0.36 min
Total fixed time	= 2.30 min

The travel time will be
Hauling, 500 ft ÷ 352 ft/min	= 1.42 min
Returning, 600 ft ÷ 396 ft/min	= 1.52 min
Total time per trip	= 5.24 min

Trips per h, 50 ÷ 5.24 = 9.5
Volume per h, 9.5 trips × 8.8 yd^3 = 83.6 yd^3

Example 4-8. Determine the probable production rate for a wheel-type tractor with a 225-hp diesel engine and a wheel-type scraper whose struck capacity is 11 yd^3. The scraper will be loaded to an average heaped capacity of 13 yd^3 loose measure. The material will be good common earth, with a swell of 25 percent. While the tractor has a maximum speed of 23 mi/h, the average speed on a construction road will be considerably less, especially for short hauls over poorly maintained earth roads. The following conditions will apply to the unit.

Haul distance, 860 ft at average speed of 9 mi/h loaded
Return distance, 940 ft at average speed of 11 mi/h
Net pay load, 13 ÷ 1.25 = 10.4 yd^3 bm
Assume a 45-min operating hour

The time per round trip will be
Loading, 80 ft ÷ 88 ft/min	= 0.91 min
Hauling, 860 ft ÷ 792 ft/min	= 1.09 min
Dumping	= 0.75 min
Turning and delays	= 0.50 min
Returning, 940 ft ÷ 968 ft/min	= 0.97 min
Total round-trip time	= 4.22 min

No. trips per h, 45 ÷ 4.22 = 10.66
Volume hauled per h, 10.66 × 10.4 = 110.9 yd^3 bm

Cost of Excavating and Hauling Earth With Tractors and Scrapers

The cost of handling earth with tractors and scrapers may be determined by assuming that the equipment will be operated 1 h for quantity and cost purposes.

If a bulldozer is used to assist severai scraper units while they are being loaded, the cost of the bulldozer should be distributed equally among the scraper units that are assisted. The number of scrapers that a bulldozer can assist will be equal to the total cycle time for a scraper to load, haul, dump, and return to the pit, divided by the time required by the bulldozer to assist in loading a scraper and to get into position to load another scraper.

Consider the scraper unit in Example 4-8:

> The round-trip time is 4.22 min
> The loading time is 0.91 min

If the round-trip time, 4.22 min, is divided by the loading time, 0.91 min, it appears that a bulldozer should be able to assist 4.63 scrapers. However, when a bulldozer finishes assisting one scraper, the bulldozer must move into position to assist the next scraper, which may require the bulldozer to travel more than 100 ft for its new position. For this reason it appears that the bulldozer cannot assist more than 5 scrapers. Thus one-fifth the cost of a bulldozer will be charged to each scraper.

The cost per h will be

Tractor-scraper, 1 h @ $68.25	=	$ 68.25
Bulldozer, 0.2 h @ $59.20	=	11.84
Scraper operator, 1 h @ $17.25	=	17.25
Bulldozer operator, 0.2 h @ $17.25	=	3.45
Total cost	=	$100.79

Volume hauled per h, 110.9 yd^3
Cost per yd^3, $100.79 ÷ 110.9 yd^3 = 0.91

Haul Distances for Scrapers

For short-haul distances, the production rate of a scraper depends primarily on the fixed time of the scraper—the time to load, accelerate, decelerate, turn, and dump. As the length of the haul distance increases, the production rate of a scraper will depend primarily on the travel time of the scraper, or the time to haul and return. Therefore, the haul distance must be known to estimate the production rate and the cost of excavation by a scraper for a particular job.

The earthwork required for a highway project is defined by the mass diagram on the plan and profile sheets for the project. Figure 4-11 is a portion of a profile and mass diagram. The amount of material that must be excavated, and the distance it must be hauled, can be determined from the mass diagram. This information enables one to estimate the production rate and the cost of excavation.

Example 4-9. Determine the total cost and the cost per cubic yard for the earthwork of the first balance point of the mass diagram in Fig. 4-11. A self-loading scraper with a struck capacity of 15 yd^3 will be used to excavate and haul the soil. The scraper has a load time of 2.3 min and a dump time of 1.8 min. Assume a 20-mi/h haul speed and a 25-mi/h return speed. The material will be common earth, with a swell of 25 percent.

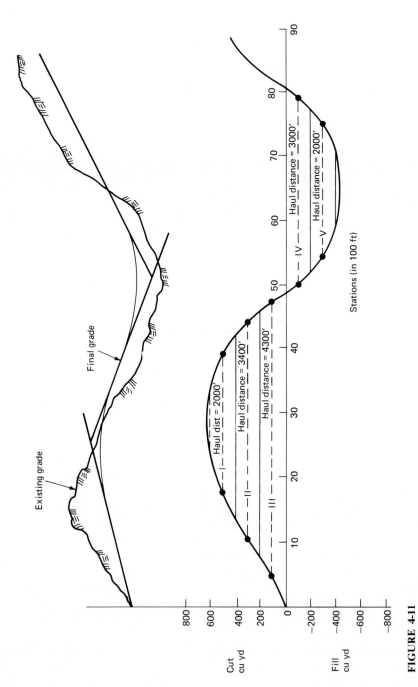

FIGURE 4-11
Profile and mass diagram for highway project.

80

The mass diagram can be arbitrarily divided into three 200-yd^3 sections. For each section the haul distance can be determined from the diagram. The haul distance, cycle time, number of trips per 45-min hour, quantity hauled per trip, production rate, and time required for each section will be as follows:

Quantity hauled per trip, $15 \div 1.25 = 12$ yd^3 bm

| Section | Quantity, yd^3 | Distance, ft | Cycle time, min | | | | | Trips per 45-min hour | Quantity hauled per trip | Production rate, yd^3/h | Total time, h |
			Load	Haul	Dump	Return	Total				
I	200	2,000	2.3	1.14	1.8	0.91	6.15	7.3	12	87.6	2.3
II	200	3,400	2.3	1.93	1.8	1.55	7.58	5.9	12	70.8	2.8
III	200	4,300	2.3	2.44	1.8	1.95	8.49	5.3	12	63.6	3.1

Total = 600   Total time = 8.2 h

The cost will be

Scraper, 8.2 h @ $68.50   = $561.70
Operator, 8.2 h @ $16.20 =   132.84
      Total cost   = $694.54
Cost per yd^3, $694.54 \div 600 =$ 1.16

Shaping and Compacting Earthwork

When earth is placed in a fill, it is necessary to spread it in uniformly thick layers and compact it to the specified density. Unless sufficient moisture is present, water should be added to produce the optimum moisture content, which will permit more effective compaction. Spreading may be accomplished with graders (Fig. 4-12) or

FIGURE 4-12
Motor grader shaping a fill area. (*Caterpillar, Inc.*)

FIGURE 4-13
Three-drum sheep's-foot roller.

bulldozers, or both, while compaction may be accomplished with tractor-pulled sheep's-foot rollers (Fig. 4-13), smooth-wheel rollers, pneumatic rollers, vibrating rollers, or other types of equipment. For some projects the best results are obtained by using more than one type of equipment. Regardless of the type of equipment selected, there should be enough units to shape, wet, and compact the earth at the rate at which it will be delivered.

Example 4-10. Estimate the total cost per cubic yard for shaping, sprinkling, and compacting earth in a fill such as a dam or a highway. The earth will be placed by tractor-scrapers at the rate of 396 yd^3/h, with the volume measured after compaction. The earth will be placed in layers not to exceed 6 in thick when compacted. The moisture content, probably averaging 8 percent when the earth is placed in the fill, must be increased to 12 percent. The water will be placed with truck-mounted sprinklers.

For the stated rate of placing the earth, the surface area placed in 1 h will be

$$\frac{396 \text{ yd}^3 \times 27}{0.5} = 21,384 \text{ ft}^2$$

If it requires three passes of a grader to smooth each layer of earth satisfactorily, the area which the grader must cover in 1 h will be $3 \times 21,384 = 64,152$ ft^2. A motor grader with a 12-ft blade should cover a strip or lane whose effective width will be about 8 ft for each pass of the grader. The grader should be able to travel at an average speed of at least 2 mi/h, allowing for stops and lost time turning around.

The area covered in 1 h by a grader should be $2 \times 5,280 \times 8 = 84,500$ ft^2. Thus one grader should be sufficient to do the job.

The earth will be compacted by sheep's foot rollers pulled by a crawler tractor. It is estimated that 10 passes will be required to attain the specified compaction. The equivalent area that the rollers must cover in 1 h will be $10 \times 21,384 = 213,840$ ft^2.

If roller drums 4 ft long are operated at an average speed of $2\frac{3}{4}$ mi/h, allowing for lost time, the area covered by one drum in 1 h will be $2\frac{3}{4} \times 5,280 \times 4 = 58,080$ ft^2.

The number of drums required will be $213,840 \div 58,080 = 3.68$. It will be necessary to use 4 drums, which will be pulled by one tractor.

The compacted earth will weigh $3,150$ lb/yd^3. The weight of water required per hour will be 4 percent of the weight of the earth placed $= 0.04 \times 396 \times 3,150 = 49,896$ lb, equal to $49,896 \div 8.33 = 5,990$ gal/h. If a 2,000-gal sprinkler truck can make one trip in 40 min, two trucks will be required.

The cost per h will be

Motor grader	= $ 27.50
Grader operator	= 17.25
Tractor and sheep's-foot roller	= 50.40
Tractor operator	= 17.25
Sprinkler trucks, 2 @ $18.60	= 37.20
Truck drivers, 2 @ $8.10	= 16.20
Foreman	= 17.30
Total cost per h	= $183.10
Cost per yd^3 of earth, $183.10 \div 396 =$	0.46

Drilling and Blasting Rock

Before rock can be excavated, it must be loosened and broken into pieces small enough to be handled by the excavating equipment. The most common method of loosening is to drill holes into which explosives are placed and detonated.

FIGURE 4-14
Gasoline-engine-operated hand drill.

FIGURE 4-15
Truck-mounted drill equipped with
a dust collector.

Holes may be drilled by one or more of several types of drills, such as jackhammers, wagon drills, drifters, churn drills, rotary drills, etc., with the selection of equipment based on the size of the job, type of rock, depth and size of holes required, production rate required, and topography at the site.

Jackhammers may be used for holes up to about $2\frac{1}{2}$ in in diameter and 15 to 18 ft deep. For deeper holes, the production rates are low and the costs are high. Wagon drills may be used for holes 2 to $4\frac{1}{2}$ in in diameter, with depths sometimes as great as 40 ft, although shallower depths are more desirable. Drifters are used to drill approximately horizontal holes, up to about 4 in in diameter, in mining and tunneling operations. Jackhammers, wagon drills, and drifters are operated by compressed air, which actuates the hammer that produces the percussion which disintegrates the rock and blows it out of the holes. Replaceable bits, which are attached to the bottoms of the hollow drill steels, are commonly used.

A churn drill disintegrates the rock by repeated blows from a heavy steel bit, which is suspended from a wire rope. Holes in excess of 12 in in diameter may be drilled several hundred feet deep with this equipment. Water, which is placed in a hole during the drilling operation, will produce a slurry with the disintegrated rock. A bailer is used to remove the slurry.

Rotary drills may be used to drill holes 3 to 8 in or more in diameter to depths in excess of 100 ft. Drilling is accomplished by a bit, which is attached to the lower end of a drill stem. Either water or compressed air may be used to remove the rock cuttings.

FIGURE 4-16
Blast hole drill.

Dynamite is frequently used as the explosive, although several other types of explosives are available. Dynamite is available in sticks of varying sizes, which are placed in the holes. The strength, which is specified as 40, 60, etc., percent, indicates the concentration of the explosive agent, which is nitroglycerin. The dynamite is usually exploded by a blasting cap, which is detonated by an electric current. At least one blasting cap is required for each hole. The charges in several holes may be shot at one time.

The amount of dynamite required to loosen rock will vary from about 0.25 to more than 1 lb/yd^3, depending on the type of rock, the spacing of holes, and the degree of breakage desired.

Cost of Operating a Drill

The items of cost in operating a jackhammer or a wagon drill will include equipment and labor. The equipment cost will include the drill, drill steel, bits, air compressor, and hose. Since each of these items may have a different life, it should be priced separately.

Drill steel is purchased by size, length, and quality of steel used, with the cost of steel based on its weight. While the life of a drill steel will vary with the class of

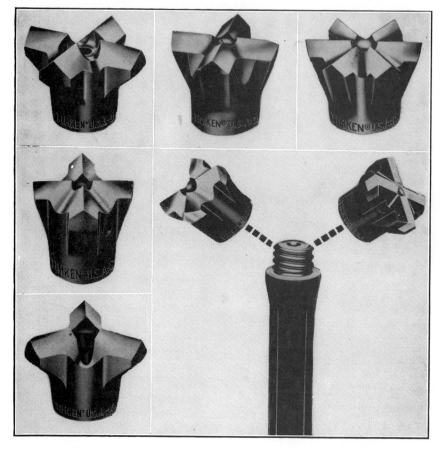

FIGURE 4-17
Multiuse rock bits.

rock drilled and the conditions under which the steel is used, records from drilling projects show consumptions varying from $\frac{1}{20}$ to $\frac{1}{10}$ lb/yd^3 of rock to be representative, with the higher consumption applicable to the harder rocks. Thus, if a drill steel for a wagon drill, 12 ft long, weighing 4.6 lb/ft, whose total weight is about 55 lb, is consumed at the rate of $\frac{1}{12}$ lb/yd^3 of rock, this steel should drill enough hole to produce about 660 yd^3 of rock.

Detachable bits are commonly used with jackhammers, drifters, and wagon drills. The depth of hole that may be drilled with a bit before it must be resharpened or discarded will vary considerably with the class of rock and the type of bit, with values ranging from less than a foot to as much as 100 ft or more for steel bits. Bits with carbide inserts will give much greater depths.

Rates of Drilling Holes

The rate of drilling rock will vary with several factors, including the type and hardness of the rock, type and size of drill used, depth of holes, spacing of holes,

TABLE 4-14
Representative rates of drilling rock with various types of drills

Size of hole, in	Class of rock	Rate of drilling, ft/h				
		Jack-hammer	Wagon drill	Churn drill	Rotary drill	Diamond drill
$1\frac{3}{4}$	Soft	15–20	30–45	—	—	5–8
	Medium	10–15	25–35	—	—	3–5
	Hard	5–10	15–30	—	—	2–4
$2\frac{3}{8}$	Soft	10–15	30–50	—	—	5–8
	Medium	7–10	20–35	—	—	3–5
	Hard	4–8	15–30	—	—	2–4
3	Soft	—	30–50	—	—	4–7
	Medium	—	15–30	—	—	3–5
	Hard	—	8–20	—	—	2–4
4	Soft	—	10–25	—	—	3–6
	Medium	—	5–15	—	—	2–4
	Hard	—	2–8	—	—	1–3
6	Soft	—	—	4–7	25–50	3–5
	Medium	—	—	2–5	10–25	2–4
	Hard	—	—	1–2	6–10	1–3

topography at the site, condition of the drilling equipment, etc. Although the rates given in Table 4-14 are based on observations, they should be used as a guide only. The rates include an allowance for lost time at the job.

Holes should be drilled 1 ft, or more, deeper than the desired effective depth of rock loosened. This is necessary because the rock usually will not break to the full depth of the holes over the entire area blasted.

Example 4-11. Estimate the cost of drilling and blasting limestone rock for the stated conditions.

Diameter of holes, 3 in
Depth of holes, 14 ft
Effective depth of holes, 13 ft
Spacing of holes, 8 ft × 8 ft
Two medium-size wagon drills used, each with 50 ft of $1\frac{1}{2}$-in hose and connections
Dynamite required, 1 lb/yd^3 of rock
Cost of bits, $12.05 each
No. times sharpened, 3
Cost of sharpening bit, $3.15 each time
Depth of hole before sharpening bit, 36 ft
Drill steel consumed, $\frac{1}{15}$ lb/yd^3 of rock
Estimated rate of drilling, 24 ft/h
Base all quantities and costs on operating 1 h
Total depth of hole drilled, 2 × 24 = 48 ft

Effective depth of hole drilled, $48 \times \frac{13}{14} = 44.6$ ft
Volume of rock produced, $44.6 \times 8 \times 8 \div 27 = 106$ yd^3
No. holes drilled, 48 ft ÷ 14 ft/hole = 3.44

The cost per h will be

Dynamite, 106 yd^3 × 1 lb = 106 lb @ $1.16	=	$122.96
Electric caps, 3.44 @ $1.08	=	3.71
Electric wire	=	0.60
Air compressor, 1 h @ $42.30	=	42.30
Hose, 2-in diameter, 1 h @ $1.08	=	1.08
Wagon drills, 2 h @ 3.92	=	7.84
Air hoses, 2 h @ 0.27	=	0.54
Drill steel, 106 yd^3 × $\frac{1}{15}$ lb = 7.1 lb @ $1.05 =		7.45

Drill bits:

Original cost, each	= $12.05
Sharpening, 3 × $3.15 =	9.45
Total cost per bit	= $21.50

Total depth of hole drilled by one bit, 4 × 36	=	144 ft
Cost of bit per ft of hole, $21.50 ÷ 144 ft	= $	0.149
Cost of bits per h 48 × $0.149	= $	7.15
Drill operators, 2 h @ $10.80	=	21.60
Helpers, 2 h @ $8.50	=	17.00
Powderman, 1 h @ $10.85	=	10.85
Helper, 1 h @ $7.25	=	7.25
Foreman, 1 h @ $18.30	=	18.30
Total cost	=	$268.63
Cost per yd^3, $268.63 ÷ 106	=	2.53

PROBLEMS

4-1. A truck will haul an average load of 12.5 yd^3, heaped capacity, loose measure per load. Determine the volume in bank measure when the material is sand, ordinary earth, dense clay, and well-blasted rock.

4-2. Estimate the total direct cost and the cost per foot for excavating a sewer trench 33 in wide, 10 ft 6 in deep, and 1,480 ft long in ordinary earth, by using a ladder-type trenching machine. There are no obstructions to delay the machine. Assume a 50-min hour.

   Use the information given in the text to estimate the digging speed of the trenching machine. It will cost $1,800.00 to transport the machine to the job, get it ready to operate, then return it to the storage yard. The following crew at the specified hourly wage rates will be required:

Trenching machine	$82.50
Trencher operator	17.25
Trencher oiler	11.25
Laborers, 3 each	7.50
Foreman	18.50

4-3. What is the maximum length of trench that can be excavated by hand economically rather than by a ladder-type trenching machine for the stated conditions?

> Width of trench, 3 ft 0 in
> Average depth, 4 ft 6 in
> Class of soil, ordinary earth

For hand excavation it will be necessary to shovel the earth back from the edge of the trench. Assume that 10 laborers plus a foreman will be used. It will cost $2,200.00 to transport a trenching machine to the job and back to the storage yard. Assume the machine will operate 50 min/h. Equipment and labor costs per hour will be

Trenching machine	$92.75
Trencher operator	17.25
Trencher oiler	11.25
Laborers, each	7.50
Foreman	18.50

4-4. In excavating for a basement of a building, a contractor will use a $\frac{3}{4}$-yd^3 crawler-type shovel, powered by a 120-hp diesel engine to load earth into 6-yd^3 gasoline-engine-powered-trucks, which will haul the earth to a dump. The earth is hard, tough clay.

The dimensions of the pit are 140 ft 0 in wide, 320 ft 0 in long, and 10 ft 0 in deep.

The earth will be hauled to a dump 3 mi from the pit at an average hauling speed of 30 mi/h and a return speed of 40 mi/h. In addition to the time required to load a truck, haul the load to the dump, and return to the pit, a truck will consume an average of 8 min each trip at the dump and waiting at the pit to be loaded.

The average angle of swing for the shovel will be 120°. Assume that all equipment will work 50 min/h. It will cost $1,200.00 to transport the shovel to the job and back to the storage yard.

Use the information available in the text to determine the total cost and the cost per cubic yard, bank measure, for excavating and hauling the earth. The labor costs per hour will be

Shovel	$72.80
Trucks	27.25
Shovel operator	17.25
Shovel oiler	11.25
Truck drivers, each	9.45
Laborers, 3 each	7.50
Foreman	18.50

4-5. A project requires 692,500 yd^3, bank measure, of dense clay earth, which will be excavated by a 2-yd^3 diesel-engine-powered crawler-type power shovel, loaded into 10-yd^3, loose measure, diesel-engine-powered dump trucks, and hauled 4 mi to the project. The trucks will average traveling 30 mi/h loaded and 40 mi/h empty.

The shovel can excavate earth at the optimum depth and at 120° average angle of swing. Assume that the shovel and the trucks will operate 45 min/h. It will cost

$2,400.00 to transport the shovel to the job, set it up to operate, and return it to the storage yard.

Use information available in the book to estimate the total cost of the job and the cost per cubic yard for excavating and hauling the earth. Labor costs per hour will be

Shovel	$93.50
Trucks, each	31.40
Shovel operator	17.25
Shovel oiler	11.25
Truck drivers, each	9.45
Laborers, 2 each	7.50
Foreman	18.30

It is estimated that the average truck time at the dump, waiting to be loaded at the shovel and for other delays, will amount to 8 min per round trip.

4-6. A $1\frac{1}{2}$-yd^3 diesel-engine-powered crawler-type dragline will be used to excavate a ditch in dense clay. The ditch will be trapezoidal in cross section, 16 ft 0 in wide at the bottom, 36 ft 0 in wide at the top, 8 ft 0 in deep, and 6,480 ft long.

There will be no obstructions to affect the rate of excavating the earth. The excavated earth will be cast along the edge or edges of the ditch. The average angle of swing will be 145°.

Estimate the total cost and the cost per cubic yard, bank measure, for excavating the ditch.

Assume a 50-min hour. It will cost $3,200.00 to transport the dragline to the job, set it up for operation, and return it to the storage yard. Labor costs per hour will be

Dragline	$74.50
Dragline operator	17.25
Dragline oiler	11.25
Laborer, 1	7.50
Foreman	18.50

4-7. Ordinary earth will be excavated and hauled by diesel-engine-powered wheel-type scrapers, whose capacities are 14 yd^3 struck and 18 yd^3 heaped, loose measure. The distance from the borrow pit to the dump site will average 2,680 ft over level earth haul road. The scrapers can average 15 mi/h loaded and 20 mi/h empty.

A 270-hp crawler tractor will be used to assist the scrapers in loading. Use the example on pages 78–79 of this book to determine the number of scrapers that one crawler tractor can serve. *Note:* Generally it should require from 50 to 70 s to load a scraper if adequate tractor power is provided. Thus a bulldozer tractor should be able to serve a scraper about every $2\frac{1}{2}$ to 3 min.

Use one bulldozer and the appropriate number of scrapers to estimate the cost per cubic yard for excavating and hauling the earth. Base your costs on bank-measure volume.

The fixed time for the scrapers, including loading, dumping, turning, waiting to load or dump, etc., should average about $2\frac{1}{2}$ min per trip. Assume a 45-min hour. Each tractor operator will be paid $17.25 per hour.

4-8. In constructing a fill, it is necessary to excavate ordinary earth and haul it 1,800 ft along an approximately level haul road. The earth will be excavated and hauled with units, each consisting of a two-wheel tractor and a two-wheel scraper. Each scraper will haul 15 yd³, loose measure, per load and will be pulled by a 180-hp tractor. One 180-hp crawler-type tractor will be required to assist the four scraper units during loading.

The scrapers will average 12 mi/h loaded and 18 mi/h empty. The fixed time, including loading, dumping, and waiting in the pit, will average $2\frac{1}{2}$ min per trip. Assume a 50-min working hour.

Determine the probable direct cost per cubic yard, bank measure, for excavating and hauling the earth. Labor and equipment costs per hour will be

Tractor-scraper	$94.50
Tractor operators, each	17.25
Foreman	18.50

4-9. Estimate the total time and cost and the cost per cubic yard for the second balance point, sections IV and V, of the mass diagram in Fig. 4-11. A self-loading scraper with 12-yd³ loose measure will be used for the excavation and hauling.

Assume an average haul speed of 15 mi/h and an average return speed of 20 mi/h. Use a 2.5-min load time and a 2.0-min dump time. Assume a 50-min hour and a swell of 18 percent. The hourly cost of equipment and labor will be

Scraper	$65.40
Operator	17.25

4-10. Estimate the total direct cost and the cost per cubic yard for drilling and blasting limestone rock, classified as medium rock, for a pit 80 ft wide, 108 ft long, and 12 ft deep. Holes $2\frac{1}{2}$ in in diameter will be spaced in patterns 6 by 8 ft, over the entire area, with the outside holes being along the sides and ends of the pit. The holes will be drilled 13 ft deep.

The holes will be drilled with two wagon drills, each with 50 ft of $1\frac{1}{2}$-in hose with fittings. The air will be supplied by a 600 ft³/min compressor powered with a diesel engine, equipped with 100 ft of 2-in hose.

It is estimated that drill steel will be consumed at a rate of 1 lb per 20 yd³ of rock. Each bit is expected to drill 32 ft of hole, after which it will be discarded. It will require $\frac{3}{4}$ lb of dynamite for each cubic yard of rock blasted. Assume a 45-min working hour.

Use the information in this book to determine the rate of drilling and the costs of equipment. Other costs and labor costs per hour will be

Transporting equipment to and from job	$360.00
Drill steel, per lb	2.25
Bits, each	2.86
Dynamite, per lb	1.16
Electric caps, each	1.08
Generator and lead wire for job	32.46
Air compressor, per h	38.65
Air hose, 2-in diameter	1.24
Wagon drills, 2 used, each	3.74

Drill operators, 2, each	10.80
Drill helpers, 2, each	8.50
Powderman	10.50
Powderman's helper	8.50
Foreman	18.30

4-11. In operating a quarry for the production of crushed limestone, two heavy wagon drills will be used to drill holes 3 in in diameter, 16 ft deep, spaced 9 ft apart each way. The effective depth of the holes will be 15 ft.

One 600-ft^3/min portable air compressor, operated by a diesel engine, will be used to supply air for the drills. Each drill will require 50 ft of $1\frac{1}{2}$-in air hose with fittings. The consumption of drill steel with be $\frac{1}{12}$ lb/yd^3 of rock produced. Each bit will drill 36 ft of hole, after which it will be resharpened for four additional uses. It will require $\frac{3}{4}$ lb of dynamite for each cubic yard of rock produced. One detonator cap will be required for each hole.

Use the information in this book to determine the direct cost per cubic yard for drilling and blasting the rock.

Since the equipment will be left at the quarry, there will be no cost for transporting it to and from the job. Use the following costs for supplies and labor.

Drill steel, per lb	$ 2.25
Bits, each	8.56
Sharpening each bit, each time	1.54
Dynamite, per lb	1.16
Detonator caps, each	1.08
Air compressor, per h	42.30
Air hose, 2 in/h	0.27
Wagon drills, each per h	3.74
Drill operators, 2, each	10.80
Drill helpers, 2, each	8.50
Compressor operator	10.80
Powderman	10.85
Powderman's helper	7.25
Foreman	18.30

REFERENCES

1. *Power Crane Applications in Industrial Plants*, Power Crane & Shovel Association, Milwaukee.
2. *Caterpillar Performance Handbook*, 18th ed., Caterpillar, Inc., Peoria, Ill.
3. *Caterpillar Performance Handbook for Hydraulic Excavators*, Caterpillar, Inc., Peoria, Ill.
4. Peurifoy, R. L. and W. B. Ledbetter: *Construction Planning, Equipment and Methods*, 4th ed., McGraw-Hill, New York, 1985.

CHAPTER
5

HIGHWAYS AND PAVEMENTS

Operations Included

Operations to be discussed in this chapter include clearing and grubbing land and placing concrete and asphalt pavements. Even though the coverage is limited to only a few of the methods used, the discussions and examples presented should illustrate how estimates can be prepared for projects constructed by other methods.

CLEARING AND GRUBBING LAND

Land-Clearing Operations

Clearing land may be divided into several operations, depending on the type of vegetation to be removed, the type and condition of the soil and topography, the amount of clearing required, and the purpose for which the clearing is done, as listed:

1. Complete removal of all trees and stumps, including tree roots
2. Removal of all vegetation above the surface of the ground only, leaving the stumps and roots in the ground
3. Disposal of the vegetation by stacking and burning it

Types of Equipment Used

Several types of equipment are available for use in clearing land, including tractor-mounted bulldozers, tractor-mounted special cutting blades, and tractor-mounted rakes.

TRACTOR-MOUNTED BULLDOZERS. Whereas bulldozers were used extensively in the past to clear land, they are now being replaced by special blades mounted on tractors. There are some objections to using bulldozers for this work. Before felling large trees, the bulldozers must excavate the earth from around the trees and cut the main roots, which leaves objectionable holes in the ground and requires considerable time. Also, when stacking the felled trees and other vegetation bulldozers transport earth to the piles of trees, which makes burning more difficult.

TRACTOR-MOUNTED SPECIAL BLADES. Two types of special blades are used to fell trees, and both are mounted on the front ends of tractors.

One is a single-angle blade with a projecting stinger on the lead side, extending ahead of the blade, so that it can be forced into and through the tree to split and weaken it. If a tree is too large to be felled in one pass, the trunk can thus be split and removed in parts. Also, the tractor can make a pass around the tree with the stinger penetrating the ground to cut the main horizontal roots of the tree. Figure 5-1 illustrates this blade in use. The unit may be used to remove stumps and to stack material for burning.

Another type of special blade is a V blade, with a protruding stinger at its lead point, as illustrated in Fig. 5-2. The sole effect of the blade permits it to slide along the surface of the ground, thereby cutting vegetation flush with the surface. However, it can be lowered below the surface of the ground to remove stumps. Also, the blade may be raised to permit the stinger to pierce the tree above the ground, as illustrated in Fig. 5-1.

TRACTOR-MOUNTED RAKES. Figure 5-3 illustrates a tractor-mounted rake which can be used to grub and pile trees, boulders, and other materials without transporting excessive amounts of earth. This can be a very effective machine for stacking materials in piles for burning.

Disposal of Brush

When brush is to be disposed of by burning, it should be piled in stacks and windrows, with a minimum amount of earth included. Shaking the rake while it is moving the brush will help remove the earth.

Because burning is usually necessary while the brush contains considerable moisture, it may be desirable to provide a continuous external source of fuel and heat to assist in burning the material. The burner illustrated in Fig. 5-4, which

FIGURE 5-1
Tractor-mounted V blade splitting a large tree.

consists of a gasoline-engine-driven pump and a propeller, is capable of discharging a liquid fuel onto the pile.

Rates of Clearing Land

As previously stated, the rates of clearing land will depend on several factors, including, but not limited to, the density of vegetation, sizes and varieties of trees, type of soil, topography, rainfall, types of equipment used, skill of equipment operators, and requirements of the specifications governing the project.

Equation (5-1) may be used as a guide in estimating the required time to fell trees only, by using a shear-type cutting blade, as illustrated in Fig. 5-1, mounted on a crawler tractor of the size indicated in Table 5-1. Before preparing an estimate, the estimator should visit the project to be cleared to obtain information needed to evaluate the variable factors in the formula. With this information, reasonably applicable values can be assigned to the factors listed in Table 5-1. Thus we have

$$T = B + M_1 N_1 + M_2 N_2 + M_3 N_3 + M_4 N_4 + DF \qquad (5\text{-}1)$$

FIGURE 5-2
Tractor-mounted V blade for clearing land.

where T = time per acre, min
    B = base time for tractor to cover 1 acre with no trees requiring
        splitting or individual treatment, min
    M = time required per tree in each diameter range, min
    N = number of trees per acre in each diameter range, obtained from
        field survey
    D = sum of diameters of all trees per acre, if any, larger
        than 6 ft in diameter at ground level, ft
    F = time required per foot of diameter to fell trees larger than 6 ft
        in diameter, min

Equation (5-1) can also be used to estimate the time required to stack felled trees into windrows spaced approximately 200 ft apart, by letting $M_1, M_2, \ldots$ represent the time required to move a tree into a windrow. Table 5-2 gives representative values for the time required to pile trees.

FIGURE 5-3
Tractor-mounted clearing rake.

FIGURE 5-4
Burning brush with forced draft and fuel oil.

TABLE 5-1
Representative times required to cut trees with tractor-mounted blades

| Size tractor, hp | B, min | Time to cut a tree†, min | | | | Time per foot for diameters above 6 ft (F) |
		1- to 2-ft diameter M_1	2- to 3-ft diameter M_2	3- to 4-ft diameter M_3	4- to 6-ft diameter M_4	
93	40	0.8	4.0	8.0	25	
130	28	0.5	2.0	4.0	12	4.0
190	21	0.3	1.5	2.5	7	2.0
320	18	0.3	0.5	1.5	4	1.2

† The listed times are for cutting trees flush with the surface of the ground. If it is necessary to remove the stumps, the time should be increased by 50 percent.

Cost of Clearing Land

Very little information on this subject has been released. However, in 1958 the Agricultural Experiment Station of Auburn University, Auburn, Alabama, conducted tests to determine the cost of clearing land by using three sizes of crawler tractors, equipped with bulldozer blades and with shearing blades, such as the one illustrated in Fig. 5-1. The results of the tests have been published in a booklet.

For test purposes, an area of 24 acres was divided into 12 plots of 2 acres each, with dimensions 198 ft wide and 440 ft long. Each size tractor cleared two plots using a bulldozer blade and two plots using a shear blade. The net time required to fell, stack, and burn the material from each plot was determined. The trees consisted of pine, oak, hickory, and gum, distributed by species, size, and

TABLE 5-2
Representative times required for stacking trees with tractor-mounted blades or rakes

| Size tractor, hp | B, min | Time to stack a tree, min | | | | Time per foot for diameters above 6 ft (F) |
		1- to 2-ft diameter M_1	2- to 3-ft diameter M_2	3- to 4-ft diameter M_3	4- to 6-ft diameter M_4	
93	35	0.3	0.6	2.5	. . .	
130	28	0.2	0.4	1.5	3.0	
190	24	0.1	0.3	1.0	2.0	0.4
320	20	0.0	0.1.	0.7	1.2	0.2

TABLE 5-3

Types of equipment used, species, sizes, and densities of trees removed

Plot no.	Blade used†	Percentage by species		Percentage by size trees		No. trees per acre
		Hardwood	Pine	To 6 in	Above 6 in	
1	B	79	21	87	13	375
2	B	98	2	74	26	285
3	B	97	3	76	24	385
4	B	56	44	87	13	585
5	B	53	47	93	7	680
6	B	78	22	87	13	755
7	S	80	20	86	14	690
8	S	29	71	98	2	1,545
9	S	72	28	82	18	445
10	S	60	40	98	2	710
11	S	89	11	72	28	410
12	S	75	25	76	24	400

† B denotes a bulldozer, S denotes a shearing blade.

density as listed in Table 5-3. The diameters of the trees were measured at breast height.

The trees were felled and then pushed along the surface of the ground and stacked in windrows not more than 198 ft apart, after which they were burned. During the burning operation the timber was pushed into stacks to increase the burning effectiveness, using a tractor-mounted blade.

Table 5-4 shows the averge time required by each size crawler tractor and type of blade to fell, stack, and dispose of 1 acre of timber. The smaller times

TABLE 5-4

Average machine time required to clear 1 acre of land based on size tractor and blade used

Operation	Time per acre, h					
	93 fwhp		130 fwhp		190 fwhp	
	B†	S	B	S	B	S
Felling	2.19	1.58	1.71	1.14	0.92	0.71
Stacking	0.52	0.55	0.56	0.60	0.48	0.46
Disposal	1.75	0.84	1.80	0.78	1.93	0.70
Total	4.46	2.97	4.07	2.52	3.33	1.87

† B denotes a bulldozer, and S denotes a shearing blade.

required to dispose of trees felled with the shearing blades were the results of smaller amounts of soil in the roots of the trees felled with this type of blade.

Example 5-1. Estimate the direct cost per acre for clearing, grubbing, and disposing of the vegetation on 46 acres of land. All trees and shrubs are to be pushed down, and all roots larger than 1 in in diameter in the top 18 in of soil are to be removed, stacked, and burned on the site.

The area is the right of way for a highway 150 ft wide. After the trees are felled, they will be pushed into windrows about 150 ft apart and will be burned.

The trees and shrubs will be felled by using tractor-mounted V blades, as illustrated in Fig. 5-2. Roots and any remaining stumps will be removed by rippers mounted on the rear of tractors. All the material will be pushed into windrows by using tractor-mounted rakes similar to the one illustrated in Fig. 5-4.

The area is covered with elm and oak trees, plus smaller shrubs, whose average count per acre is as follows:

> Elm trees, 24 to 36 in in diameter, 6 per acre
> Elm trees, 12 to 24 in in diameter, 20 per acre
> Oak trees, 24 to 36 in in diameter, 8 per acre
> Oak trees, 12 to 24 in in diameter, 18 per acre
> Smaller trees, 124 per acre

The soil is sandy clay, reasonably well drained, with no standing water or ponds of water. The work will be done during the summer, when rainfall should average 3 to 4 in per month. As the material is burned, the tractor-mounted rake will push and restack it into tighter piles to provide better burning. Fuel oil will be applied as needed to ensure good burning.

Tractors having the following sizes will be used:

> For felling trees, 180 hp
> For the rake, 180 hp
> For removing stumps and roots, 180 hp

From Table 5-1 and Eq. (5-1), the time required to cut and fell the trees on an average acre should be

$$T = 21 + 0.3 \times 38 + 1.5 \times 14$$
$$= 21 + 11.4 + 21.0$$
$$= 53.4 \text{ min, or } 0.89 \text{ h}$$

With a tractor-mounted rooter with four teeth to cut a swath 6 ft wide each pass, it will require 35 passes to cover a square acre, 208 ft by 208 ft. The total distance traveled by the rooter will be $35 \times 208 = 7{,}280$ ft. At an average speed of 1 mi/h it will require 0.92 h/acre to remove the stumps and roots.

From the information in Table 5-2 and Eq. (5-1), the time required to stack the trees from an acre of land will be

$$T = 28 + 0.2 \times 38 + 0.4 \times 14$$
$$= 28 + 7.6 + 5.6$$
$$= 41.2 \text{ min, or } 0.68 \text{ h}$$

Additional stacking of the material by a tractor-mounted rake during the burning will require at least one-half as much time as the initial stacking. Assume that this time will be 0.35 h/acre.

A summary of times for the several operations, based on a 60-min hour, will be

Operation	Time
Cutting trees	0.89
Removing roots and stumps	0.92
Stacking material	0.68
Restacking material	0.35

Because of the nature of the work performed in clearing land, it is probable that the equipment will work no more than 40 min/h. If the time per operation is adjusted to reflect this operating condition, the adjusted time per operation will be

Cutting trees, $0.89 \times \frac{60}{40}$ = 1.34 h
Removing roots and stumps, $0.92 \times \frac{60}{40}$ = 1.38 h
Stacking trees, $0.68 \times \frac{60}{40}$ = 1.02 h
Restacking trees, $0.35 \times \frac{60}{40}$ = 0.53 h

The operations of stacking and restacking will use the same equipment, namely, a tractor-mounted rake. The time required to stack and restack the material will be 1.55 h/acre. This should be the governing time for progress on the project, and this rate will be used for cutting and felling the trees and removing the roots and stumps.

The cost per acre, based on 1.55 h/acre, will be

Moving in and out, $860.00 ÷ 46 acres =	$ 18.70
Tractors, 3 × 1.55 = 4.65 h @ $36.84 =	171.31
V blade, 1.55 h @ $5.86 =	9.08
Rake, 1.55 h @ $6.84 =	10.60
Ripper, 1.55 h @ $6.26 =	9.70
Ripper teeth, 1.55 h @ $3.14 =	4.87
Hand tools =	9.54
Fuel for burning, 30 gal @ $0.96 =	28.80
Tractor operators, 4.65 h @ $15.80 =	73.47
Laborers, 4 × 1.55 = 6.20 h @ $6.90 =	42.78
Foreman, 1.55 h @ $18.60 =	28.83
Pickup truck, 1.55 h @ $5.96 =	9.24
Total cost =	$416.92

CONCRETE PAVEMENT

General Information

The cost of concrete pavement in place includes the cost of fine-grading the subgrade; side forms; steel reinforcing, if required; aggregate; cement; mixing,

FIGURE 5-5
Autograde trimmer trimming subgrade for pavement. (*CMI Corporation.*)

placing, spreading, finishing, and curing concrete; expansion-joint material; and shaping the shoulders adjacent to the slab. If the subgrade is dry, it may be necessary to wet it before placing the concrete.

If the pavement is not uniformly thick, the average thickness may be determined in order to find the area of the cross section. This area multiplied by the length will give the volume of concrete required, usually expressed in cubic yards. Payment for concrete pavement usually is at an agreed price per square yard of area.

Construction Methods Used

At least two methods are used in placing concrete pavement.

The older method, which is still used, is to bring the base under the pavement to the specified density, grade, and shape and then to install side forms, to confine the concrete until it sets and to control the thickness of the slab.

A newer but very satisfactory method is to use a slip-form paver to shape the slab. The two side forms, which confine the outer edges of the freshly placed concrete, are mounted on a self-propelled paver, which spreads, vibrates, and

FIGURE 5-6
Slip-form paver paving concrete highway. (*CMI Corporation.*)

screeds the concrete to the specified thickness and surface shape as it moves along the job. Figure 5-6 illustrates a slip-form paver on a project.

Preparing the Subgrade for Concrete Pavement

If the subgrade has been previously constructed, compacted, and shaped to the approximate final evaluation, the preparation for the placing of concrete will usually include the installation of the side forms, fine-grading the subgrade to the exact shape and elevation, and possibly wetting and compacting after the fine-grading is complete.

The forms most commonly used are made of steel whose height is equal to the thickness of the concrete adjacent to the forms. Forms are manufactured in sections 10 ft long, with three holes per section for pins, which are driven into the ground to maintain alignment and stability. Since the tops of the forms are used to control the elevations of the subgrade and the concrete slab, they must be set to the exact elevation required.

Forms usually are left in place 8 to 12 h after the concrete is placed, after which they are removed, cleaned, oiled, and reused. Handling may be done by hand or with a motor crane and a truck.

Fine-grading is usually accomplished with a self-propelled mechanical subgrader which rides on the forms. This machine loosens the earth to the required depth and removes it with conveyer blades which deposit it outside the forms.

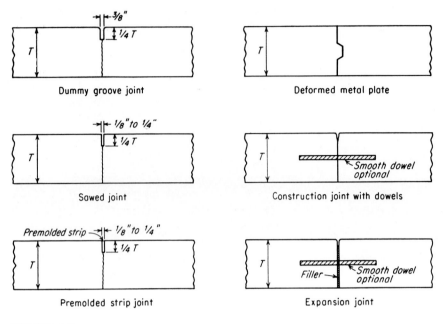

FIGURE 5-7
Representative details of joints in concrete pavement.

Self-propelled rollers or vibrating equipment may be used to compact the subgrade after the fine-grading is completed, with water added if necessary.

Handling, Batching, and Hauling Materials

A suitable site is needed for the storage and batching of materials. The equipment will include bulk cement bins, overhead bins for aggregate storage, equipped with weight batchers, and one or more clamshells to handle the aggregate. A cost study should be made prior to locating this plant to determine if more than one location is desirable. For a long paving project, more than one setup may be justified to reduce the length of haul for the batched materials.

Table 5-5 gives recommended sizes of roadbuilder's bins for storing and batching aggregate and sizes of clamshells for handling the aggregate.

If the concrete is to be mixed in a paver, dump trucks are used to haul the materials from the batching plant to the paver. Such trucks will haul two or more batches per load, depending on the size of the truck and the size of the batch. A batch includes the cement and aggregate required for one operation of the mixer. The weight of batched material will run about 3,850 lb/yd^3. If a size 34E paver operating at 110 percent capacity is used, the weight of a batch will be

$$\frac{34 \times 1.10 \times 3,850}{27} = 5,350 \text{ lb}$$

TABLE 5-5
Recommended size of bins and clamshells for handling and storing aggregate

Size mixer	Min. size bin, tons	Size clamshell bucket, yd^3	Size crane, yd^3	Boom length, ft	Operating radius, ft
1 27E single-drum	75	$\frac{3}{4}$	$\frac{3}{4}$	45	40
1 34E single-drum	75	1	1	45	40
1 16E dual-drum	50	$\frac{1}{2}$	$\frac{1}{2}$	40	35
1 34E dual-drum	100	$1\frac{3}{4}$	$1\frac{3}{4}$	50	42
2 34E dual-drums	190	3	$2\frac{1}{2}$	60	50

For three batches, the total weight on the truck chassis will be about

$$\begin{aligned}
\text{Weight of 3 batches, } 3 \times 5{,}350 &= 16{,}050 \text{ lb} \\
\text{Weight of truck body and hoist} &= \underline{\ 4{,}000 \text{ lb}} \\
\text{Total weight} \qquad\qquad &= 20{,}050 \text{ lb}
\end{aligned}$$

The number of trucks required to supply the aggregate may be obtained from the formula

$$N = \frac{60}{Kt}\left(\frac{60L}{S} + T\right) \tag{5-2}$$

where N = no. of trucks
K = no. of batches per truck
t = cycle time per batch, s
L = round-trip distance, mi
S = average speed of trucks, mi/h
T = average time per truck at batching plant and paver, plus reasonable delays, min

Mixing and Placing Concrete

Concrete for pavements is usually mixed in paving mixers. The Mixer Manufacturer's Bureau of the Associated General Contractors of America lists three sizes as standard, namely, 16-, 27-, and 34-ft^3 nominal capacities, when the mixers are operated on slopes not greater than 6 percent. Sizes 16E and 34E are available as single- and dual-drum units. The use of a dual-drum unit, which has a separate premixing and final mixing drum, will permit a mixer to produce a batch in less than a minute under favorable conditions.

Table 5-6 gives representative production rates for paving mixers based on the mixing time of 60 s when they are operating at 110 percent capacity. The rates

TABLE 5-6
Representative production rates for paving mixers for 60-s mixing time

	Type mixer			
	Single-drum		Dual-drum	
Size mixer	27E	34E	16E	34E
No. batches per hour	42	42	65	65
Vol. concrete, yd^3/h	47	59	42	90
Vol. concrete per 8 h, yd^3	376	472	336	720

are based on an operating factor of 0.85 for single-drum mixers and 0.75 for dual-drum mixers. For other operating conditions, the rates given in the table should be adjusted.

After a batch is mixed, it is discharged into a bucket, which moves along a boom attached to the mixer, to permit the batch to be distributed over the base. After concrete is placed across the full width of the pavement, it is spread and vibrated by a self-propelled spreader, which travels on the forms. Subsequent operations include screeding, belting, and finishing, all of which are accomplished by self-propelled equipment.

A modified paving method has been developed which uses slip forms attached to the sides of a self-propelled machine, mounted on crawler tracks, which spreads, vibrates, and finishes the concrete pavement. The use of this machine is reported to reduce the cost of pavement $0.75 to $1.00 per square yard.

Joints

To reduce the danger of irregular and unsightly joints across and along concrete pavement, it is common practice to install joints at regular intervals. Three types are installed: construction, transverse, and longitudinal.

Transverse joints are installed across the pavement at spacings varying from 15 to 30 ft, or sometimes more. Longitudinal center joints are installed along the length of the pavement, usually when the width exceeds 12 ft and the thickness is less than 12 in.

Joints may be constructed by grooving the freshly placed concrete, by installing premolded joint material or wood planks as the concrete is placed, or by sawing the grooves after the concrete has been placed. Self-propelled or hand-pushed machines, using abrasive or diamond blades, are used for this operation. The width of joints specified may vary from $\frac{1}{8}$ to $\frac{3}{8}$ in. Cutting speeds should vary from 2 to 10 ft/min depending on the width and depth of the joint, type of blade used, and kind of aggregate used. Figure 5-7 shows sections through several types of joints.

FIGURE 5-8
Saw for cutting joints in concrete pavement.

The grooved or sawed joints should be filled with a sealing compound, usually an asphaltic or rubber-base material, applied either hot or cold.

Steel tie bars or rods may be installed across all joints, primarily to transfer shearing forces.

Curing Concrete Pavement

Curing is accomplished by covering the fresh slab with burlap, cotton mats, waterproof paper, or an impervious membrane-producing compound, which is sprayed on the surface and sides soon after the concrete is placed.

The burlap and mats must be kept wet for the specified time, while the paper and membrane prevent or reduce the evaporation of initial water in the concrete. The membrane-producing compounds, frequently called *curing compounds*, will cover 30 to 50 yd^2/gal.

> **Example 5-2.** Estimate the total cost and the cost per square yard, for bid purposes, for placing a concrete pavement 30 ft wide, with 9-in average thickness, and 5.78 mi long.
>
> The mix design specifies 524 lb of cement, 1,278 lb of fine aggregate, 2,096 lb of coarse aggregate, and 32.5 gal of water per cubic yard of concrete. A 34E dual-drum mixer operating at 110 percent capacity will be used to mix the concrete.

Transverse joints $\frac{1}{8}$ in wide and 2 in deep will be installed at 20-ft intervals across the slab. A longitudinal joint $\frac{1}{8}$ in wide and 2 in deep will be installed along the centerline of the slab. All joints will be sawed within 24 h after the concrete is placed. The joints will be sealed with hot asphaltic compound, applied with a pressure sealer.

No steel reinforcing or dowels will be placed in the pavement.

The batching plant for the aggregate will be set up near the midpoint of the project. The plant will include a bulk-cement silo of 500 barrels (bbl), equipped with a screw conveyer and a bucket elevator, a three-compartment 100-ton aggregate bin, and weight batchers, arranged for single-stop delivery to batch trucks. Trucks with a capacity of three batches each will haul the aggregate an average distance of 2.5 mi at an average speed of 30 mi/h. A $1\frac{1}{2}$-yd^3 crawler-mounted clamshell will be used to charge the aggregate into the bins.

Water will be obtained from a private pond, requiring a 3-in gasoline-engine-operated pump having a capacity of 20,000 gal/h. The water will be hauled an average distance of 3 mi by trucks whose capacities are 2,500 gal each. The average haul speed is estimated to be 25 mi/h. It will require 10 min to fill a truck and 10 min to empty a truck. The truck driver will operate the pump at the pond and the pump to empty the truck tank into a wheel-mounted truck tank that will move along with the paver.

The calculations will be as follows:

Volume of concrete, $\dfrac{5.78 \times 5,280 \times 30 \times 9}{27 \times 12} = 25,432 \text{ yd}^3$

$$\begin{array}{ll} \text{Add for waste and overrun, } 2\% \times 25,432 & = \underline{509 \text{ yd}^3} \\ \text{Total volume} & = 25,941 \text{ yd}^3 \end{array}$$

Area of pavement, $(5.78 \times 5,280 \times 30)/9$ = 101,728 yd^2
Mixer capacity, $(34 \times 1.10)/27$ = 1.39 yd^3 per batch
Mixer output, 65×1.39 = 90 yd^3/h
Assume a 50-min working hour
Probable output of mixer, $90 \times 50/60$ = 75 yd^3/h
Probable total time, $25,941 \div 75$ = 346 h
The costs will be based on 346 h

Use Eq. (5-2) to determine the number of trucks needed to haul the aggregate.

$$N = \frac{60}{Kt}\left(\frac{60L}{S} \times T\right) = \frac{60}{3 \times 55}\left(\frac{60 \times 5}{30} + 12\right)$$

$$= 8 \text{ trucks}$$

It may be desirable to have an extra truck available on a standby basis.

Quantity of water required for concrete only = 90 yd$^3 \times 32.5$ gal = 2,925 gal/h

Note that the quantity of water is based on the maximum output of the mixer, because at times this amount of water will be needed.

Cycle time for water truck:
 Filling tank, 10 min   = 0.167 h
 Emptying tank, 10 min   = 0.167 h
 Traveling, 6 min $\div$ 25 mi/h   = 0.240 h
  Total time   = 0.574 h

Capacity of water truck, 2,500 gal ÷ 0.574 = 4,360 gal/h
One water truck will be sufficient

The costs will be
  Materials:
    Cement, 25,941 yd^3 × (524/376) = 36,151 bbl @ $16.80 = $ 607,337
    Sand, 25,941 × (1,278/2,000) = 16,576 tons @ $7.92   =   131,282
    Gravel, 25,941 × (2,096/2,000) = 27,186 tons @ $9.11 =   247,664
    Water, 25,941 × 32.5 gal = 843 kgal @ $0.42   =   354
  Water, for general use, 843 kgal @ $0.42   =   354
    Curing compound, 101,728 yd^2 ÷ 45 yd^2/gal =
      2,261 gal @ $6.18   =   13,971
    Sealing compound for joints, 8.5 tons @ $92.60   =   787
      Subtotal cost of materials   = $1,001,749
  Equipment:
    Moving to and from project, excluding labor   = $ 4,620
    Batching plant, 346 h @ $24.82   =   8,588
    Clamshell, 346 h @ $71.24   =   24,649
    Mixer, 346 h @ $69.72   =   24,123
    Dump trucks, 8 × 346 = 2,768 h @ $22.36   =   61,892
    Side forms, 2 × 5.78 × 5,280 = 61,036 lin ft @
      $46.80 per 1,000 ft   =   2,856
    Air compressor and hammer to drive form pins,
      346 h @ $6.10   =   2,111
    Fine-grader, 346 h @ $18.94   =   6,553
    Earth roller, smooth-wheel, 8 tons, 346 h @ 3.10   =   1,072
    Spreader vibrator, 346 h @ $22.68   =   7,847
    Finisher, 346 h @ $11.94   =   4,131
    Truck for handling forms, 346 h @ $8.14   =   2,816
    Water truck, 346 h @ $11.56   =   3,999
    Water tank, 346 h @ $6.94   =   2,401
    Water pump, 346 h @ $3.72   =   1,287
    Motor grader, 346 h @ $18.52   =   6,408
    Concrete saws and blades, 2 × 346 = 692 h @ $17.86   =   12,359
    Asphalt heater, 346 h @ $9.75   =   3,374
    Air compressor for applying joint sealer, 346 h @ $5.13 =   1,775
    Pickup trucks, 2 × 346 = 692 h @ $4.96   =   3,432
  Sundry equipment and tools   =   785
      Subtotal cost of equipment   = $ 187,078
  Labor:
    Setting up and dismantling plant   = $ 2,496
    Batching plant operators, 2 × 346 = 692 h @ $10.90   =   7,543
    Clamshell operator, 346 h @ $17.28   =   5,979
    Clamshell oiler, 346 h @ $11.25   =   3,892
    Truck drivers, 10 × 346 = 3,460 h @ $8.10   =   28,026
    Form workers, 4 × 346 = 1,384 h @ $7.84   =   10,850
    Fine-grader operator, 346 h @ $15.80   =   5,467
    Roller operator, 346 h @ $11.56   =   4,000
    Mixer operator, 346 h @ $14.64   =   5,065

Spreader operator, 346 h @ $13.58	=	4,698
Finisher operator, 346 h @ $13.58	=	4,698
Grader operator, 346 h @ $15.80	=	5,466
Saw operators, 2 × 346 = 692 h @ $15.60	=	10,795
Applying joint sealer, 2 workers, 692 h @ $12.40	=	8,581
Laborers, 6 × 346 h = 2,076 h @ $8.10	=	16,815
Foreman, 346 h @ $16.40	=	5,674
Subtotal labor costs	= $	130,045
Workers' compensation insurance,		
$130,045 × $4.37 per $100.00	= $	5,683
Social security tax, $130,045 × 7.51%	=	9,766
Unemployment tax, $130,045 × 3%	=	3,901
Subtotal labor taxes	= $	19,350

Summary of costs:

Materials	=	$1,001,749
Equipment	=	187,078
Labor	=	130,045
Labor taxes	=	19,350
		1,338,222
General overhead, 4% of $1,338,222	=	53,529
		$1,391,751
Profit, 6% of $1,391,751	=	83,505
Performance bond:		
$500,000 at $12.00 per $1,000	=	6,000
$975,256 at $7.50 per $1,000	=	7,314
Total cost, bid price		$1,488,570
Cost per yd^2, $1,488,570 ÷ 101,728	=	14.63

ASPHALT PAVEMENT

Asphalt pavements are obtained by mixing and placing one or more types of mineral aggregate and bitumen or asphaltic binders. It is beyond the scope of this book to cover all the methods used.

Aggregates

The aggregates commonly used include sand, gravel, limestone, crushed iron ore, or other suitable crushed stone. Local materials should be used if they are of good quality. To produce the desired density, the aggregates should be proportioned by size from the largest to the smallest particles specified. Specifications usually designate the percentages passing and retained on screens having given size openings.

Asphalts

Asphalts used for paving construction may be classified as road oils, or slow-curing (SC) asphalts; cutback asphalts, medium-curing (MC) asphalts; and rapid-curing

(RC) asphalts; asphalt cements, either petroleum or natural asphalts; emulsified asphalts; and powdered asphalts. In addition to asphalts, coal tars and water-gas tars sometimes are used for paving construction.

Cold-Mix Asphaltic-Concrete Pavement

If a cold-mix method is used to construct the pavement, the aggregate is mixed with the specified quantity of asphalt. The asphalt may or may not be heated, depending on the type of aggregate and asphalt used. Mixing is usually accomplished in a pugmill mixer for a specified time, about 1 min for most jobs. After the batch is thoroughly mixed, it is discharged into a truck and hauled to the job, where it is spread by hand or by a mechanical spreader in layers of specified thickness. Compaction to the desired density is obtained by rolling with smooth-wheel or pneumatic rollers, producing a designated pressure. It may be necessary to place the material in two or more layers, with each layer rolled, to obtain uniform compaction throughout the full depth of the pavement.

If a travel plant is used to mix the material, the aggregate is placed in windrows along the road to be paved. The mixer picks up the aggregate with a bucket conveyer, mixes it with the asphalt, and deposits it in layers at the rear of the mixer. This method eliminates trucks for hauling the mixture and reduces spreading costs.

Hot-Mix Asphaltic-Concrete Pavement

If a hot-mix method is used to construct the pavement, the aggregate is heated in a rotary kiln or other suitable apparatus, permitting constant agitation, to 300 to 375°F. After it is heated, it is screened and recombined, to give the specified size grading, and then discharged into a pugmill type of mixer. The asphalt cement is heated to the specified temperature, usually 275 to 375°F. The aggregate and the asphalt cement are mixed in proper proportions in the pugmill for the specified time or until a uniform mixture is obtained. The quantity of asphalt cement varies from 4 to 9 percent of the weight of the finished product. Both batch- and continuous-type mixers are used.

After the material is mixed, it is discharged into trucks and hauled to the job, where it is spread in uniformly thick layers by a mechanical spreader or by some other suitable method. After each layer is deposited, it is compacted by a wheel-type roller to the desired density. It may be necessary to cover the material with a canvas or tarpaulin during transit to prevent excessive loss of heat between the mixing plant and the job.

Cost of Hot-Mix Asphaltic-Concrete Pavement

The initial cost of the plant for producing hot-mix asphaltic-concrete pavement is relatively high. The cost of moving a plant to a location and setting it up for operation can run as high as $8,000 to $15,000 or more. Consequently, a single setup is generally made for a given job.

The cost of hot-mix asphaltic-concrete pavement will include the cost of aggregate, asphalt, moving the plant to the location, setting it up, operating the plant, hauling the material to the job, spreading, and rolling. The cost of operating the plant will include depreciation, interest, insurance and taxes, maintenance and repairs, and fuel. Additional costs may include the rental of a site for the plant and a railroad spur track for the delivery of aggregate and asphalt.

Equipment for Hot-Mix Asphaltic-Concrete Pavement

The total equipment for mixing and placing hot-mix asphaltic-concrete pavement will include such items as a clamshell to handle the aggregate, cold-aggregate storage and feeder bins, a cold-aggregate elevator, an aggregate drier, a dust collector, a hot-aggregate elevator, aggregate screens, hot-aggregate storage bins and batcher, a mixer, asphalt storage tanks, fuel-oil storage tank, steam boiler or hot oil heater to heat the asphalt, trucks to haul the asphalt mixture, an asphalt distributor, a mechanical spreader, and rollers to compact the pavement. If the subgrade or subbase is not already prepared, additional equipment will be required to do this work.

Example 5-3. Estimate the total direct cost and the cost per ton for mixing and placing hot-mix asphaltic-concrete pavement 18 mi long, 24 ft wide, and 5 in thick on a previously prepared subbase. Prior to placing a 3-in-thick base, an asphalt prime coat will be applied to the top of the subbase at the rate of 0.3 gal/yd^2. A 2-in-thick wearing course will be placed on top of the base with no tack coat required.

The paving mixtures shall meet the requirements for grading and mix composition shown in the accompanying table.

Course Thickness, in	Base 3	Wearing 2
Sieve size	Combined aggregate including filler; percent passing, by weight	
2 in	100	
1½ in	95–100	
1 in	—	100
¾ in	70–85	95–100
½ in	—	75–90
No. 4	35–50	45–60
No. 10	25–37	35–47
No. 40	15–25	23–33
No. 80	6–16	16–24
No. 200	2–6	6–12
Asphalt cement, percentage of combined weight	6.0	6.0

The plant will require 60 percent coarse aggregate, larger than no. 10 sieve, and 40 percent fine aggregate and filler. Assume that 5 percent of the coarse aggregate, 10 percent of the fine aggregate, and 2 percent of the asphalt will be lost through waste or for other reasons. The combined material will weigh about 3,600 lb/yd^3 when compacted.

The material will be mixed in a continuous mixer whose maximum capacity is 150 tons/h. Based on a 50-min hour, the average output will be $0.833 \times 150 = 125$ tons/h.

The mixing plant will include the units listed below with costs as indicated.

1 four-compartment bin and feeder	
1 fuel-oil-fired drier	
1 dust collector	
1 hot elevator	
1 screening unit	
1 mixer, 150 tons/h continuous type	
Cost of above-listed equipment =	$348,620
2 asphalt storage tanks of 10,000 gal =	26,840
1 fuel-oil storage tank of 2,000 gal =	2,925
1 hot-oil heater =	22,385
1 asphalt pump with engine =	4,845
1 set of pipes for asphalt =	12,960
900 gal of heat-exchange oil =	1,170
1 clamshell =	146,875
Subtotal cost	$566,620

Assume that this equipment has a useful life of 5 yr with no salvage value at the end of its life and that it will be used an average of 1,400 h/yr.

The average value will be $0.6 \times \$566,620 = \$339,972$
The annual costs will be:

Depreciation, $566,620 ÷ 5	= $113,324
Maintenance and repairs, 10% of $566,620	= 56,662
Investment, 14% × $339,972	= 47,596
Total annual fixed cost	$217,582

The hourly costs will be:

Fixed cost, $217,582 ÷ 1,400	= 155.42
Hot-oil heater fuel, 16 gal @ $0.81	= 12.96
Drier fuel, 360 gal @ $0.78	= 280.80
Diesel fuel for mixer engine, 4.6 gal @ $0.89	= 4.09
Diesel fuel for drier engine, 8.2 gal @ $0.89	= 7.30
Lubricating oil, 0.5 qt @ $1.51	= 0.76
Other lubricants and grease	= 0.26
Total cost per hour	$461.59
Cost per ton, $461.59 ÷ 125 tons/h	3.69

The total quantities will be

Area, 18.04 mi × 5,280 ft × 24 ft/9 = 254,000 yd²
Material for base course,

$$\frac{254,000 \times 3 \times 3,600}{36 \times 2,000} = 38,200 \text{ tons}$$

Material for wearing course,

$$\frac{254,000 \times 2 \times 3,600}{36 \times 2,000} = 25,400 \text{ tons}$$

Total weight = 63,600 tons
Total effective weight, 0.94 × 63,600 = 59,800 tons

Assume that the cost of moving the plant will be charged to this project and that the cost of moving it out will be charged to the next project. If this is not possible, the entire cost of moving to and from this project should be charged to this project.

The costs will be:

Materials:

Coarse aggregate, 0.6 × 59,800 × 1.05 = 37,674 tons @ $9.12	= $	343,587
Fine aggregate, 0.4 × 59,800 × 1.10 = 26,312 tons @ $7.92	=	208,391
Asphalt, 0.06 × 63,600 × 1.02 = 3,892 tons @ $82.80	=	322,258
Priming oil, 254,000 yd² × 0.3 gal = 76,200 gal @ $0.42		32,004
Total cost of materials	= $	906,240
Cost per ton, $906,240 ÷ 63,600	=	14.25

Equipment at mixing plant:

Moving to and setting up plant	= $	8,560
Plant cost, 63,600 tons ÷ 125 tons/h = 508 h @ $461.59 =		234,488
Total cost at mixing plant		$ 243,048

Labor at mixing plant:

Foreman, 508 h @ $19.60	= $	9,957
Crane operator, 508 h @ $17.26	=	8,768
Crane oiler, 508 h @ $11.25	=	5,715
Weighers, 2 × 508 = 1,016 h @ $11.40	=	11,582
Batch plant operator, 508 h @ $17.85	=	9,068
Mechanic, 508 h @ $13.75	=	6,985
Laborers, 3 × 508 = 1,524 h @ $8.10	=	12,344
Total labor cost		$ 64,419

Hauling mixture from mixing plant to roadway, subcontract,

63,600 tons @ $1.19	= $	75,684

Equipment placing the pavement:
Asphalt distributor, 1,500-gal cap at 60% of 508 =

305 h @ 21.75	= $	6,634
Lay-down paver, 125 tons/h, 508 h @ $20.85	=	10,591
Roller, smooth-wheel, 12 tons, 508 h @ $11.18	=	5,679
Pneumatic roller, self-propelled, 508 h @ $15.75	=	8,001
Total cost	$	30,905

Labor placing pavement:

Foreman, 508 h @ $16.40	= $	8,331
Distributor operator, 305 h @ $12.84	=	3,916
Helper, 305 h @ $7.90	=	2,409
Paver operator, 508 h @ $15.85	=	8,052
Roller operators, 2 × 508 h = 1,016 h @ $11.56	=	11,745
Laborers, 3 × 508 = 1,524 h @ $8.10	=	12,344
Total cost	$	46,797

Other direct costs:

Land site rental, 4 mo × $160	= $	640
Plant and road tools	=	1,240
Service trucks, 2 × 508 h = 1,016 @ $5.51	=	5,598
Barricades, flares, signals	=	1,480
Truck driver, 508 h @ $8.10	=	4,115
Helper, 508 h @ $8.10	=	4,115
Superintendent, 4 mo @ $3,120	=	12,480
Time clerk, 4 mo @ $1,344	=	5,376
Total costs	$	35,044

Summary of direct costs:

Materials	= $	906,240
Equipment	=	243,048
Labor at mixing plant	=	64,419
Hauling	=	75,684
Equipment placing pavement	=	30,905
Labor placing pavement	=	46,797
Other direct costs	=	35,044
Total direct costs		$1,402,137
Cost per ton, $1,402,137 ÷ 63,600		22.05

PROBLEMS

5-1. Estimate the total cost and the cost per acre for clearing and grubbing 64 acres of sandy land. All trees and shrubs and all roots larger than 1 in in diameter in the top 16 in of soil are to be removed, stacked, and burned on the site. There are no obstructions to delay progress during the operations.

Use the information in this book as a guide in determining the production rates and the cost of wages and equipment.

The average tree count per acre is as follows:

Elm trees, 22 to 30 in in diameter	6
Elm trees, 12 to 22 in in diameter	16
Oak trees, 24 to 30 in in diameter	12
Oak trees, 12 to 24 in in diameter	14
Smaller trees	56

The cost of moving to the job and back to storage will be $684.00.

5-2. An area of 180 acres is to be cleared of all trees and shrubs above the surface of the ground. Stumps and roots below the surface of the ground do not need to be removed.

The trees and shrubs will be cut off even with the surface of the ground by one or more tractor-mounted V blades and then stacked for burning on the site, using one or more tractor-mounted rakes such as the one illustrated in Fig. 5-2. The soil is sandy clay and relatively dry.

Use the wage rates and equipment costs from the example beginning on page 100 of this book.

The average tree count per acre will be

Hardwood trees, 20 to 30 in in diameter	18
Hardwood trees, 10 to 20 in in diameter	56
Small trees	48

The cost of moving to the job and back to storage will be $976.00.

5-3. Estimate the total cost and the cost per square yard of concrete, for bid purposes, for furnishing the materials, equipment, and labor for constructing a pavement 26 ft wide, 9 in thick, and 7.24 mi long. Use the same types of equipment, equipment costs, wage rates, material costs, and overhead costs as those in the example beginning on page 107 of this book.

The batched aggregate will be hauled to the mixer in trucks that hold three batches each. The average haul distance for the aggregate will be 2 mi. Use the same batch mix as in the example.

No reinforcing steel or dowels will be placed in the concrete. Joints will be sawed as specified in the example.

The cost of moving the plant to the project and back to the storage yard will be $10,780, excluding the cost of labor in setting it up and taking it down.

5-4. Estimate the total direct cost and the cost per ton for furnishing materials and for mixing and placing hot-mix asphaltic-concrete pavement 14.64 mi long, 28 ft wide, and 4 in thick on a previously prepared base. The pavement will be placed in two layers, each 2 in thick, after compaction.

Use the same methods, material costs, equipment, equipment costs, wage rates, and production rates as those used in the example beginning on page 112 of this book. Change the asphalt cement to 7 percent of the combined weight of the concrete.

CHAPTER
6

PILING AND BRACING

When a project requires excavation into earth which is so unstable that the walls must be supported to prevent them from caving into the pit or trench, it will be necessary to install a system of shores, braces, or solid sheeting along the walls to hold the earth in position. If groundwater is present, it may be necessary to install semiwatertight sheeting around the walls to exclude or reduce the flow of water into the pit. Timber and steel are used for braces and sheeting.

For large excavation projects where the water table is near the surface of the area to be excavated, a well-point dewatering system may be required. A series of wells are drilled around the excavation area. Pumps are installed to draw the water table down so that the excavation can be performed in a dry condition. Well-point systems are generally installed by contractors who specialize in dewatering techniques. The cost will depend on the depth of the water table, type of soil, and size of the excavation area.

Piles are installed under a structure to transmit the loads from the structure into a deeper soil which has sufficient strength to support the loads. The loads are transferred by skin friction or end load bearing or both. Timber, concrete, or steel piles are installed by the driving action from the hammer of pile-driving equipment. Piles are driven from depths that range from 10 to 200 ft, depending on the loads that must be transmitted and the strength of the soil.

Drilled shaft foundations are installed by drilling holes to depths in the soil which have sufficient strength to support the loads from the structure above the foundation. A belling tool can be used to widen the bottom of the drilled hole, to increase the load-bearing capacity of the foundation. For sandy soils a steel casing may be required to stablize the soil during the drilling operation. Reinforced concrete is installed after the shaft has been drilled and belled. Drilled shafts can be installed with diameters from 10 to 96 in and at depths from 5 to 100 ft.

Shoring Trenches

If a trench is excavated deeper than 5 ft into reasonably firm earth, shoring is generally required (Fig. 6-1). Rough lumber, such as 2- by 10-in or 2- by 12-in planks (as long as the deepest portion of the trench), may be installed on opposite sides of the trench in a vertical position to prevent the earth from caving in. The spacing of shores should be about 6 to 8 ft along the trench. Wood or metal trench braces, spaced about 3 to 4 ft apart, one above the other, may be used to hold the shores against the walls of the trench. The shores and braces are removed as the trench is backfilled. The shores may be used several times, and the metal trench braces should last several years. A truck will be needed to haul the shores and braces to different locations along the trenches. Two or more workers will be required to install and remove the shores and braces. A method of estimating the cost of shoring a trench is illustrated in the following example.

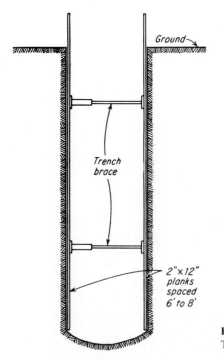

FIGURE 6-1
Trench braces and shores in a sewer trench.

Example 6-1. Estimate the cost of providing shores for 100 ft of trench with depth varying from 8 to 10 ft, using 2- by 12-in planks and trench braces. The planks will be spaced 6 ft apart along the trench. Trench braces will be placed not more than 2 ft above the bottom of the trench and spaced not more than 3 ft apart vertically to within 2 ft below the surface of the ground.

Assume that the shore lumber will be used 30 times.

The estimated cost will be determined as follows.

No. of shores, $100 \div 6 = 17$
No. of planks, $2 \times 17 = 34$

Lumber, 34 pc, 2 in $\times$ 12 in $\times$ 10 ft $= \dfrac{680 \text{ fbm} \times \$540/1{,}000 \text{ fbm}}{30 \text{ uses}}$ = \$ 12.24

Trench braces, 3 per shore = $3 \times 17 = 51$ @ $0.19	=	9.69
Labor installing, 17×0.40 h = 6.8 h @ $7.50	=	51.00
Labor removing, 17×0.3 h = 5.1 h @ $7.50	=	38.25
Truck, part time, 3 h @ $14.75	=	44.25
Total cost	=	\$155.43

Sheeting Trenches

If the earth is so unstable that it must be restrained for the full areas of the wall of a trench, it will be necessary to install solid sheeting to the full depth of the trench, as illustrated in Fig. 6-2. For depths up to about 12 ft, 2-in-thick lumber may be used

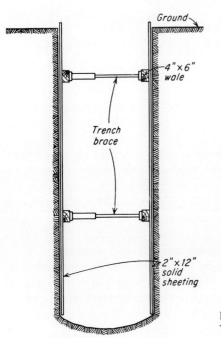

FIGURE 6-2
Trench braces, wales, and solid bracing.

for sheeting with 4- by 6-in lumber wales and braces; for depths of 12 to 20 ft, 3-in-thick lumber should be used for sheeting with 6- by 8-in or 8- by 8-in lumber for wales and braces. The lumber may be rough sawed or S4S.

Depending on the stability of the earth, it may be necessary to drive the sheeting and install some of the braces ahead of excavating. The sheeting may be driven with a maul or with a pneumatic hammer, such as a paving breaker equipped with a suitable driving head.

Example 6-2. Estimate the cost of installing and removing solid sheeting and braces for 100 ft of trench 7 ft deep. Use 2- by 12-in lumber 8 ft 0 in long for sheeting with two horizontal rows of 4- by 6-in wales on each side of the trench for the full length of the trench. Trench braces will be placed 4 ft apart along each row of wales.

Assume that the sheeting can be used 4 times and the wales 10 times. The sheeting will be driven one plank at a time by using a 75-ft^3/min air compressor and a pneumatic hammer.

The estimated cost will be determined as follows.

Sheeting, $\dfrac{200 \text{ ft} \times 12}{11.25} = 214 \text{ pc} = \dfrac{214 \times 2 \times 12 \times 8}{12 \times 1,000} = 3.424$ M fbm

@ \$135.00 per use ....= \$ 462.24

Wales, $\dfrac{200 \text{ ft} \times 4 \times 6}{12 \times 1,000} = 0.4$ M fbm @ \$54.00 per use = 21.60

Trench braces, $2 \times 100 \div 4 = 50$ @ \$0.18 .... = 9.00

Labor-driving sheeting, 2 men, 214 pc ÷ 5 pc per hr =

42.8 hr × 2 = 85.6 hr @ \$7.50 ....= 642.00

Labor-installing wales and braces, 50 each ÷ 5 per hr =

10 hr × 2 men = 20 hr @ \$7.50 ....= 150.00

Labor removing sheeting, wales, and braces,

$\frac{1}{4}$ of time to install = 26.4 hr @ \$7.50 ....= 198.00

Air compressor, 85.6 ÷ 2 = 42.8 hr @ \$6.25 ....= 267.50

Air hammer, hose, etc., 42.8 hr @ \$0.85 ....= 36.38

Total cost ....= \$1,786.72

Pile-Driving Equipment

Equipment used to drive piles on land usually consists of a skid rig or a truck-mounted or a crawler-mounted crane, leads, a hammer, and a source of steam or compressed air to drive the hammer. When piles are driven in water, the driving rig is usually mounted on a barge. The actual driving of piles usually is accomplished with hammers. Several types are available, including the drop hammer, single-acting steam hammer, double-acting steam hammer, differential-acting steam hammer, self-contained diesel-operated hammer, and vibratory hammer.

The size of a drop hammer is indicated by the weight of the hammer, whereas the size of a steam or diesel hammer is indicated by the theoretical foot-pounds of energy delivered by each blow. Steam hammers can be operated by compressed air at pressures of 80 lb to 150 lb/in^2.

FIGURE 6-3
A single-acting steam hammer driving a steel shell
for a cast-in-place concrete pile.

Table 6-1 gives recommended sizes of hammers for different types and sizes of piles and driving conditions. Table 6-2 gives information for various sizes and types of hammers.

Sheet Piling

Sheet piles are used to form a continuous wall by installing sheets of steel that interlock. The cost of sheet piling in place will include the cost of the piling, driving equipment, and labor. If the piling is to be salvaged, there will be an additional cost for extracting it. It is common practice to drive two piles simultaneously.

Table 6-3 gives information for sheet piles manufactured by steel companies in the United States. Table 6-4 gives representative rates for driving steel-sheet piles when they are driven in pairs with the size hammer recommended in Table 6-1.

TABLE 6-1
Recommended sizes of hammers for driving piles, in theoretical foot-pounds of energy per blow

Length of piles, ft	Depth of penetration	Weight of piles, lb/lin ft									
		Steel sheet†			Timber			Concrete		Steel	
		20	30	40	30	60	150	400	40	80	120
		**Driving through ordinary earth, moist clay, and loose gravel; normal frictional resistance**									
25	$\frac{1}{2}$	2,000	2,000	3,600	3,600	7,000	7,500	15,000	3,600	7,000	7,500
	Full	3,600	3,600	6,000	3,600	7,000	7,500	15,000	4,000	7,500	7,500
50	$\frac{1}{2}$	6,000	6,000	7,000	7,000	7,500	15,000	20,000	7,000	7,500	12,000
	Full	7,000	7,000	7,500	7,500	12,000	15,000	20,000	7,500	12,000	15,000
75	$\frac{1}{2}$	—	7,000	7,500	—	15,000	—	30,000	7,500	15,000	15,000
	Full	—	—	12,000	—	15,000	—	30,000	12,000	15,000	20,000
		**Driving through stiff clay, compacted sand, and gravel; high frictional resistance**									
25	$\frac{1}{2}$	3,600	3,600	3,600	7,500	7,500	7,500	15,000	5,000	9,000	12,000
	Full	3,600	7,000	7,000	7,500	7,500	12,000	15,000	7,000	10,000	12,000
50	$\frac{1}{2}$	7,000	7,500	7,500	12,000	12,000	15,000	25,000	9,000	15,000	15,000
	Full	—	7,500	7,500	—	15,000	—	30,000	12,000	15,000	20,000
75	$\frac{1}{2}$	—	7,500	12,000	—	15,000	—	36,000	12,000	20,000	25,000
	Full	—	—	15,000	—	20,000	—	50,000	15,000	20,000	30,000

† The indicated energy is based on driving two steel-sheet piles, simultaneously. When single piles are driven, use approximately two-thirds of the indicated energy.

TABLE 6-2
Data on pile-driving hammers

Hammer	Model	Ram weight, lb	Blows per minute	Rated energy, ft·lb
Vulcan single-action	2	3,000	70	7,260
	1	5,000	60	15,000
	8	8,000	50	26,000
	10	10,000	50	32,500
	12	12,000	55	39,000
	24	24,000	60	72,000
	340	40,000	60	120,000
	360	60,000	62	180,000
McKiernan-Terry single-action	S5	5,000	60	16,250
	S8	8,000	55	26,000
	S10	10,000	55	32,500
	S14	14,000	60	37,500
	S20	20,000	60	60,000
	S40	40,000	55	120,000
	S60	60,000	55	180,000
Raymond International single-action	1	5,000	60	15,000
	1S	6,500	58	19,500
	0	7,500	52	24,375
	2/0	10,000	50	32,500
	3/0	12,500	48	40,600
	4/0	15,000	46	48,750
	5/0	17,500	44	56,875
	8/0	25,000	35	81,250
Vulcan differential action	30C	3,000	133	7,260
	50C	5,000	120	15,100
	80C	8,000	111	24,450
	120C	12,000	108	32,000
	140C	14,000	103	36,000
	200C	20,000	98	50,200
	400C	40,000	100	113,488
	600C	60,000	100	164,507
McKiernan-Terry double-action	9B3	1,600	145	8,750
	10B3	3,000	105	13,100
	11B3	5,000	95	19,150
Raymond International differential action	65C	6,500	110	19,500
	80C	8,000	100	24,450
	150C	15,000	100	48,750

TABLE 6-3
Properties of steel-sheet piles

Section No.	Width, in	Web thickness, in	Weight lb/lin ft of pile	Weight lb/ft² of wall
PSX32	$16\frac{1}{2}$	$\frac{29}{64}$	44.0	32.0
PS32	15	$\frac{1}{2}$	40.0	32.0
PS28	15	$\frac{3}{8}$	35.0	28.0
PSA28	16	$\frac{1}{2}$	37.3	28.0
PSA23	16	$\frac{3}{8}$	30.7	23.0
PDA27	16	$\frac{3}{8}$	36.0	27.0
PMA22	$19\frac{5}{8}$	$\frac{3}{8}$	36.0	22.0
PZ27	18	$\frac{3}{8}$	40.5	27.0
PZ32	21	$\frac{3}{8}$	56.0	32.0
PZ38	18	$\frac{3}{8}$	57.0	38.0

Example 6-3. Estimate the cost of furnishing and driving steel-sheet piling for a cofferdam to enclose a rectangular area 60 by 100 ft. The piles will be 16 in wide, weighing 30.7 lb/lin ft, section no. PSA23, Table 6-3, 24 ft long. The piles will be driven in pairs to full penetration into soil having normal frictional resistance.

A hammer delivering approximately 6,000 ft·lb of energy per blow should be used. A suitable steel cap should be placed on top of the piles to protect them from

TABLE 6-4
Representative number of steel-sheet piles driven per hour

Length of pile, ft	Depth of penetration	Weight of pile, lb/lin ft 20	Weight of pile, lb/lin ft 30	Weight of pile, lb/lin ft 40
20	$\frac{1}{2}$	6	$5\frac{3}{4}$	$5\frac{1}{2}$
	Full	$5\frac{1}{2}$	$5\frac{1}{4}$	$5\frac{1}{4}$
25	$\frac{1}{2}$	5	$4\frac{3}{4}$	$4\frac{1}{2}$
	Full	$4\frac{1}{4}$	4	4
30	$\frac{1}{2}$	$4\frac{1}{2}$	$4\frac{1}{4}$	4
	Full	4	$3\frac{3}{4}$	$3\frac{1}{2}$
35	$\frac{1}{2}$	4	$3\frac{3}{4}$	$3\frac{1}{2}$
	Full	$3\frac{1}{2}$	$3\frac{1}{4}$	3
40	$\frac{1}{2}$	—	$3\frac{1}{4}$	3
	Full	—	3	$2\frac{3}{4}$
45	$\frac{1}{2}$	—	—	$2\frac{3}{4}$
	Full	—	—	$2\frac{1}{2}$
50	$\frac{1}{2}$	—	—	$2\frac{1}{2}$
	Full	—	—	$2\frac{1}{4}$

damage during driving. The hammer will be suspended from an 8-ton 12-ft-radius gasoline-engine-powered crawler-mounted crane. Steam will be supplied by a 25-hp boiler, fired with fuel oil.

The cost will be determined as follows.

$$\text{No. piles required,} \quad \frac{320 \text{ ft} \times 12 \text{ in}}{16 \text{ in}} = 240$$

Corner piles will be of the same section, bent at right angles.
The cost will be:

Piling, 240 × 24 × 30.7 = 176,832 lb @ $0.32 = $56,586.24
Moving equipment to job and back to storage =   1,800.00
Crane, 240 piles ÷ 4 per h = 60 h @ $41.50   =   2,490.00
Hammer, 60 h @ $13.90   =   834.00
Boiler, fuel, and water, 60 h @ $10.80   =   648.00
Other equipment and supplies, 60 h @ $6.75   =   405.00

Allow 8 h for crew to set up and take down equipment

Foreman, 60 + 8 = 68 h @ $18.30   =   1,244.40
Crane operator, 68 h @ $17.25   =   1,173.00
Crane oiler, 68 h @ $11.25   =   765.00
Firer, 68 h @ $12.50   =   850.00
Worker on hammer, 68 h @ $9.45   =   642.60
Helpers, 2 × 68 = 136 h @ $7.50   =   1,020.00
      Total cost   = $68,458.24

Wood piles

The cost of wood load-bearing piles in place will include the cost of the piles delivered to the job; the cost of moving the pile-driving equipment to the job, setting it up, taking it down, and moving out; and the cost of equipment and labor driving the piles. Since the cost of moving in, setting up, taking down, and moving out is the same regardless of the number of piles driven, the cost per pile will be lower for a greater number of piles.

The cost of wood piles is usually based on the length, size, quality, treatment, and location of the job. Wood piles are tapered from an endpoint diameter of 6 to 8 in to a butt diameter of 12 to 14 in. Lengths vary from 30 to 50 ft. For piles that are to be driven over 50 ft in depth, a splice of two sections of piles is required.

A preservative treatment of creosote or pentachlorophenol is generally applied to wood piles. Treatments reduce the tendency of decay but often increase the brittlement of the wood, which increases the possibility of breakage during the driving process. A steel boot can be placed on the pile point to reduce the tendency of breakage during driving of the pile.

Because piles are sometimes broken during the driving process, a reasonable allowance for extra piles should be included in the estimate. If the contractor is paid for the total number of linear feet of piles driven, any piles which cannot be driven to full penetration will result in some wastage. When it is necessary to cut off the tops of piles to a fixed elevation, the estimate should include the cost of cutting.

The estimator should thoroughly evaluate each project to determine the number and length of piles, number of splices, desired steel boots for pile points, potential for breakage, and waste for cutting the tops of piles to match the elevation of the pile caps.

Driving Wood Piles

A relatively fixed amount of energy is required to drive a pile of a given length into a given soil, regardless of the frequency of the blows. Consequently, a hammer delivering a given amount of energy per blow will reduce the driving time if it strikes more blows per minute. This is particularly true in driving piles into materials where skin friction and not point resistance is to be overcome, for the skin friction will not have as much time to develop between blows when the blows are struck more frequently.

Table 6-5 gives the approximate number of wood piles that should be driven per hour for various lengths and driving conditions. The rates are based on using a hammer of the proper size, as indicated in Table 6-1.

> **Example 6-4.** Estimate the cost of furnishing and driving 160 wood piles 36 ft long into a soil having normal frictional resistance. The piles will be approximately 14 in in diameter at the butt and 7 in at the tip. The piles will be treated with creosote preservative at the rate of 8 lb/ft^3.
>
> The average weight of the piles will be about 37 lb/lin ft. Reference to Table 6-1 indicates that the hammer should deliver approximately 7,000 theoretical ft·lb of energy per blow. Use a Vulcan size 2 single-acting hammer. The driving rate is estimated to be 3 piles per hour.
>
> Include 5 extra piles for possible breakage or damage.

TABLE 6-5
Representative number of wood piles driven per hour, full penetration

Length of piles, ft	Piles driven per hour	
	Normal friction	High friction
20	5	4
28	$4\frac{1}{4}$	3
32	$3\frac{1}{2}$	$2\frac{1}{2}$
36	3	$2\frac{1}{4}$
40	$2\frac{3}{4}$	2
50	$2\frac{1}{2}$	$1\frac{3}{4}$
60	2	$1\frac{1}{2}$

The cost will be

Piles for the job, 165 × 36 = 5,940 lin ft @ $3.54 = $21,027.60
Moving equipment to the job and back to storage = 1,200.00
Crane, 8 ton, 12 ft-radius, 160 ÷ 3 = 53 hr @ $57.80 = 3,063.40
Pile hammer, 53 hr @ $15.60 = 826.80
Boiler, fuel, water, etc., 53 hr @ $10.80 = 572.40
Leads, hose, accessories, 53 hr @ $5.20 = 275.60

Allow 16 hr for the crew to set up and dismantle the equipment

Foreman, 53 + 16 = 69 hr @ $18.30 = 1,262.70
Crane operator, 69 hr @ $17.25 = 1,190.25
Crane oiler, 69 hr @ $11.25 = 776.25
Fireman, 69 hr @ $12.50 = 862.50
Man on hammer, 69 hr @ $9.45 = 652.05
Helpers, 2 men × 69 hr = 138 hr @ $7.50 = 1,035.00

Total cost = $32,744.55
Cost per pile, $32,744.55 ÷ 160 = 204.65
Cost per lin ft, $204.65 ÷ 36 = 5.68

Concrete Piles

Concrete piles are generally fabricated as prestressed or post-tensioned concrete members by a concrete supplier who specializes in precast concrete. The cost will depend upon the length, diameter, strength and the distance from the fabrication yard of the supplier to the job site.

The weight of concrete piles is high, therefore the cost of transporting the piles to the job can be significant. The amount of steel, strength of concrete, and amount of prestressing depend on the specifications prepared by the designer.

Because of the varied conditions that can affect the costs, it is not possible to prepare a general estimate that will apply to a particular project without the information that must be obtained from a concrete supplier that is to furnish the piles. To estimate the cost, the estimator must obtain price quotes from the supplier.

Cast-in-Place Concrete Piles

Several methods of casting concrete piles in place are used. In general, they involve driving tapered steel shells or steel pipes which are later filled with concrete. Although the shells are left in place, the pipes sometimes are withdrawn as the concrete is deposited.

Cast-in-place piles are especially suitable for use on projects where the soil conditions are such that the depth of penetration is not known in advance and the depth varies among the piles driven.

Monotube Piles

The monotube pile is obtained by driving a fluted tapered-steel tube, closed at the tip, to the desired penetration. Additional sections may be welded to the original

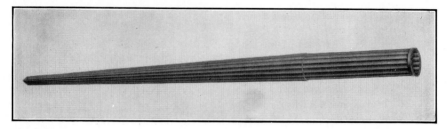

FIGURE 6-4
Monotube pile.

tube as driving progresses to increase the length, if required. After driving is completed in a given area, the condition of each tube is examined by lowering a light into the tube. A damaged tube can be removed and replaced. Concrete is then poured into the tubes. No additional reinforcing is required. If steel dowels are specified, they are driven into the tops of the piles before the concrete hardens (Fig. 6-4).

Several advantages are claimed for this type of pile. Damaged or broken piles will not result from driving. Driving subsequent piles will not damage piles previously driven because the concrete is not poured until driving has progressed beyond the area affected by driving. There is no loss of materials, since any excess tube length can be cut off and used on another tube. Cutting off of piles that cannot be driven to the expected penetration is eliminated.

Table 6-6 gives dimensions and other information on type-F tubes.

TABLE 6-6

Data on Monotube piles, type F, taper 1 in in 7 ft, tip diameter 8 in

Length, ft	Butt diam., in	Weight of shells, lb			Volume of concrete per pile, ft^3
		No. 11 gauge	No. 9 gauge	No. 7 gauge	
10	9.4	137	166	195	3.5
15	10.1	202	248	293	5.8
20	10.8	271	334	396	8.4
25	11.5	348	430	511	11.4
30	12.2	425	526	627	14.3
35	12.9	512	635	757	18.1
40	13.6	600	744	988	22.1
45	13.9	700	865	1,030	26.6
50	14.6	794	983	1,172	31.0
55	15.3	894	1,111	1,328	36.5
60	16.0	1,004	1,244	1,438	42.2
70	17.4	1,231	1,529	1,826	55.5
75	18.1	1,356	1,685	2,013	62.8

Raymond Step-Tapered Concrete Piles

This pile is obtained by driving a corrugated-steel shell, made by joining sections 4, 8, 12, or 16 ft long to produce a total length to fit the particular need of a project. A shell may be assembled that uses more than one length, if desired. Figure 6-5 illustrates a pile shell after the sections are assembled. The lower end of the bottom shell is closed with a steel plate or a hemispherical boot. If shells having tip diameters larger than the one shown in Fig. 6-5 are desired, the sections below the desired diameter are omitted.

A steel mandrel, which has stepped outer surfaces to correspond with the inside diameters of the shell sections, is inserted into the shell before it is driven. This mandrel bears on each of the stepped shoulders of the shell to distribute the driving energy throughout the shell. When the desired penetration is obtained, the mandrel is withdrawn and the shell is inspected for possible damage, after which it is filled with concrete.

Information furnished by the manufacturer gives the volume of concrete required to fill the shells.

Cost of Cast-in-Place Concrete Piles

In estimating the cost of cast-in-place concrete piles, it is necessary to determine the cost of the shells delivered to the project, the cost of equipment and labor required to drive the shells, and the cost of the concrete placed in the shells. All these costs except the cost of driving the shells are easy to obtain. The rate of driving varies with the length of the piles, class of soil into which they are driven, type of driving equipment used, spacing of the piles, topography of the site, and weather conditions. If a project requires the driving of a substantial number of piles, it may be desirable to make subsoil tests, to obtain reliable information related to the ease or difficulty of driving piles and the lengths of piles required (Fig. 6-6).

Steel Piles

Steel piles are used when they seem more suitable or economical than wood or concrete piles. They are especially suited to projects where it is necessary to drive piles through considerable depths of poor soil to reach solid rock or other formation having high load-supporting properties. Steel H sections or wide-flange beams are most frequently used, although fabricated sections are sometimes used for special conditions. Rolled sections are less expensive than fabricated sections.

Driving operations are similar to those for other piles. Steel piles do not require the care in handling that must be observed in handling precast-concrete piles. The danger of damaging a pile is reduced by using a driving cap which fits the particular pile.

Table 6-7 gives the weights in pounds per linear foot for various steel piles, and Table 6-8 gives approximate rates for driving steel piles when one is using a hammer of the recommended size.

Raymond steel-encased concrete
piles are installed by driving
the required length of steel shell
and internal steel mandrel;
withdrawing the mandrel, leaving
the steel shell in place; inspecting
the driven shell internally; and
filling the shell with concrete.
Numbered shell sections are made
in 4-, 8-, 12-, and 16-ft. lengths.
Longer lengths can be furnished
for special conditions. The point
section is closed at the bottom
by a flat steel plate welded to the
boot ring. Shell sections are screw-
connected.

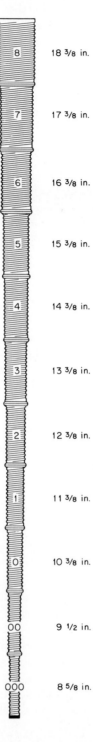

8   18 ³/₈ in.

7   17 ³/₈ in.

6   16 ³/₈ in.

5   15 ³/₈ in.

4   14 ³/₈ in.

3   13 ³/₈ in.

2   12 ³/₈ in.

1   11 ³/₈ in.

0   10 ³/₈ in.

00   9 ¹/₂ in.

000   8 ⁵/₈ in.

FIGURE 6-5
Raymond step-taper pile.

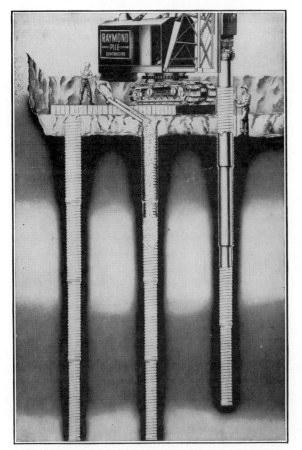

FIGURE 6-6
Driving Raymond step-taper piles.

TABLE 6-7
Data on selected steel-pile sections

Designation	Size, in	Area in² H sections	Weight, lb/lin ft
HP14	$14 \times 14\frac{1}{2}$	34.44	117
		30.01	102
		26.19	89
		21.46	73
HP12	12×12	21.76	74
		15.58	53
HP10	10×10	16.76	57
		12.35	42
HP8	8×8	10.60	36

TABLE 6-8

Approximate number of steel piles driven to full penetration† per hour

Length of pile, ft	Weight, lb/lin ft			
	102	74	57	36
30	3	$3\frac{1}{4}$	$3\frac{1}{2}$	$3\frac{1}{2}$
35	$2\frac{1}{2}$	$2\frac{3}{4}$	3	3
40	$2\frac{1}{4}$	$2\frac{1}{4}$	$2\frac{1}{2}$	$2\frac{1}{2}$
45	$1\frac{1}{2}$	$1\frac{3}{4}$	$1\frac{3}{4}$	
50	$1\frac{1}{2}$	$1\frac{1}{2}$	$1\frac{3}{4}$	
55	$1\frac{1}{4}$	$1\frac{1}{2}$		
60	1	1		
70	$\frac{3}{4}$			
80	$\frac{1}{2}$			

† The rates are based on piles being delivered to the job in lengths up to 40 ft. For lengths greater than 40 ft, a field weld will be required for each pile.

Example 6-5. Estimate the cost for bid purposes for furnishing and driving 180 H-section steel piles 12 by 12 in, 40 ft long, and weighing 74 lb/lin ft. The piles will be driven to full penetration into soil having high frictional resistance.

Reference to Table 6-1 indicates that a 15,000-ft·lb single-acting steam hammer will be satisfactory.

A diesel-engine-powered crawler crane will be used to lift the piles into position, support the leads, and lift the hammer. The size crane required may be determined as follows.

Item	Weight on crane, lb
Piles, 40 ft × 74 lb	2,960
Hammer	9,600
Leads	5,000
Pile cap	440
Total weight	18,000 lb, or 9 tons

Assume that during the driving of the piles the maximum radius required will be 25 ft for the crane. This radius and the load on the crane will give a product equal to $9 \times 25 = 225$ ton·ft. Use a 20-ton 12-ft-radius crane to give a load-radius product of 240 ton·ft.

Reference to Table 6-8 indicates a probable driving rate of $2\frac{1}{4}$ piles per hour. The time required to drive the piles should be about $180 \div 2\frac{1}{4} = 80$ h. This time will be used.

The cost will be:

Materials:
  Piles, 180 × 40 × 74 = 532,800 lb @ $0.31 = $165,168.00
Equipment:
  Moving to and from job   = $ 3,100.00
  Crane, 80 h @ $63.73   =   5,098.40
  Hammer, 80 h @ $16.47   =   1,317.60
  Boiler, fuel, water, hose, etc., 80 h @ $12.74 = 1,019.20
  Leads, 80 h @ $4.80   =   384.00
  Pickup truck, $\frac{3}{4}$-ton, 80 h @ $11.50 =   920.00
  Utility truck, 2-ton, 80 h @ $14.25   =   1,140.00
  Tools and supplies, 80 h @ $5.70   =   456.00
    Subtotal cost of equipment   = $ 13,435.20

Labor:
  Allow 16 h to set up and dismantle equipment
  Foreman, 80 + 16 = 96 h @ $18.30   = $ 1,756.80
  Crane operator, 96 h @ $17.25   =   1,656.00
  Crane oiler, 96 h @ $11.25   =   1,080.00
  Firer, 96 h @ $9.45   =   907.20
  Guiders, 2 × 96 = 192 h @ $8.10   =   1,555.20
  Laborers, 2 × 96 = 192 h @ $7.50   =   1,440.00
  Superintendent, 0.75 month @ $2,913.33 = 2,185.00
    Subtotal cost of labor   = $10,580.20

Summary of direct costs:
  Materials   = $165,168.00
  Equipment = 13,435.20
  Labor   = 10,580.20
    Subtotal = $189,183.40

Workers' compensation insurance,
$10,580.20 @ $8.14 per $100.00   =   861.23
Social security tax, 7.51% of $10,580.20 = 794.57
Unemployment tax, 3% of $10,580.20 =   317.41
                                    = $191,156.61
General overhead, 5% of $191,156.61 =   9,557.83
                                    = $200,714.44
Profit, 8% of $200,714.44   =   16,057.16
                                    = $216,771.60
Performance bond, $216,771.60
@ $12.00 per $1,000   2,601.26
  Total cost, amount of bid   = $219,372.86
Cost per lin ft, $219,372.86 ÷ 7,200 =   30.47

Jetting Piles into Position

If the formation into which the piles are driven contains considerable sand, the rate of driving may be increased by jetting with water. To accomplish this with concrete piles, steel pipes may be placed in the piles when they are cast. The pipe, $1\frac{1}{2}$ to 2 in in diameter, should extend from the tip to within about 2 ft of the top, where it protrudes from the side by the use of an ell. It may require 100 to 400 gal/min of water or more to jet a concrete pile into position. Required pressures will be 100 to 200 lb/in^2. For suitable soil conditions this method will sink piles much more quickly than pile-driving hammers will.

Drilled Shafts and Underreamed Footings

Drilled-shaft foundations are installed by placing reinforced concrete in a hole that has been drilled into the soil. Truck-mounted rotary drilling rigs can drill holes with diameters of 10 to 96 in. Depths up to approximately 100 ft are possible; however, most drilled shafts range from 6 to 30 ft. The bottom of the drilled hole is underreamed with a belling tool to provide an increased bearing capacity of the foundation.

Two types of belling tools can be used to underream a drilled shaft; one opens out from the top, as shown in Fig. 6-7a, and one opens out from the bottom, as shown in Fig. 6-7b. Equation (6-1) can be used to determine the volume of the shaft for the underream of the type of hole illustrated in Fig. 6-7b.

Equation 6-1 gives the volume of concrete required to fill the hole for the shaft.

$$V_c = \frac{\pi d_s^2}{4}(D_f - h) + \frac{\pi h}{12}(d_s^2 + d_s d_B + d_B^2) \tag{6-1}$$

Reference Fig. 6-7b for definition of variables.

Cohesive soils, clays, generally have sufficient strength to provide natural stability of the sides of the drilled shaft. Cohesionless soils, sands, may require the installation of a cylindrical steel casing during the drilling process to hold the sides of the shaft until the reinforcing steel and concrete are placed. The casing is pulled from the hole during placement of the concrete.

Table 6-9 provides representative rates of drilling various diameters of shafts. These rates can vary, depending upon the moisture and hardness of the soil and the size of the rotary drilling rig. A thorough evaluation of the soil test boring logs is necessary in order to determine the rate of drilling.

Vertical reinforcing steel is tied together by horizontal ties or spiral hoops and lowered into position in the drilled shaft. Lugs, attached to the reinforcing cage, hold them in position at the center of the hole until concrete is placed.

The cost of drilled-shaft foundations will include mobilization of equipment, drilling of the shaft, casing (if required), fabrication and placement of the reinforcing steel, and concrete material and placement.

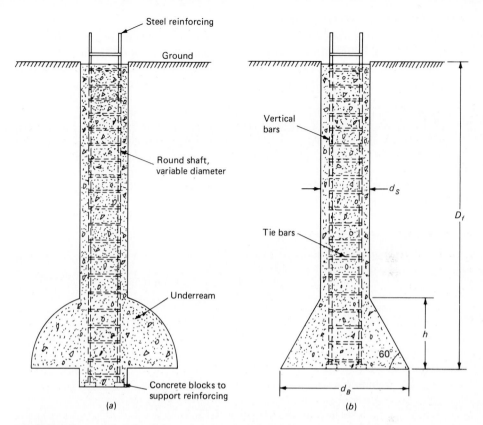

FIGURE 6-7
Drilled hole for cast-in-place concrete piles; underream made by a belling tool that opens out (*a*) from the bottom and (*b*) from the top.

TABLE 6-9
Representative rates of drilling shafts

Diameter, in	Rate of drilling in clays, ft/h		
	Soft	Firm	Hard
18	26.0	20.0	6.0
24	22.0	15.0	5.0
30	19.0	11.0	4.0
36	16.0	8.0	3.0
42	14.0	7.0	2.0
48	13.0	6.0	1.7
54	12.0	5.0	1.5
60	11.0	4.0	1.2
72	10.0	3.0	0.7
84	9.0	2.0	0.5
96	8.0	1.0	0.3

Example 6-6. A 15-mi-long electric transmission line requires 64 steel truss towers to support the conductors and shield wires. There are four drilled-shaft foundations for each tower, one for each leg of each tower, similiar to the type shown in Fig. 6-7b.

Each foundation is to be installed with a 24-in diameter at a depth of 9 ft in firm clay. A 42-in-diameter underream is required for each foundation.

Reinforcing steel consists of eight vertical bars 9 ft long with $\frac{3}{4}$-in diameter. Horizontal ties will be placed at 12 in on centers. Tie bars will have $\frac{3}{8}$-in diameters and will have a 3-in overlap on their ends. A 3-in clear cover is required on all reinforcing steel.

Ready-mixed concrete will be used, with the delivery trucks discharging directly into the holes.

Estimate the total cost, cost per linear foot, and cost per cubic yard of concrete in place.

The quantities will be:
 Number of drilled shafts, 64 × 4 = 256
 Total depth of drilled shaft, 256 × 9 = 2,304 lin ft
 Concrete volume, from Eq. (6-1):
  Volume per shaft

$$\frac{3.14 \times 2.0^2}{4}(9.0 - 1.3) + \frac{3.14(1.3)}{12}(2.0^2 + 2.0 \times 3.5 + 3.5^2) = 32.1 \text{ ft}^3$$

  Total concrete, (32.1 ÷ 27) × 256   = 304.4 yd³

 Reinforcing steel:
  Vertical bars
   Total length, 256 × 8 × 9 = 18,432 lin ft
   Total weight, 18,432 ft @ 1.502 lb/ft = 27,684.9 lb
  Horizontal tie bars
   Centerline diameter,

$$24 - (2 \times 3) - \tfrac{3}{8} = 17.625 \text{ in}$$

  Circumference,

$$[(3.14 \times 17.625) + 3] \div 12 = 4.86 \text{ ft}$$

   Total length, 256 × 9 × 4.86 = 11,197 lin ft
   Total weight, 11,197 ft @ 0.376 lb/ft = 4,210.1 lb

The costs will be:
 Moving equipment to and from job   = $ 1,800.00
 Drilling, 2,304 lin ft ÷ 15 ft/h = 153.6 h
  Drilling rig, 153.6 h @ $68.50   = 10,521.60
  Driller, 153.6 h @ $17.25   = 2,649.60
  Laborers, 2 × 153.6 = 307.2 h @ $8.10   = 2,488.32
  Foreman, 153.6 h @ $18.30   = 2,810.88
 Reinforcing steel:
  Vertical bars in place, 27,684.9 lb @ $0.31 = 8,582.32
  Tie bars in place, 4,210.1 lb @ $0.34   = 1,431.43
 Concrete in place, 304.4 yd³ @ $51.85   = 15,783.14
  Total cost   = $46,067.29
  Cost per lin ft, $46,067.29 ÷ 2,304   =   19.99
  Cost per yd³, $46,067.29 ÷ 304.4   =   151.34

PROBLEMS

6-1. Estimate the total direct cost and the cost per linear foot for furnishing and driving 96 creosote-treated wood piles to full penetration into soil having high frictional resistance. The piles will be 40 ft long, with 14-in minimum diameter at the butt and 6-in minimum diameter at the tip. The tip of each pile will be fitted with a metal point to facilitate driving.

The crane used on the job will be a 8-ton 12-ft-radius diesel-engine-powered crawler-mounted unit. The cost of moving the equipment to the job and back to the storage yard will be $1,950.

Use the information appearing in the example beginning on page 126 to determine the cost of the job. The piles will cost $4.10 per linear foot delivered to the job. The pile points will cost $9.45 each. It will not be necessary to cut off the tops of the piles.

6-2. Estimate the total direct cost and the cost per linear foot for furnishing and driving the piles of Prob. 6-1 to full penetration into soil having normal frictional resistance. Use a metal point on the tip of each pile.

The cost of moving to the job and back to the storage yard will be $1,400.

6-3. Estimate the total direct cost and the cost per linear foot for furnishing and driving 84 steel piles 12 in by 12 in weighing 53 lb/lin ft to full penetration into soil having high frictional resistance. The piles will be 40 ft long.

The piles will cost $0.34 per pound delivered to the job.

The cost of moving the equipment to the job and back to the storage yard will be $1,540. Use a 20-ton 12-ft-radius diesel-engine-powered crawler-mounted crane to handle the piles and hammer. Use cost information appearing in the example on page 132 to estimate the cost of the project.

6-4. Estimate the total direct cost and the cost per linear foot for furnishing and driving 66 steel piles 14 by 14 in weighing 117 lb/lin ft into soil having high frictional resistance. The piles will be driven to full penetration. The piles will be 40 ft long.

Assume that the leads will weigh 6,000 lb. Select the size hammer suitable for driving the piles, and then select the size crane required to handle the piles, hammer, and leads. Assume a 12-ft radius for the crane.

Material costs; wage rates, and other conditions will be the same as for the example on page 132.

6-5. Estimate the total cost and the cost per cubic yard for installing 95 drilled-shaft foundations in hard clay soil. Each foundation is to be installed, similar to the type shown in Fig. 6-7b, with a 30-in-diameter shaft and a 54-in-diameter underream at a depth of 14 ft.

Vertical reinforcing steel consists of twelve $\frac{7}{8}$-in-diameter bars 14 ft long. Horizontal ties are to be placed at 15 in on centers and lap 6 in. Ties are $\frac{1}{2}$ in in diameter.

Material costs, wage rates, and other conditions will be the same as the example on page 136.

CHAPTER
7

CONCRETE STRUCTURES

Costs of Concrete Structures

The items which govern the total costs of concrete structures include the following:

1. Forms
2. Reinforcing steel
3. Concrete
4. Finishing, if required
5. Curing

When one is preparing an estimate, it is suggested that the cost of each of these items be determined separately and that a code number be assigned to each item. The cost of an item should include all materials, equipment, and labor required for that item. Following this procedure will simplify the preparation for an estimate.

FORMS

Forms for Concrete Structures

Concrete structures may be constructed in any shape for which it is possible to build forms. However, the cost of forms for complicated shapes is considerably greater than for simple shapes because of the extra material and labor costs required to build them and the low salvage value after they are used.

Since the major cost of a concrete structure frequently results from the cost of forms, the designer of a structure should consider the effect which the shape of a structure will have on the cost of forms.

Forms for concrete are fabricated from lumber, plywood, steel, aluminum, and various composition materials, either separately or in combination. If the form material will be used only a few times, lumber will usually be more economical than steel or aluminum. However, if the forms can be fabricated into panel sections or other shapes, such as round column forms, that will be used many times, then the greater number of uses obtainable with steel and aluminum may produce a lower cost per use than with lumber.

That material should be selected for forms which will give the lowest total cost of a structure, considering the cost of the forms plus the cost of finishing the surface of the concrete, if required, after the forms are removed. The use of wood planks usually will leave form marks which may have to be removed at considerable cost, whereas the use of plywood, pressed wood, or metal forms might eliminate the cost of removing form marks. It is good practice to spend an extra 5 ¢/ft^2 on forms if by so doing it is possible to eliminate a finishing cost of 10 ¢/ft^2.

The cost of forms will include the cost of materials, such as lumber, nails, bolts, form ties, and the cost of labor making, erecting, and removing the forms. Frequently there will be a cost for power equipment, such as saws and drills, and hand tools. If it is possible to reuse the forms, an appropriate allowance should be made for salvage value. If the forms are treated with oil prior to each use, the cost of the oil should be included.

Materials for Forms

Forms for concrete structures are generally fabricated from standard-dimension lumber and plywood sheets. Because the lumber must be cut and fastened for the particular shape of the concrete member, considerable waste often occurs. The cost of materials for forms should include an allowance for waste.

Plywood Used for Forms

Plywood sheets are made in various thicknesses with waterproof glue. Those commonly used are $\frac{1}{4}$, $\frac{3}{8}$, $\frac{1}{2}$, $\frac{5}{8}$, and $\frac{3}{4}$ in. It is available in sheets 4 ft wide and 8, 10, or 12 ft long. Other dimensions may be obtained on special order. It is necessary to install the plywood with the outer layers perpendicular to the studs.

Dimension Lumber Used for Forms

The net dimensions of lumber are given in Chap. 11. Because the net dimensions are less than the nominal dimensions, it is necessary to use the former to determine the quantity required. Standard lengths are available in multiples of 2 ft, with the longer lengths carrying an extra charge. If an odd length, such as 9 ft, is desired, it may be obtained by cutting an 18-ft-long plank into two equal lengths or by cutting 1 ft off a 10-ft-long plank, with a resulting waste.

Nails Required for Forms

The quantity of nails required for forms usually will vary from 10 to 20 lb per 1,000 fbm of lumber for the first use of the lumber and from 5 to 10 lb per 1,000 fbm for additional uses if the forms can be reused without completely refabricating them.

Form Oil

When form oil is used to treat the surfaces that will come in contact with concrete, it should be applied at a rate of 300 to 500 ft^2/gal. It may be applied with mops, brushes or pressure sprayers.

Labor Required to Build Forms

The factors which determine the amount of labor required to build forms for concrete structures include the following:

1. Size of the forms.
2. Kind of materials used. Large sheets of plywood require less labor than planks.
3. Shape of the structure. Complicated shapes require more labor than simple shapes.
4. Location of the forms. Forms built above the ground require more labor than forms built on the ground or on a floor.
5. The extent to which prefabricated form panels or sections may be used.
6. Rigidity of dimension requirements.
7. The extent to which power equipment is used to fabricate the forms.

If forms are prefabricated into panels or sections and then assembled, used, removed, and reused, it is desirable to estimate separately the labor required to make, assemble, and remove them. Because making is required only once, for additional uses it is only necessary to assemble and remove the forms.

The production rates given in this book are based on using power saws and other equipment as much as possible. If the fabricating is done with hand tools, the labor required should be increased.

The tables on production rates give the minimum and maximum number of labor-hours required to do a specified amount of work, for both capenters and helpers. If union regulations specify the ratio of helpers to carpenters for certain locations, it may be necessary to transfer some of the helper time to carpenter time. The tables include an allowance for lost time by using a 45- to 50-min hour.

Forms for Footings and Foundation Walls

Foundation walls include grade beam footings, basement walls for buildings, retaining walls, concrete cores for earth-filled dams, vertical walls for water reservoirs, etc. The forms for such walls consist of $\frac{3}{4}$-in-thick plywood sheathing, 2- by 4- or 2- by 6-in studs, double 2- by 4- or 2- by 6-in wales, and 2- by 4- or 2- by 6-in braces. The sheathing, studs, and wales are erected on both sides of the wall, while the braces are spaced about 6 to 8 ft apart on one side of the wall only.

Design of Forms for Foundation Walls

To estimate correctly the quantity of materials for wall forms, it is necessary to design the forms for the particular wall. Proper design will ensure adequate safety with economy.

The factors which affect the design are

1. The consistency and proportions of the concrete
2. The rate of filling the form
3. The temperature of the concrete
4. The method of placing the concrete
5. The depth of drop and distribution of reinforcing steel

These factors affect the pressure at any given depth below the surface of the freshly placed concrete. The rate of filling and the temperature determine the effective depth of liquid concrete, that is, the depth which has not set long enough to possess internal rigidity. As the concrete sets, it reduces the pressure on the forms. Concrete sets more slowly at low temperatures than at high ones. Lower maximum pressures will be exerted by concrete having a low slump than one having a high slump.

The maximum pressure of concrete on wall and column forms may be determined from the following formulas, which were developed by the American Concrete Institute and published in 1971 [1].

For walls:

$$P_m = 150 + \frac{9,000R}{T} \qquad \text{for } R \text{ up to 7 ft/h} \tag{7-1}$$

$$P_m = 150 + \frac{43,400}{T} + \frac{2,800R}{T} \qquad \text{for } R \text{ more than 7 ft/h} \tag{7-2}$$

For columns:

$$P_m = 150 + \frac{9,000R}{T} \tag{7-3}$$

TABLE 7-1
Relation among the rate of filling wall forms, maximum pressure, and temperature (ACI)

Rate of filling forms, ft/h	Maximum concrete pressure, lb/ft²						
	Temperature, °F						
	40	50	60	70	80	90	100
1	375	330	300	279	262	250	240
2	600	510	450	409	375	350	330
3	825	690	600	536	487	450	420
4	1,050	870	750	664	600	550	510
5	1,275	1,050	900	793	712	650	600
6	1,500	1,230	1,050	921	825	750	690
7	1,725	1,410	1,200	1,050	933	850	780
8	1,793	1,466	1,246	1,090	972	877	808
9	1,865	1,522	1,293	1,130	1,007	912	836
10	1,935	1,578	1,340	1,170	1,042	943	864
15	2,185†	1,858	1,573	1,370	1,217	1,099	1,004
20	2,635†	2,138†	1,806	1,570	1,392	1,254	1,144

† These values are limited to 2,000 lb/ft².

where P_m = maximum pressure on forms, lb/ft²
$\quad\quad\ R$ = rate of filling forms, ft/h
$\quad\quad\ T$ = temperature of concrete, °F

The maximum pressure obtained from Eq. (7-2) is limited to 2,000 lb/ft², and for Eq. (7-3) it is limited to 3,000 lb/ft². The pressures are for concrete that is consolidated by internal vibration as it is placed.

Table 7-1 gives the maximum pressures resulting from placing concrete in wall forms. Table 7-2 gives the maximum pressures resulting from placing concrete in column forms.

Regardless of the pressures given in Tables 7-1 and 7-2, the maximum pressures on wall and column forms will not exceed the weight of 1 ft³ of concrete times the depth to the given area in feet.

The pressures given in Table 7-1 and Table 7-2 are effective for concrete weighing 150 lb/ft³. If forms are filled with concrete whose weight is more or less than 150 lb/ft³, the maximum pressures appearing in the two tables should be multiplied by a factor equal to the weight of 1 ft³ of the concrete in question divided by 150 [2].

Design of Forms for Concrete Walls Using Tables

A number of publications contain tables that may be used in designing or building forms for concrete walls. In general, these tables have been useful and satisfactory.

TABLE 7-2

Relation among the rate of filling column forms, maximum pressure, and temperature (ACI)

Rate of filling forms, ft/h	Maximum concrete pressure, lb/ft^2						
	Temperature, °F						
	40	50	60	70	80	90	100
1	375	330	300	279	262	250	240
2	600	510	450	409	375	350	330
3	825	690	600	536	487	450	420
4	1,050	870	750	664	600	550	510
5	1,275	1,050	900	793	712	650	600
6	1,500	1,230	1,050	921	825	750	690
7	1,725	1,410	1,200	1,050	937	850	780
8	1,950	1,590	1,350	1,179	1,050	950	870
9	2,175	1,770	1,500	1,307	1,162	1,050	960
10	2,400	1,950	1,650	1,436	1,275	1,150	1,050
12	2,850	2,310	1,950	1,693	1,500	1,350	1,230
15	3,525†	2,850	2,400	2,093	1,837	1,650	1,500
20	4,650†	3,750†	3,150†	2,721	2,400	2,150	1,950

† These values are limited to 3,000 lb/ft^2.

Table 7-3 contains information which should enable one to select a design that is suitable for most projects and conditions. The following conditions apply to the information contained in the table.

1. All lumber used is S4S stress-grade pine or fir as specified by the National Forest Products Association publication 06, National Design Specifications for Stress-Grade Lumber and Its Fastenings, 1973 edition, Washington, D.C.
2. Current dimensions of lumber, effective on January 1, 1988, were used in determining the values appearing in the table.
3. The listed maximum pressures of concrete were determined by using the ACI formulas [Eqs. (7-1) and (7-3)]. These formulas provide for the increased pressures resulting from compacting the concrete with interval vibrators as it is placed.
4. The listed maximum spacings for form ties are based on limiting the stresses in the lumber to safe values. When these spacings are used, the designer or form builder should be sure to select form ties that will be strong enough to resist the total stesses in the ties. Or, if it is desirable to use ties having lower strengths, they may be used with safety if the spacing of the ties is reduced sufficiently. For example, if a given design from Table 7-3 permits a maximum safe spacing of 48 in but requires that 6,000-lb form ties be used for this spacing, the spacing may be reduced to 24 in, which will permit the use of 3,000-lb ties.

TABLE 7-3
Design of forms for concrete walls

Minimum temperature of concrete, °F	50			70			90		
Rate of filling forms, ft/h	2	4	6	2	4	6	2	4	6
Maximum pressure, lb/ft²	510	870	1,230	409	664	921	350	550	750
Maximum spacing of studs for safe value of sheathing, in									
For 1-in sheathing	22	17	14	24	19	16	26	21	18
For 2-in sheathing	38	29	24	42	33	28	45	36	31
Maximum spacing of wales for safe value of studs, in									
2 × 4 studs 1-in sheathing	26	23	21	28	25	23	29	26	24
4 × 4 studs 1-in sheathing	40	35	33	43	38	35	45	40	37
2 × 6 studs 1-in sheathing	41	36	33	44	39	36	46	41	38
2 × 6 studs 2-in sheathing	31	27	25	33	29	27	35	31	29
4 × 4 studs 2-in sheathing	31	27	25	33	29	27	34	30	28
3 × 6 studs 2-in sheathing	41	36	33	43	38	35	45	41	37
Maximum spacing of form ties for safe values of wales, in									
Double 2 × 4 wale 2 × 4 stud 1S	34	28	24	37	31	27	39	33	29
Double 2 × 4 wale 4 × 4 stud 1S	30	24	21	32	27	24	34	29	26
Double 2 × 4 wale 2 × 6 stud 1S	27	22	20	29	24	22	31	26	23
Double 2 × 6 wale 2 × 6 stud 1S	43	35	31	46	38	34	49	41	37
Double 2 × 6 wale 3 × 6 stud 2S	43	35	31	48	39	35	50	41	37

5. The values appearing in Table 7-3 limit the deflections of all members, including sheathing, to not more than $\frac{1}{360}$.

6. Most of the values appearing in Table 7-3 are limited by the allowable unit fiber stress in bending, which, because of the short duration, permits stresses up to 1,800 lb/in² in stress-grade lumber.

7. If concrete is placed in forms at rates other than those listed in Table 7-3, or if it is placed at different temperatures from those listed, the table may still be used to design forms. For example, assume that the concrete is placed at 60°F and the forms are filled at a rate of 7 ft/h. Table 7-1 indicates a maximum pressure of 1,200 lb/ft². Table 7-3 indicates that if forms are filled at a rate of 6 ft/h at a temperature of 50°F, the maximum pressure will be 1,230 lb/ft². If the forms are designed to resist this pressure, they should be safe for the pressure of 1,200 lb/ft².

Materials Required for Forms for Foundation Walls

Wall forms are frequently built as indicated in Fig. 7-1. Assume that the rate of filling the forms will be 4 ft/h at a temperature of 70°F.

The sheathing will be $\frac{3}{4}$-in plywood. Studs will be 2 by 4 in, and wales will be double 2- by 4-in planks.

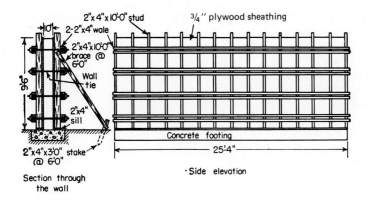

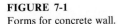

FIGURE 7-1
Forms for concrete wall.

Reference to Table 7-3 indicates the following pressure and spacings:

Maximum pressure, 664 lb/ft^2
Maximum spacing of studs, 19 in, use 18 in
Maximum spacing of wales, 25 in, use 24 in
Maximum spacing of form ties, 31 in

The wall is 9 ft 6 in high and 25 ft 4 in long. The $\frac{3}{4}$-in plywood sheathing will be placed with the 4-ft width in the vertical direction and the 8-ft length in the horizontal direction.

The total quantity of sheathing will be:
    No. sheets in vertical direction, 9 ft 6 in ÷ 4 ft/sheet = 2.37, use 3 sheets
    No. sheets in horizontal direction, 25 ft 4 in ÷ 8 ft/sheet = 3.16, use 4 sheets
    No. sheets required per side, 3 × 4 = 12
    No. sheets required for wall, 12 × 2 = 24
Studs required:
    Length of wall, (25 × 12) + 4 = 304 in
    Spacing of studs, 18 in
    No. studs required per side, (304 ÷ 18) + 1 = 18
    No. studs required for wall, 2 × 18 = 36
    Lumber required, 36 pc, 2 × 4 in × 10 ft 0 in = 240 fbm
Wales required:
    Height of wall, 114 in
    Spacing of wales, 24 in
    No. required per side, 114 ÷ 24 = 4.75, use 5 wales
    For each wale use 2 pc of 2- × 4-in × 12-ft 0-in and 2 pc of 2- × 4-in × 14-ft 0-in
        lumber
Lumber required:
        20 pc, 2 × 4 in × 12 ft 0 in   =   160 fbm
        20 pc, 2 × 4 in × 14 ft 0 in   =   187 fbm

Sills required:

2 pc, 2 × 4 in × 12 ft 0 in	=	16 fbm
2 pc, 2 × 4 in × 14 ft 0 in	=	18 fbm

Scab splices required:
  No. required per side 2 × 5 = 10
  Total number, 2 × 10 = 20
  Lumber required:

20 pc, 2 × 4 in × 2 ft 0 in	=	27 fbm

Braces required, one side only:
  Spacing, 6 ft 0 in, with one at each end
  No. required, 5

Lumber, 5 pc, 2 × 4 in × 10 ft 0 in =		33 fbm

Stakes:

5 pc, 2 × 4 in × 3 ft 0 in	=	10 fbm
Total quantity of lumber	=	691 fbm

Area of the wall, $2 \times 9.5 \times 25.33 = 481$ ft^2
Quantity of nails required, 691 fbm × 10 lb/1000 fbm = 6.9 lb
Form ties required:
  No. per wale, 304 in ÷ 31 in = 10
  No. per side, 5 rows @ 10 per row = 50
  Maximum load on a tie, $(25 \times 31) \div 144 \times 664 = 3{,}431$ lb
    Use 4,000-lb ties, or if 3,000-lb ties are to be used, the spacing should be reduced
to $31(3{,}000 \div 3{,}431) = 27$ in
    No. of ties $[(304 \div 27) + 1]5$ rows = 61

Summary of materials to build forms:
  Plywood required = 24 sheets
  Lumber required = 691 fbm
  Nails required = 6.9 lb
  Ties required = 61

Form Ties

To resist the internal pressure resulting from the concrete, form ties are placed between the forms as shown in Fig. 7-2. A great many varieties are available. Most serve two purposes: They hold the forms apart prior to placing the concrete, and they resist the bursting pressure of the concrete after it is placed.

The types available consist of narrow steel bands, plain rods, rods with hooks or buttons on the ends, and threaded rods, with suitable clamps or nuts to hold them in position. While most ties are designed to break off inside the concrete, there are some which can be pulled out of the wall after the forms are removed and the holes filled with concrete. Manufacturers specify the safe working stresses for their ties.

In ordering form ties it is necessary to specify the thickness of the wall, sheathing, studs, and wales. As a guide to the cost of form ties, a popular type,

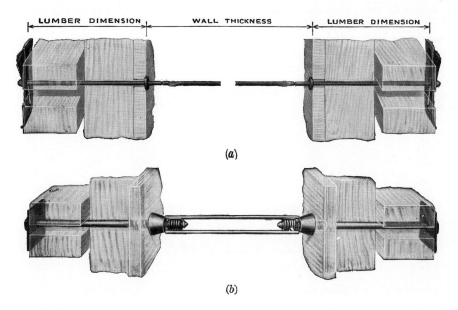

FIGURE 7-2
Form ties for concrete wall: (*a*) snap ties and (*b*) coil ties.

having a safe working strength of 3,000 lb, is priced as shown in the accompanying table.

For wall thickness, in	Cost per 100
From 4 to 12	$37.50
Over 12 to 18	41.50
Over 18 to 24	49.50
Over 24 to 30	52.22
Over 30 to 36	57.17

Two end clamps are required for each tie. The cost will be about $48.00 per 100 clamps. Clamps can also be rented for a specified time, usually about 4 months.

Labor Building Wood Forms for Foundation Walls

For walls up to about 12 ft high, 2- by 4-in studs and double 2- by 4-in wales are adequate. For greater heights, up to approximately 20 ft, 2- by 6-in studs and double 2- by 6-in wales may be desirable. Double rows of braces should be used for walls over 12 ft high, one near the middle and the other near the top of the wall.

TABLE 7-4
**Quantities of lumber and form ties and the
labor-hours required for 100 ft² of wall forms**

Height of wall, ft	Lumber, fbm	Form ties†	Labor-hours building and removing	
			Carpenter	Helper
4	175	16	3.0–4.0	2.0–3.0
6	200	14	3.5–5.5	2.0–3.0
8	225	14	4.0–5.0	2.5–3.5
10	250	16	4.5–5.5	3.0–4.0
12	275	16	5.0–6.0	3.5–4.5
14	325	10	5.5–6.5	4.0–5.0
16	350	10	6.5–7.5	4.5–5.5
18	375	10	7.0–8.0	5.0–6.0

† The form ties must be strong enough to withstand the stresses for
the particular uses. The lumber and form ties are based on filling the
forms at a rate of 4 ft/h at a temperature of 70°F.

Table 7-4 gives the quantities of lumber and form ties and the labor-hours for erecting forms for 100 ft² of wall for various heights. The quantity of form ties should be based on the area of only one side of the wall.

Plywood Forms

Plywood for forms is made in various thicknesses with waterproof glue. Those commonly used are $\frac{1}{4}$, $\frac{3}{8}$, $\frac{1}{2}$, $\frac{5}{8}$, and $\frac{3}{4}$ in. It is necessary to install the plywood with the outer layers perpendicular to the studs.

Plywood is available in 4- by 8-ft, 4- by 10-ft, and 4- by 12-ft sizes. Studs should be spaced so that there will be a stud at the end of each plywood sheet. To keep the damage to plywood to a minimum, a standard pattern for holes for wall ties should be adopted. The adoption of such a practice will reduce the need for drilling additional holes for subsequent uses. The use of small nails will reduce the damage during removal. With reasonable care, plywood can be used many times.

Table 7-5 gives the quantities of plywood, lumber, and form ties and the labor-hours required for 100 ft² of concrete wall using $\frac{3}{4}$-in-thick plywood for various heights. For wall heights other than multiples of 2 ft, there will be side wastage. Also door and window openings, wall columns, and irregular-shaped walls will usually cause end wastage. The estimator should consider the extent of wastage from the nature of the structure.

Prefabricated Form Panels

Contractors usually make prefabricated form panels using 2- by 4-in lumber for frames and 1-in-thick planks or $\frac{3}{4}$-in-thick plywood for sheathing. Because plywood

TABLE 7-5

Quantities of $\frac{3}{4}$-in plywood, lumber, and form ties and the labor-hours required for 100 ft^2 of wall

Height of wall, ft	Plywood, ft^2	Lumber, fbm	Form ties†	Labor-hours building, erecting, and removing	
				Carpenter	Helper
4	100	100	16	3.0–3.5	1.5–2.0
6	100	125	14	3.5–4.0	1.5–2.0
8	100	160	14	3.5–4.5	2.0–2.5
10	100	200	16	4.0–5.0	2.5–3.0
12	100	240	16	4.5–5.5	3.5–4.0
14	100	300	10	6.0–7.0	4.0–4.5
16	100	325	10	6.5–7.5	4.5–5.0
18	100	350	10	7.0–8.0	5.0–5.5

† The form ties must be strong enough to resist the stress for the particular use. The lumber and form ties are based on a rate of fill of approximately 4 ft per hr at a temperature of 70 deg.

sheathing is used, it is good practice to make panels 2 ft 0 in wide and 4, 6, 8, 10, or 12 ft long. An assortment of lengths will permit considerable flexibility in fitting the lengths to variable-length walls. When the panels are erected, one on top of the other, the 2-ft width will permit the use of form ties at the top and bottom of adjacent panels, thus eliminating the need for holes through the plywood.

Figure 7-3 shows a method of constructing a typical panel, using $\frac{3}{4}$-in-thick plywood for the sheathing and 2- by 4-in planks for the frame. The total weight of this panel will be about 140 lb, which will require two people to handle it.

The quantity of lumber required for the panel in Fig. 7-3 will be

Plywood, $2 \times 8 = 16$ ft^2
2 pc, 2×4 in $\times$ 8 ft 0 in   = 11 fbm
7 pc, 2×4 in $\times$ 1 ft 8 in   = 8 fbm
    Total lumber   = 19 fbm
Quantity of lumber per ft^2, $19 \div 16 = 1.2$ fbm

The labor required to build the panel should be about as follows:

Carpenter building frame, 19 fbm $\div$ 75 fbm/h = 0.25 h
Carpenter ripping and installing plywood   = 0.15 h
    Total carpenter time   = 0.40 h
Helper   = 0.20 h
    Carpenter time per 100 ft^2, $\frac{100}{16} \times 0.4$ = 2.5 h

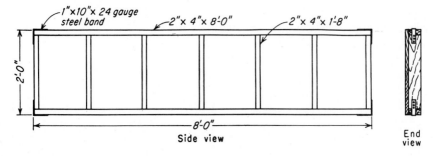

FIGURE 7-3
Prefabricated panel form.

Figure 7-4 shows a typical set of forms for a foundation wall using prefabricated panels. The bottoms of the forms are held together by steel tie bands, which pass under the concrete wall and are nailed to the bottom members of the panel frames. The tops of the forms may be held together with planks, as shown in Fig. 7-4, or with steel tie bands. The top and bottom edges of the panels may be grooved slightly to receive the form ties.

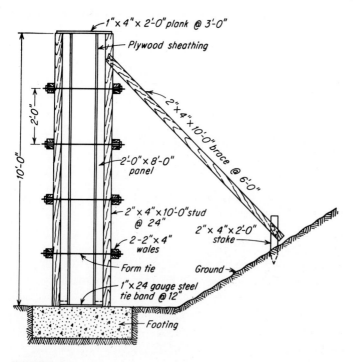

FIGURE 7-4
Form for concrete wall.

TABLE 7-6

Quantities of plywood, lumber, and form ties and the labor-hours required for 100 ft² of wall area, using prefabricated panels

Height of wall, ft	Plywood ft²	Lumber, fbm	Form ties†	Labor-hours making panels		Labor-hours erecting and removing forms	
				Carpenter	Helper	Carpenter	Helper
4	100–110	200–225	11	2.0–2.5	1.5–2.0	2.5–3.0	2.0–2.5
6	100–110	210–230	15	2.0–2.5	1.5–2.0	3.0–3.5	2.5–3.0
8	100–110	220–240	16	2.0–2.5	1.5–2.0	3.5–4.0	3.0–3.5
10	100–110	225–250	17	2.0–2.5	1.5–2.0	4.0–4.5	3.5–4.0
12	100–110	230–260	18	2.0–2.5	1.5–2.0	4.5–5.0	4.0–4.5

† The form ties must be strong enough to resist the stress for the particular use.

Table 7-6 gives the quantities of plywood, lumber, and form ties and the labor-hours required for 100 ft² of contact area for concrete walls using prefabricated panels, based on a rate of fill not exceeding 4 ft/h at a temperature of 70°F.

The increase in the number of uses of the lumber, and the reduction in the labor costs after the panels are made, will frequently permit a saving in form costs, compared with other types of forms.

Commercial Prefabricated Forms

Plywood sheathing attached to steel or aluminum frames in various sizes may be purchased or rented from the manufacturers. Figure 7-5 illustrates a set of wall forms in place.

Forms consisting of steel sheets attached to steel frames may also be purchased or rented. All these forms can be used with straight or circular walls or for walls which vary in thickness, either uniformly or in steps.

Forms for Concrete Columns

Concrete columns may be round, square, or rectangular. Forms for round columns are made of fiber tube, fiberglass, or steel. They are generally patented by manufacturers and may be purchased from building suppliers or may be rented for use on a particular job. Square or rectangular column forms are generally made with $\frac{3}{4}$-in plywood sheathing and either wood yokes and bolts or steel column clamp to hold the plywood sheathing.

Round fiber-tube forms are fabricated by numerous manufacturers. They can be used only once. After placement and proper curing of the concrete, the forms are peeled, or torn, from the column. The cost for this type of form will vary, depending on the diameter, length, and number of forms purchased. The estimator must

FIGURE 7-5
Universal wall forms.

obtain the price from a building supplier before estimating the cost for a project. The accompanying table provides representative costs of round fiber-tube forms.

Diameter, in	Cost of round fiber-tube forms per linear foot
12	$ 3.60
18	7.97
24	12.26
30	19.14
36	22.61
42	33.96
48	55.51

Round fiberglass column forms may be reused as many as 4 times. If a project has a sufficient number of columns that the column forms can be reused 3 or 4 times, then it may be economical to purchase fiberglass forms; otherwise, they may be rented. The estimator must determine the number of reuses before deciding whether it is more economical to purchase or rent round fiberglass column forms.

Round steel column forms are generally rented by the contractor for use on a particular project. Although they may be purchased and reused many times, there may be a substantial length of time between reuse, creating extra time for transporting, storing, and general handling of the forms.

Forms for square or rectangular concrete columns are generally made with $\frac{3}{4}$-in plywood sheathing with wood yokes and bolts or steel column clamps. They are generally fabricated by the contractor at the yard of the home office or at the job site.

Forms should be designed to resist the high pressures which result from quick filling with concrete. If the forms are filled in 30 min or less, the concrete will exert the full hydrostatic pressure based on a weight of 150 lb/ft^3.

If the forms are to be reused, they should be designed so that they can be removed with the least possible dismantling, usually by removal in two symmetric sections. For square or rectangular columns, the sides are made of tongue-and-groove planks, prefabricated to the correct widths and lengths, or of plywood. The planks for a side are fastened together with 1- by 4-in or 1- by 6-in wood cleats, spaced 3 to 4 ft apart along the form sides. The sides are assembled for the form. To facilitate use and reuse, forms should be identified with numbers or letters written on them.

Clamps are installed around the forms prior to filling them with concrete. The clamps are made of lumber, lumber and bolts, or steel. The steel clamps are adjustable for a wide range of sizes and may be reused several hundred times. The spacing should be such that the column forms will safely resist the maximum possible internal pressure.

For small columns, approximately 12 in wide, 1- by 4-in wood clamps, with 4-in side perpendicular to the form, are sometimes used. Each clamp will require four pieces, which are lapped where they contact each other. Each lap should be nailed with five to six 8d nails. The clamps should be spaced about 12 to 16 in apart.

For columns larger than 12 in, clamps should consist of 4- by 4-in yokes and bolts, with two washers per bolt. The yokes are drilled for the bolts. Figure 7-6 shows the details of this clamp.

If it is expected that column clamps will be needed for many uses, it will be more economical and satisfactory to purchase or rent adjustable steel clamps. Each clamp is made of four arms, about $\frac{5}{16}$ in thick and varying in width from 2 to 3 in, depending on the length. The arms are fastened in pairs with rivet hinges. Slots spaced along the arms permit their use on columns varying in size from approximately 10 by 10 in to 48 by 48 in. Figure 7-7 shows how the clamps are assembled around the forms. For most columns using 1-in sheathing, it is sufficiently acccurate to assume one steel clamp for each foot of height.

Table 7-7 shows the approximate safe spacing of steel clamps and yoke and bolt clamps for various column heights.

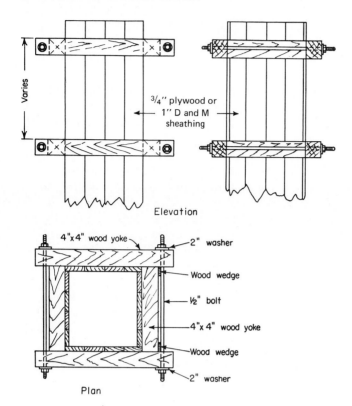

FIGURE 7-6
Column clamps made of wood and bolts.

FIGURE 7-7
Adjustable column clamps.

TABLE 7-7

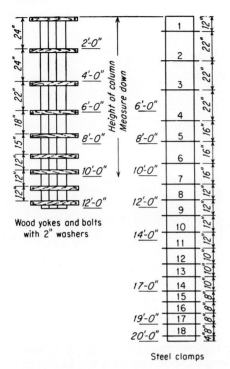

Wood yokes and bolts
with 2" washers

Steel clamps

Materials Required for Forms for Concrete Columns

The sheathing or sides for forms may be 1-in-thick D and M lumber or plywood. S4S lumber is available in net widths of $3\frac{1}{2}$, $5\frac{1}{2}$, $7\frac{1}{4}$, $9\frac{1}{4}$, and $11\frac{1}{4}$ in. D and M lumber, frequently called *tongue-and-groove lumber*, is available in net widths of $3\frac{1}{8}$, $5\frac{1}{8}$, and $8\frac{7}{8}$ in. For quantity purposes these widths are designated as 4, 6, 8, and 10 in wide.

Example 7-1. Determine the quantity of lumber required for forms for a concrete column 16 by 16 in by 11 ft 6 in long, using 1-in-thick D and M lumber, 1- by 6-in wood cleats, and steel-column clamps.

The sheathing for two sides will be 16 in wide; for the other two sides, it will be $17\frac{1}{2}$ in wide. The 16-in sides can be made with the following pieces:

Item	Net width, in
2 pc, $5\frac{1}{8}$ in	$10\frac{1}{4}$
2 pc, $3\frac{1}{8}$ in	$6\frac{1}{4}$
Total	$16\frac{1}{2}$

The $17\frac{1}{2}$-in side can be made with the following pieces:

Item	Net width, in
3 pc, $5\frac{1}{8}$ in	$15\frac{3}{8}$
1 pc, $3\frac{1}{8}$ in	$3\frac{1}{8}$
Total	$18\frac{1}{2}$

It will be necessary to rip a small quantity of lumber from the side of each panel. Five 1- by 6-in cleats will be used for each side panel.

The quantity of lumber required will be

For side panels:
$$\begin{aligned}
\text{10 pc, } 1 \times 6 \text{ in} \times 12 \text{ ft 0 in} &= 60 \text{ fbm} \\
\text{6 pc, } 1 \times 4 \text{ in} \times 12 \text{ ft 0 in} &= 24 \text{ fbm}
\end{aligned}$$

$$\begin{aligned}
\text{For cleats, 20 pc, } 1 \times 6 \text{ in} \times 1 \text{ ft 6 in} &= 15 \text{ fbm} \\
\text{For floor template, 4 pc, } 2 \times 4 \text{ in} \times 3 \text{ ft 0 in} &= \underline{8 \text{ fbm}} \\
\text{Total quantity} &= 107 \text{ fbm}
\end{aligned}$$

Area of column surface, $4 \times \frac{16}{12} \times 11.5 = 61.3 \text{ ft}^2$

Quantity of lumber per ft^2 of area, $107 \div 61.3 = 1.75 \text{ fbm}$

If S4S lumber is used for the sheathing, the quantity will be reduced to about 1.6 fbm/ft^2 of area.

Because the lumber used as braces for column forms can be removed, generally without damage, and reused elsewhere, it should not be necessary to charge all, if any, of this lumber to the column forms.

Cost of Lumber for Forms

The cost per use for lumber for forms should be based on its initial cost at the job divided by a realistic number of times that it can be used. For example, in this book the specified cost of common-grade lumber is $480.00 per 1,000 fbm. For some special grades and sizes the cost may be higher, as it is in actual practice. If form lumber can be used 5 times, the cost per use will be $96.00 per 1,000 fbm.

TABLE 7-8
Materials required for column forms per ft^2

	Kind of clamps	
Kind of lumber	Yokes and bolts	Steel clamps
D and M	3.0 fbm	1.7 fbm
S4S	2.75 fbm	1.5 fbm
Bolts and washers	0.43 lb	None

The quantity of plywood sheathing for column forms will depend on the shape, cross section, length, and number of columns. Because plywood sheets are available only in 4-ft widths, the estimator must determine the arrangement and number of sheets, including waste, necessary to build the forms for a particular job.

Quantity of Nails Required for Forms

If *D* and *M* or S4S lumber is used for forms, the quantity of nails required should be about 10 to 15 lb per 1,000 fbm of lumber. If plywood is used for sheathing, the quantity of nails should be less, usually about 7 to 12 lb per 1,000 fbm of lumber, including the plywood.

Cost of Adjustable Steel Column Clamps

Steel clamps should cost from $0.70 to $1.40 per use, the cost varying with the size of the columns.

Labor Making and Erecting Forms for Concrete Columns

Column forms are usually prefabricated at the job prior to erecting in place. The lumber is cut and ripped to the proper sizes by power saws, the sides are drawn together on a table by using carpenter's clamps, and the cleats are attached with 8d common nails, which are clinched after they are driven. The four sides are assembled and nailed together or clamped together by two clamps. After the forms are assembled, they are carried to the place of use and set into templates that have been accurately and securely located. It may be necessary to support the forms above the floor or column base until the column reinforcing is fastened to the dowels. Some contractors prefer to place the column reinforcing first and assemble the forms around it.

If it is not necessary to rebuild the forms for each use, the estimator should separate the labor into two operations, making the forms and erecting and removing them. For subsequent uses, if the sizes are not altered, only the labor for erecting and removing the forms should be considered.

TABLE 7-9
Approximate labor-hours required to make, erect, and remove 100 ft² of wood column forms

Labor	Wood yokes and bolts	Adjustable steel clamps
Carpenter making	3.5–4.5	2.5–3.5
Carpenter erecting	6.0–7.0	5.5–6.5
Helper making	1.0–1.5	0.5–1.0
Helper erecting and removing	3.5–4.5	4.0–5.0

Example 7-2. Estimate the cost of forms for 20 columns 16 by 16 in by 11 ft 6 in long. The sheathing will be 1-in D and M lumber. Adjustable steel clamps will be used with the columns. Assume that the lumber will be used 5 times. The lumber will cost $480.00 per 1,000 fbm at the job.

The quantity of material will be

Contact area per column, $(16 \div 12) \times 4 \times 11.5 = 61.33$ ft^2
Total area of column forms, $61.33 \times 20 = 1,227$ ft^2
Area of forms that must be fabricated $= 1,227 \div 5 = 245.4$ ft^2

The cost will be:

Materials:

Lumber, $1,227 \times 1.8 = 2,209$ @ $0.48 ÷ 5 uses	=	\$ 212.06
Nails, $2,209 \times 15$ per $1,000 = 33$ lb @ \$0.90	=	29.70
Form oil, $1,227 \div 400 = 3.1$ gal @ \$7.85	=	24.34
Power saws, $2,209 \div 400$ ft^2/h $= 5.5$ h @ \$3.75	=	20.63
Column clamps, $20 \times 8 = 160$ @ 0.85	=	136.00

Labor:

Carpenter making forms,		
$\quad$ 245.4 ft^2 $\times$ 3 h/100 ft^2 $= 7.4$ h @ \$16.70	=	123.58
Helper making forms,		
$\quad$ 245.4 ft^2 $\times$ 1 h/100 ft^2 $= 2.5$ h @ \$12.10	=	30.25
Carpenter erecting and removing forms,		
$\quad$ 1,227 ft^2 $\times$ 6 h/100 ft^2 $= 73.6$ h @ 16.70	=	1,229.12
Helper erecting and removing forms,		
$\quad$ 1,227 ft^2 $\times$ 4.5 h/100 ft^2 $= 55.2$ h @ \$12.10	=	667.92
Foreman, based on 5 carpenters, 22 h @ \$18.30		
$\quad$ $(7.4 + 73.6) \div 5 = 16.2$ h @ \$18.30	=	296.46
$\quad$ Total cost	=	\$2,770.06
$\quad$ Cost per ft^2, \$2,770.06 ÷ 1,227	=	2.26

Without reusing the forms the cost would be:

Lumber, $1,227 \times 1.8 = 2,209$ fbm @ \$04.8	=	\$1,060.32
Nails, 33 lb @ \$0.90	=	29.70
Form oil, 3.1 gal @ \$7.85	=	24.34
Power saws, 5.5 h @ \$3.75	=	20.63
Column clamps, 160 @ \$0.85	=	136.00
Carpenter making forms, 36.8 h @ \$16.70	=	614.56
Helper making forms, 12.3 h @ \$12.10	=	148.83
Carpenter erecting forms, 73.6 h @ \$16.70	=	1,229.12
Helper erecting forms, 55.2 h @ \$12.10	=	667.92
Foreman, based on 5 carpenters,		
$\quad$ $(36.8 + 73.6) \div 5 = 22.1$ h @ \$18.30	=	404.43
$\quad$ Total cost	=	\$4,335.85
$\quad$ Cost per ft^2, \$4,335.85 ÷ 1,227	=	3.53

Economy of Reducing the Size of Concrete Columns

Example 7-2 demonstrated the economy affected by reusing the same-size column forms. As noted, the cost per square foot of surface area for the additional uses of the forms, without remaking to different sizes, was reduced from $3.53 to $2.26 for a saving of $1.27 per square foot. The saving on forms for a column in Example 7-2 was

$$\frac{\$4,335.85 - \$2,770.06}{20} = \$78.29$$

If these columns are erected in a multistory building whose load conditions will permit the sizes of the columns for higher floors to be reduced to 14 by 14 in, thereby requiring that different forms be made or that the forms from the larger columns be remade for use as forms for the smaller columns, then the additional cost of making or remaking the smaller forms may exceed the saving in the cost of concrete required for the smaller columns. Thus, reducing the sizes of concrete columns to reduce the cost of concrete may not be economical. Also, if a larger-size column is used for the higher floors, it is probable that the quantity of reinforcing steel required in a column will be less than would be required in a smaller column. This practice could result in additional savings.

Shores

Shores are members of form work whose function is to support loads from the forms for concrete beams, girders, slabs, etc. Shores may be made from commercial lumber, or they may be obtained from commercial manufacturers, by purchase or rental.

Wood Shores

Wood shores are frequently made at the project from lumber in sizes 4 by 4 in, 4 by 6 in, 6 by 6 in, or larger. Either rough-sawed or S4S lumber may be used. Figure 7-10 illustrates typical shores in use.

If the vertical posts of wood shores are adequately braced to prevent horizontal movement or buckling under loading conditions, the allowable compressive stress may be as high as $1,600\ lb/in^2$. However, if a wood T head or other wood member rests on the top of the post, the maximum safe load may be limited by the allowable unit stress between the post and the member resting on the post. The maximum compressive stress acting perpendicular to the grain of wood should be limited to $500\ lb/in^2$.

Consider a shore consisting of a 4- by 4-in S4S post and a T head of the same size and material. The actual sizes of the lumber will be $3\frac{1}{2}$ by $3\frac{1}{2}$ in. The area of contact between the top of the post and the T head will be $12.25\ in^2$. The maximum

TABLE 7-10
Maximum safe loads, in pounds, from T head or stringers on wood shores†

Size shore, in	Size T head or stringer, in		
	4 × 4 rough	4 × 4 S4S	2-in S4S
4 × 4 rough	8,000	7,000	3,000
4 × 4 S4S	7,000	6,125	2,625

† The loads are limited to 500 lb/in² of contact area between the members.

safe load on the post will be $500 \times 12.25 = 6,125$ lb. Even though the post may be capable of withstanding a load of $1,600 \times 12.25 = 19,600$ lb, the maximum load for this design should not exceed 6,125 lb.

Table 7-10 gives the safe loads that may be transmitted to the tops of wood shores from horizontal wood members resting on the tops of the shores.

Adjustable Shores

Figure 7-8 illustrates an adjustable shore made by one manufacturer. Table 7-11 gives information for this shore, and Fig. 7-9 shows them in use.

Quantity of Lumber Required for Concrete Beams Using Wood Shores

Figure 7-10 illustrates the details of constructing wood forms and shores for a typical beam-and-slab floor. Two plans for constructing beam forms are shown. Plan A requires that the 4- by 4-in T heads be long enough to permit one 2- by 4-in ledger to be nailed on each side to resist lateral movement of 2- by 4-in studs. For

TABLE 7-11
Load capacities for Symons shores, lb†

Length, ft		Height extended to, ft						
From	To	7.5	9.0	10.0	11.0	12.0	13.0	14.0
4.5	7.5	6,000						
6.5	11.5	6,000	6,000	5,500	4,500	—	—	—
7.25	13.0	6,000	6,000	6,000	5,500	3,500	3,000	—
8.25	15.0	—	6,000	6,000	6,000	4,000	3,500	2,500

† These loads may be increased 100 percent if two-way horizontal braces are installed at the midpoints, provided the maximum load does not exceed 6,000 lb.

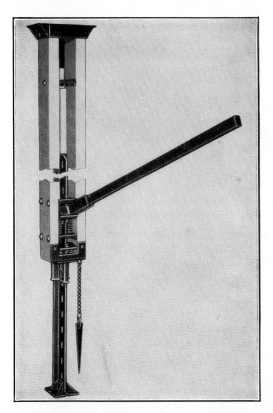

FIGURE 7-8
Adjustable shore.

FIGURE 7-9
Adjustable shores supporting floor beams.

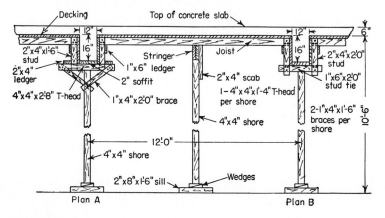

FIGURE 7-10
Woods forms and shores for concrete beams.

plan B the bottoms of the studs extend below the beam soffit about 6 in to permit a 1- by 6-in tie to be installed for each pair of studs. Plan B permits the use of shorter T heads than those required for plan A.

If the concrete beams of Fig. 7-10 are spaced 12 ft 0 in apart, with an intermediate stringer midway between the beams, as shown, the beam shores will support the weight of the beam plus a slab 6 ft wide. The volume of concrete per linear foot of beam will be $6 \times 1 \times 0.5 + 1 \times 1.33 = 4.33$ ft^3. The weight will be 650 lb. Add 40 lb/ft^2 for live load, for the storage of form lumber and other materials during construction, equal to $6 \times 40 = 240$ lb/lin ft. The total load will be 890 lb/lin ft. If 4- by 4-in S4S shores and T heads are used, the maximum safe load per shore will be 6,125 lb. The maximum safe spacing of shores will be $6,125 \div 890 = 6.9$ ft. To eliminate the danger of objectionable deflection between shores, the spacing should not exceed 5 ft.

TABLE 7-12
Quantity of lumber required for forms for concrete beams

	Lumber required, fbm/ft^2 of surface	
	Plan in Fig. 7-10	
Type of shores	A	B
Wood to 9 ft long	3.50–4.0	3.2–3.7
Adjustable	—	2.3–2.8

Example 7-3. Determine the quantity of lumber required for a beam 12 in wide, 16 in deep, and 18 ft 10 in long, based on plan A of Fig. 7-10. With a shore at each end of the beam, five shores will be required.

The quantity of lumber will be
Soffit:

1 pc, 2 × 8 in × 20 ft 0 in	=	27 fbm
1 pc, 2 × 6 in × 20 ft 0 in	=	20 fbm

Sides:
Use $3\frac{1}{2}$ pc, 1 × 6 in per side

7 pc, 1 × 6 in × 20 ft 0 in	=	70 fbm
Studs, 22 pc, 2 × 4 in × 1 ft 6 in	=	22 fbm
Ledgers, 2 pc, 2 × 4 in × 20 ft 0 in	=	27 fbm

Shores:

5 pc, 4 × 4 in × 9 ft 0 in	=	60 fbm
5 pc, 4 × 4 in × 3 ft 0 in	=	20 fbm
10 pc, 1 × 4 in × 2 ft 0 in	=	7 fbm
Sills, 5 pc, 2 × 8 in × 1 ft 6 in	=	10 fbm
Total quantity of lumber	=	263 fbm

Area of the beam, 69 ft^2
Quantity of lumber per ft^2, 263 ÷ 69 =   3.8 fbm

If the method shown in plan B is used,

The quantity of lumber will be:
Soffit:

1 pc, 2 × 8 in × 20 ft 0 in	=	27 fbm
1 pc, 2 × 6 in × 20 ft 0 in	=	20 fbm
Sides, 7 pc, 1 × 6 in × 20 ft 0 in	=	70 fbm
Studs, 22 pc, 2 × 4 in × 2 ft 0 in	=	30 fbm
Cross braces, 11 pc, 1 × 6 in × 2 ft 0 in	=	11 fbm

Shores:

5 pc, 4 × 4 in × 9 ft 0 in	=	60 fbm
5 pc, 4 × 4 in × 1 ft 4 in	=	9 fbm
10 pc, 1 × 4 in × 1 ft 6 in	=	5 fbm
Sills, 5 pc, 2 × 8 in × 1 ft 6 in	=	10 fbm
Total quantity of lumber	=	242 fbm

Quantity of lumber per ft^2, 242 ÷ 69 =   3.5 fbm

Quantity of Lumber Required for Forms for Concrete Beams Using Adjustable Shores

If adjustable shores instead of wood shores are used, the maximum safe load will be 6,000 lb. The maximum safe spacing, based on the safe load, will be about 6 ft 6 in. However, the spacing will be limited to 4 ft 0 in, with no horizontal braces required.

The quantity of lumber required with plan B of Fig. 7-10 and adjustable shores will be 242 − 84 = 158 fbm. The quantities are determined from Example 7-2.

The quantity of lumber per square foot will be 158 ÷ 69 = 2.3 fbm.

Labor Required to Build Forms for Concrete Beams

In building and erecting forms for a concrete beam, it is customary to make the soffit and then, starting at a column form, erect the soffit on the shores. The shores are wedged to the correct elevation and temporarily braced. The side forms for the beams are made and cleated with the 2- by 4-in studs, to the full length of the beam. Then they are lifted to position, nailed to the soffit, and secured against lateral failure by the method shown in plan A or B (Fig. 7-10). After the tops of the forms are brought to a straight line, the joists for the decking are installed, and then the decking is installed.

If the side forms are to be used on other beams, they should be assembled and erected in a manner which will permit them to be removed intact. The estimator should show separately the labor required to make and to erect the forms, since this practice will permit him or her to include only erection costs for additional uses.

Table 7-13 gives the labor hours required to build forms for concrete beams.

Example 7-4. Estimate the cost of forms for 12 inside concrete beams, 13 in wide, 19 in deep, and 19 ft 2 in long, using 4- by 4-in wood shores 9 ft 0 in long. The forms will be used 6 times on the job.

The area of beams will be $\dfrac{12 \times (19 + 19 + 13)}{12} \times 19$ ft 2 in $= 980$ ft^2

Assume that lumber will cost $480.00 per 1,000 fbm at job

The cost will be

Lumber, 980 ft^2 × 3.5 fbm/ft^2 = 3,430 fbm @ $0.08	=	$ 274.40
Nails, 3,430 fbm × 15 lb/1,000 fbm = 51 lb @ $0.95	=	48.45
Form oil, 980 ft^2 ÷ 400 ft^2/gal = 2.5 gal @ $7.50	=	18.75
Power saws, 3,430 fbm ÷ 400 fbm/h = 8.6 h @ $2.80	=	24.08
Carpenters, 980 ft^2 × 10 h/100 ft^2 = 98 h @ $16.70	=	1,636.60
Helpers, 980 ft^2 × 5.75 h/100 ft^2 = 56.3 h @ $12.10	=	681.23
Foreman, based on using 6 carpenters,		
98.0 ÷ 6 = 16.3 @ $18.30	=	298.29
Total cost	=	$2,981.80
Cost per ft^2, $2,981.80 ÷ 980	=	3.04

The cost for additional uses will be

Lumber, 3,430 fbm @ $0.08	=	$ 274.40
Nails, 3,430 fbm × 7 lb/1,000 fbm = 24 lb @ $0.95	=	22.80
Form oil, 2.5 gal @ $7.50	=	18.75
Power saws, no cost		
Carpenters, 980 ft^2 × 6 h/100 ft^2 = 58.8 h @ $16.70	=	981.96
Helpers, 980 ft^2 × 4.25 h per 100 ft^2 = 41.7 h @ $12.10	=	504.57
Foreman, based on using 6 carpenters,		
58.8 ÷ 6 = 9.8 h @ $18.30	=	179.34
Total cost	=	$1,981.82
Cost per ft^2, $1,981.82 ÷ 980	=	2.02

TABLE 7-13
Labor-hours required to build 100 ft² of forms for concrete beams

Operation	Using 4- × 4-in wood shores	Using adjustable shores
For inside beams and girders		
Making soffits, sides, and wood shores:		
Carpenter	3.5–4.5	3.0–3.5
Helper	1.0–1.5	1.0–1.5
Erecting soffits, sides, and shores:		
Carpenter	5.0–6.0	5.0–5.5
Helper	1.5–2.0	1.5–2.0
Removing forms:		
Helper	1.5–2.0	1.5–2.0
For outside beams and girders		
Making soffits, sides, and wood shores:		
Carpenter	3.5–4.5	3.0–4.0
Helper	1.0–1.5	
Erecting soffits, sides, and shores:		
Carpenter	7.0–8.0	6.7–7.0
Helper	3.0–3.5	3.0–3.5
Removing forms:		
Helper	3.0–3.5	3.0–3.5

Forms for Flat-Slab-Type Concrete Floors

The flat-slab-type concrete floor is sometimes used when column heads support the floors, without beams or girders. Since there are no beam forms to assist in supporting the slab forms, it is necessary to classify all forms as slab forms. The cost of forms will vary with the thickness of the slab and the height of the slab above the lower floor.

Figure 7-11 illustrates the method commonly used in building forms for this type of slab.

To determine the quantity of lumber required, it is necessary to know the thickness of the slab, the probable live load on the slab while the forms are in place, and the height of the slab above the supporting floor. The live load includes form lumber, brick, tile, and other materials temporarily stored on the slab. The live load will usually amount to about 40 to 50 lb/ft².

The form builder has considerable freedom in selecting the size and spacing of joists and stringers. Joist sizes commonly used are 2 by 4 in, 2 by 6 in, 2 by 8 in, and 2 by 10 in S4S. Stringer sizes are 2 by 8 in, 2 by 10 in, 4 by 4 in, and 4 by 6 in S4S. The 2-in-thick stringers give good economy, but the bearing stresses where they rest on the shores may be excessive and should be checked before the final design is selected. The exercise of care in designing forms will result in economical forms, considering the cost of materials and labor.

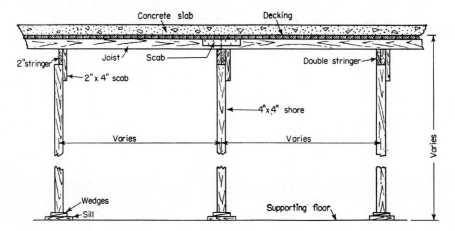

FIGURE 7-11
Wood forms for flat-slab concrete floors.

Design of Forms for Flat-Slab-Type Concrete Floors

Table 7-14 gives information that may be used to design forms for flat-slab concrete floors. The following example illustrates how the information in Table 7-14 may be used to design forms.

Example 7-5. Determine the quantity of lumber per square foot of floor area required for a 6-in-thick concrete slab whose clear height above the lower floor is 11 ft 2 in. Assume a live load of 40 lb/ft². The decking will be 1 by 6 in S4S, whose net width is $5\frac{1}{2}$ in.

The joists will be spaced 2 ft 0 in on centers.

If the selected spacings for joists, stringers, and shores will permit the use of commercial lumber lengths, which vary in 2-ft steps, with little or no end wastage, this will generally reduce the quantity and the cost of form materials. For example, if the spacing of shores is 4 ft, the stringers should be 12, 16, or 20 ft long.

Reference to Table 7-14 indicates that 2- by 8-in joists will permit a safe span of 8 ft 0 in. Thus 2- by 8-in joists 16 ft 0 in long, spaced 2 ft 0 in on centers, will be used.

Investigate the desirability of using 4- by 6-in S4S or 2- by 10-in S4S stringers.

If 4- by 6-in stringers are used, the maximum safe span will be 5 ft 0 in. The area supported by a shore will be 5 ft × 8 ft = 40 ft². The total weight on a shore from this area will be 40 × 115 = 4,600 lb. The area in compression between a stringer and the top of a 4- by 4-in shore will be 3.5 × 3.5 = 12.25 in². The unit compressive stress on the stringer will be 4,600 ÷ 12.25 = 375 lb/in², which is satisfactory.

If 2- by 10-in stringers are used, the maximum safe span will be 4 ft 0 in, which will be the maximum safe spacing of shores. The area supported by a shore will be 4 ft × 8 ft = 32 ft². The total weight on a shore will be 32 × 115 = 3,680 lb. The area in compression between a stringer and the top of a shore will be 1.5 × 3.5 = 5.25 in². The unit compressive stress on the stringer will be 3,680 ÷ 5.25 = 700 lb/in². This stress exceeds the allowable unit stress. If 2- by 10-in stringers are used, the spacing of

TABLE 7-14

Safe spans for forms for concrete slabs based on using 1-in decking and joists spaced 2 ft 0 in on centers†

Thickness of slab, in		4	5	6	8	10
Total load, lb/in²		90	103	115	140	165
Size joist, in		Safe span for joists				
2 × 4		4 ft 9 in	4 ft 9 in	4 ft 3 in	3 ft 9 in	3 ft 9 in
2 × 6		6 ft 6 in	6 ft 6 in	6 ft 0 in	5 ft 6 in	5 ft 0 in
2 × 8		8 ft 9 in	8 ft 3 in	8 ft 0 in	7 ft 6 in	7 ft 0 in
2 × 10		10 ft 0 in	10 ft 0 in	9 ft 6 in	8 ft 9 in	8 ft 3 in
Size stringer, in	Stringer spacing	Safe span for stringers				
2 × 8	6 ft 0 in	5 ft 6 in	4 ft 6 in	4 ft 0 in	3 ft 6 in	3 ft 0 in
2 × 10	7 ft 0 in	5 ft 6 in	4 ft 6 in	4 ft 0 in	3 ft 6 in	3 ft 0 in
4 × 4	5 ft 0 in	4 ft 3 in	3 ft 9 in	3 ft 9 in	3 ft 3 in	
4 × 6	6 ft 0 in	5 ft 6 in	5 ft 6 in	5 ft 0 in	4 ft 9 in	4 ft 3 in

† The spans are based on a fiber stress in bending of 1,800 lb/in² of net size for the joist or stringer. Wood capable of withstanding this stress should be used, or the spans should be reduced. All lumber is S4S.

the shores must be reduced to less than 4 ft, to limit the stress on the stringer to not more than 500 lb/in². The maximum safe spacing will be $4 \times 500/700 = 2.86$ ft.

Use 4- by 6-in S4S stringers, with shores spaced 4 ft 6 in apart to limit deflection. The forms will consist of the following materials:

Decking, 1 × 6 in S4S, 16 ft 0 in long
Joists, 2 × 8 in S4S, 16 ft 0 in long
Stringers, 4 × 6 in S4S, 18 ft 0 in long
Shores, 4 × 4 in S4S, 10 ft 0 in long

To determine the quantity of lumber required per square foot of slab, consider an area of slab 8 ft 0 in wide and 18 ft 0 in long = 144 ft².

The quantity of lumber required will be
Decking, 8 × 12 ÷ 5.5 = 17.5 pc of 1 × 6 in × 18 ft 0 in = 157 fbm
Joists, 18 ÷ 2 = 9 pc of 2 × 8 in × 8 ft 0 in   = 96 fbm
Stringer, 1 pc of 4 × 6 in × 18 ft 0 in   = 36 fbm
Shores, 18 ÷ 4.5 = 4 pc of 4 × 4 in × 10 ft 0 in   = 53 fbm
Scabs, 4 pc of 2 × 4 in × 2 ft 0 in   = 6 fbm
Sills for shores, 4 pc of 2 × 8 in × 1 ft 6 in   = 8 fbm
Two-way braces for shores, 4 × 8 ft 0 in = 32 ft + 18 ft
  = 50 ft, 1 × 6 in   = 25 fbm
    Total quantity   = 381 fbm
        Quantity of lumber per ft², 381 ÷ 144 ft²   = 2.7 fbm

If adjustable shores are used instead of wood shores, the lumber for shores and scabs will not be needed. This will reduce the quantity of lumber to $381 - 59 = 322$ fbm. The quantity of lumber per square foot will be $322 \div 144 = 2.24$ fbm.

Quantity of Lumber Required for Forms for Flat-Slab-Type Concrete Floors

Table 7-15 gives the approximate quantities of lumber required per square foot of slab area for various thicknesses of slab and heights of ceilings.

Labor Required to Build Forms for Flat-Slab-Type Concrete Floors

Table 7-16 gives the number of labor-hours required to build and remove forms for flat-slab concrete floors.

Example 7-6. Estimate the direct cost of furnishing, building, and removing forms for a flat-slab concrete floor whose size is 56 ft 8 in by 88 ft 6 in. The slab will be 6 in thick. There will be 16 column heads 6 by 6 ft each. The ceiling height is 11 ft 8 in. Assume that the lumber will be used 4 times. The lumber will cost $480.00 per 1,000 fbm at the job.

The quantities are:

Gross area of slab, 56.67×88.5	$= 5{,}015 \text{ ft}^2$
Area of column heads, 16×36	$= -576 \text{ ft}^2$
Net area of slab	$= 4{,}439 \text{ ft}^2$

TABLE 7-15
Quantities of lumber required per square foot of flat-slab concrete floor, fbm†

Ceiling height, ft	Thickness of slab, in					
	6	8	10	6	8	10
	4- × 4-in wood shores			Adjustable shores		
10	2.7	2.8	3.0	2.3	2.4	2.7
12	2.8	2.9	3.1	2.3	2.4	2.7
14	3.0	3.1	3.4	2.5	2.6	3.0
16	3.1	3.2	3.6	2.5	2.6	3.0

† The values in the table are based on a live load of 40 lb/ft² on the slab. The values do not include lumber for column heads, stairs, elevator shafts, etc.

TABLE 7-16
Labor-hours required to build and remove 100 ft² of forms for flat-slab concrete floors and column heads

Ceiling height, ft	Laborer	4- × 4-in wood shores	Adjustable shores
10–14	Carpenter	5.0–6.0	4.5–5.5
	Helper	3.5–4.0	3.0–3.5
14–16	Carpenter	5.5–6.5	5.0–6.0
	Helper	4.0–4.5	3.5–4.0

If adjustable shores are used, the cost will be:
Lumber for slab, 4,439 ft² × 2.3 fbm/ft² = 10,210 fbm
  10,210 fbm @ $0.12 = $1,225.20
Lumber for column heads, 576 ft² × 4.0 fbm/ft² = 2,304 fbm
  2,304 fbm @ $0.12 = 276.48
Nails, 12,514 ft² × 12 lb/1,000 fbm = 150 lb
  150 lb @ $0.95 = 142.50
Form oil, 5,015 ft² ÷ 400 ft²/gal = 12.5 gal
  12.5 gal @ $7.50 = 93.75
Shores, 4,439 ft² ÷ 24 ft²/shore = 185
  185 shores @ $2.25 = 416.25
Power saws, 12,514 fbm ÷ 400 fbm/h = 31.3 h
  31.3 h @ $2.88 = 90.14
Carpenters, 5,015 ft² × 5.0 h/100 ft² = 250.7 h
  250.7 h @ $16.70 = 4,186.69
Helpers, 5,015 ft² × 3.0 h/100 ft² = 150.5 h
  150.5 h @ $12.10 = 1,821.05
Foreman, based on using 5 carpenters, 250.7 ÷ 5 = 50.1 h
  50.1 h @ $18.30 = 916.83
    Total cost = $9,168.89
    Cost per ft², $9,168.89 ÷ 5,015 = 1.83

If 4- × 4-in wood shores are used, the cost will be:
Lumber for slab, 4,439 ft² × 2.7 fbm/ft² = 11,985 fbm
  11,985 fbm @ $0.12 = $1,438.20
Lumber for column heads, 576 ft² × 4.25 fbm/ft² = 2,448 fbm
  2,448 fbm @ $0.12 = 293.76
Nails, 14,433 fbm × 12 lb/1,000 fbm = 173 lb
  173 lb @ $0.95 = 164.35
Form oil, 5,015 ft² ÷ 400 ft²/gal = 12.5 gal
  12.5 gal @ $7.50 = 93.75
Power saws, 14,433 fbm ÷ 400 fbm/h = 36.0 h
  36.0 h @ $2.85 = 102.60

Carpenters, 5,015 ft^2 × 5.5 h/100 ft^2 = 275.8 h
  275.8 h @ $16.70   = 4,605.86
Helpers, 5,015 ft^2 × 3.5 h/100 ft^2 = 175.5 h
  175.5 h @ $12.10   = 2,123.55
Foreman, based on using 5 carpenters, 275.8 h ÷ 5 = 55.2 h
  55.2 h @ $18.30   = 1,010.16
    Total cost   = $9,832.23
    Cost per ft^2, $9,832.23 ÷ 5,015   =   1.96

Thus the use of adjustable shores is cheaper than using wood shores.

Forms for Slabs for Beam-and-Slab Type of Concrete Floors

The forms for the slab for beam-and-slab-type concrete floors consist of the decking, joists, stringers, shores, and braces, if required. The ends of the joists are usually supported by the beam forms. If the beams are spaced more than approximately 8 ft 0 in on centers, it will usually be economical to install a stringer between the beams to assist in supporting the decking. The use of a stringer will reduce the load on the beam forms and will permit the use of smaller joists.

Figure 7-12 illustrates the types of forms used for slab floors. Plan A does not require stringers to support the forms, while plan B requires stringers and shores.

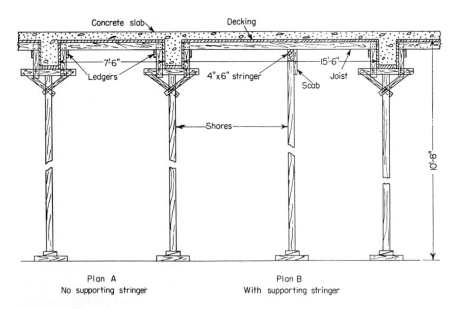

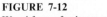

FIGURE 7-12
Wood forms for beam-and-slab concrete floors.

Quantity of Lumber Required for Forms for Beam-and-Slab Type of Concrete Floors

This determination is based on using plan A of Fig. 7-12. If the concrete slab is 6 in thick and the clear span between the beam faces is 7 ft 6 in, Table 7-14 indicates that it will be necessary to use 2- by 8-in S4S lumber for joists, spaced 2 ft 0 in on centers. Decking will be 1- by 6-in S4S lumber, whose actual width is $5\frac{1}{2}$ in. Ledgers will be 1- by 6-in S4S lumber. Assume that the bay will be 24 ft 6 in long.

The quantities are determined as follows:
Area of slab, $7.5 \times 24.5 = 184$ ft^2
No. pieces of decking required for 7.5 ft of width, $7.5 \times 12 \div 5.5 = 16.35$ (use 17 pc)
For length of 24.5 ft use
   1 pc, 14 ft 0 in long
   1 pc, 12 ft 0 in long

The quantity of lumber will be
Decking, 17 pc, 1 × 6 in × 12 ft 0in	= 102 fbm
17 pc, 1 × 6 in × 14 ft 0 in	= 119 fbm
Joists, no., 24.5 ÷ 2 = 12.25 + 1 = 14 pc, 2 × 8 in × 8 ft 0 in	= 149 fbm
Ledgers, 2 pc, 1 × 6 in × 12 ft 0 in	= 12 fbm
2 pc, 1 × 6 in × 14 ft 0 in	= 14 fbm
Total quantity	= 396 fbm
Quantity per ft^2, 396 ÷ 184	= 2.15 fbm

Quantity of Lumber Required for Forms for Beam-and-Slab Type Concrete Floors

This determination is based on using plan B of Fig. 7-12. If the concrete slab is 6 in thick and the clear span between the beam faces is 15 ft 6 in, the joists should be supported at their midpoints with a stringer, as indicated in Fig. 7-12. The stringer may be supported with either wood or adjustable shores. Use 4- by 4-in wood shores.

Reference to Table 7-14 indicates the use of 2- by 8-in S4S joists spaced 2 ft 0 in on centers. Stringers will be 4- by 6-in S4S with shores spaced not more than 5 ft 0 in apart.

The quantities are determined as follows:
Length of bay, 24 ft 6 in
Area of bay, $15.5 \times 24.5 = 380$ ft^2
No. pieces of decking required for 15.5 ft of width, $15.5 \times 12 \div 5.5 = 33.9$ (use 34 pc)

The quantity of lumber will be:
Decking, 34 pc, 1 × 6 in × 12 ft 0 in	= 204 fbm
34 pc, 1 × 6 in × 14 ft 0 in	= 238 fbm
Joists, no., 24.5 ÷ 2 = 12.25 + 1 = 14 pc, 2 × 8 in × 16 ft 0 in	= 299 fbm

Stringer, 1 pc, 4 × 6 in × 12 ft 0 in		=	24 fbm
1 pc, 4 × 6 in × 14 ft 0 in		=	28 fbm

Shores, use 4-ft 0-in spacing under the 12-ft 0-in length, and 4-ft 8-in under the 14-ft 0-in length of stringer for a total of 7 shores,

7 pc, 4 × 4 in × 10 ft 0 in	=	93 fbm
Scabs, 7 pc, 2 × 4 in × 2 ft 0 in	=	10 fbm
Sills, 7 pc, 2 × 8 in × 1 ft 6 in	=	14 fbm
Horizontal braces:		
2 pc, 1 × 6 in × 14 ft 0 in	=	14 fbm
7 pc, 1 × 6 in × 8 ft 0 in	=	28 fbm
Total quantity	=	952 fbm
Quantity per ft², 952 ÷ 380	=	2.5 fbm

Table 7-17 gives the quantities of lumber required for forms for the slab only of beam-and-slab type of concrete floors. The width of the slab is the clear distance between the faces of the beams.

Labor Required to Build Forms for Beam-and Slab Type of Concrete Floors

The labor required to build forms for the slab only for beam-and-slab type of concrete floors will be slightly less for plan A than for plan B, since stringers and shores are not installed. The beam forms will be in place when the slab forms are

TABLE 7-17

Lumber required for forms for the slab only of beam-and-slab type concrete floors, fbm/ft²

	With no supporting stringers			
	Thickness of slab, in			
Height of ceiling, ft	4	6	8	10
All heights	2.0	2.2	2.3	2.4

	With supporting stringers							
	4- × 4-in wood shores				Adjustable shores			
	Thickness of slab, in				Thickness of slab, in			
Height of ceiling, ft	4	6	8	10	4	6	8	10
10	2.3	2.4	2.5	2.6	2.1	2.2	2.3	2.4
12	2.4	2.5	2.6	2.7	2.2	2.3	2.4	2.5
14	2.6	2.7	2.8	2.9	2.4	2.5	2.6	2.7
16	2.7	2.8	2.9	3.0	2.5	2.6	2.7	2.8

TABLE 7-18

Labor-hours required to build and remove the slab forms for 100 ft² of beam-and-slab type of concrete floors

With no supporting stringers		
Height of ceiling, ft	**Laborer**	**Labor-hours**
All heights	Carpenter	3.0–4.0
	Helper	2.0–2.5

With supporting stringers			
		Labor-hours	
Height of ceiling, ft	**Laborer**	**Wood shores**	**Adjustable shores**
10–14	Carpenter	4.0–5.0	4.0–4.5
	Helper	3.0–3.5	2.5–3.0
14–16	Carpenter	4.5–5.5	4.0–4.5
	Helper	3.5–4.0	3.5–4.0

started. Ordinarily the 1- by 6-in ledgers on which the joists rest will be nailed to the beam forms before erecting them. The joists will be cut to length, sized if necessary, and nailed in place. Then the decking will be installed.

If supporting stringers and shores are required, they will be installed after some of or all the joists are in place, but prior to installing the decking.

Table 7-18 gives the labor-hours required to build and remove the forms for 100 ft² of beam-and-slab type of concrete floors.

Example 7-8. Estimate the cost of building and removing forms for a beam-and-slab type of concrete floor using adjustable shores. The slab will be 15 ft 4 in wide, 38 ft 6 in long, and 7 in thick. The ceiling height will be 12 ft 2 in. Assume that the lumber will cost \$480.00 fbm at the job and that it will be used 4 times.

The area of the slab will be $15.33 \times 38.5 = 590$ ft²

The cost will be:

Lumber, 590 ft² × 2.4 fbm/ft² = 1,416 fbm @ \$0.12	= \$169.92
Nails, 1,416 fbm × 12 lb/1,000 fbm = 17 lb @ \$0.90	= 15.30
Form oil, 590 ft² ÷ 400 ft²/gal = 1.5 gal @ \$7.50	= 11.25
Shores, 11 @ \$2.25	= 24.75
Power saws, 1,416 fbm ÷ 400 fbm/h = 3.5 h @ \$2.85	= 9.98
Carpenter, 590 ft² × 4.2 h/100 ft² = 24.8 h @ \$16.70	= 414.16
Helper, 590 ft² × 3.0 h/100 ft² = 17.7 h @ \$12.10	= 214.17
Foreman, based on using 5 carpenters, 24.8 ÷ 5 = 5 h @ \$18.30 =	91.50
Total cost	= \$951.03
Cost per ft², \$951.03 ÷ 590 ft²	= 1.61

Forms for Metal-Pan and Concrete-Joist Type of Concrete Floors

There is an increasing use of metal-pan and concrete-joist type of concrete floors. This type of construction has a number of advantages over other types of forms for concrete floors. The total quantity of concrete is less than for slab floors, which reduces the cost of materials and also permits the use of smaller beams, columns, and footings. If the pans are used enough times, the cost per square foot of use, including the labor cost of installing and removing, should compare favorably with forms built entirely of lumber. Some types of pans permit removal within about a week.

This type of construction has several disadvantages. Few building contractors can afford to purchase the pans, since they cannot be reused unless a similar project is constructed. The structure must be designed for the use of a designated size and type of pan. It is difficult to form around irregular structural units and openings.

Several plans are used for the installation of the metal pans. Contractors can rent the pans at an agreed price per square foot per use, with a minimum number of uses. Also they can subcontract the pans only in place and removed, or they can subcontract the furnishing of the pans and supporting forms completely.

Figure 7-13 gives details of the widths and depths of flange and adjustable types of Meyer Steelforms. Figure 7-14 shows plan views of the installation of flange and adjustable types of Meyer Steelforms. Figure 7-15 shows the details of supporting forms for the flange and adjustable types of metal-pan forms. The stringers may be supported by wood or adjustable shores. Figure 7-16 shows various stages in the installation and use of metal-pan forms.

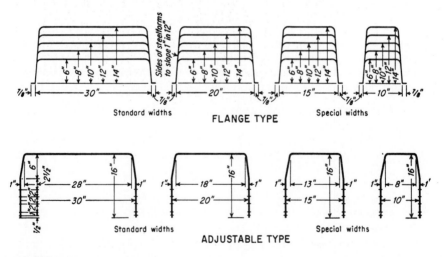

FIGURE 7-13
Dimensions of Meyer Steelforms.

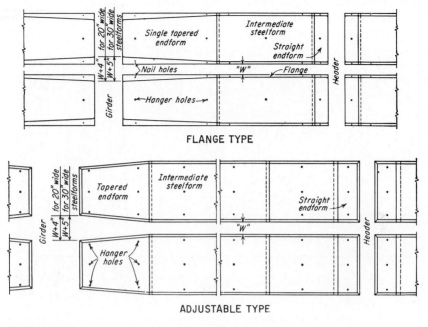

FIGURE 7-14
Plan view of Meyer Steelforms.

Metal pans are available which are designed to be self-supporting for spans up to 12 ft or more. These forms do not require intermediate stringers and shores.

Table 7-19 gives the quantities of concrete required for Meyer Steelform construction. Manufacturers of other forms will furnish similar information for their forms.

Lumber Required for Metal-Pan and Concrete-Joist Construction

The lumber required for metal-pan and concrete-joist construction consists of centering strips or planks 2 by 6 in or 2 by 8 in, stringers, shores, and braces. For irregularities around structural units or openings, lumber is sometimes used instead of metal pans.

The spacing of the centering strips will equal the width of the pan plus the thickness of the joist. By using pans of special widths it is usually possible to install pans for any width floor. Also, the pans are available in various lengths, usually in 1-ft intervals, which permits the pans to be used for any length floor.

Stringers may be 2 by 8 in, 2 by 10 in, or 4 by 6 in. They should be spaced, insofar as possible, an exact number of feet apart to reduce the end wastage of lumber in centering strips. A spacing of 4 ft 0 in is frequently used, which permits the use of centering strips in lengths 8 ft 0 in, 12 ft 0 in, 16 ft 0 in, or 20 ft 0 in. A 2-in

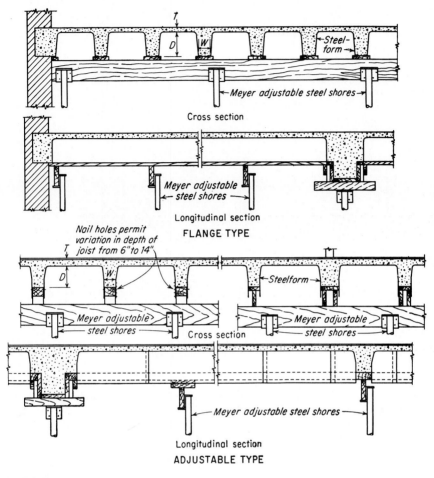

FIGURE 7-15
Supporting forms for metal pans.

stringer is not thick enough to permit safe end-butting joints of centering strips. If butt joints of centering strips are necessary, the thickness of 2-in stringers should be increased by nailing a lumber strip to the side of each stringer.

Wood or adjustable shores may be used to support the stringers. The spacing should be 4, 5, 6, 7, or 8 ft, depending on the load and the length of the shores. The use of such spacing will permit the use of commercial lengths of stringers, with a minimum end wastage. Spacings 6, 7, and 8 ft are generally possible.

Example 7-9. Determine the quantity of lumber required for 1 ft^2 of floor, using metal pans of the flange type. The thickness of the floor will be $12\frac{1}{2}$ in. This floor thickness is obtained with a 10-in-deep pan and a slab thickness of $2\frac{1}{2}$ in over the pan. Joists will be 6 in thick, and the pans will be 30 in wide. The size of the bay will be 16 ft 0 in by 24 ft 0 in. The height of the ceiling will be 12 ft 3 in.

FIGURE 7-16
Installing metal pans.

  The volume of concrete per square foot of floor will average 0.372 ft^3, the weight of which is 56 lb. Add 40 lb/ft^2 for temporary additional load. This gives a total average load of 96 lb/ft^2 on the forms.

  The 6-in-wide joist requires 2- by 8-in centering strips, 3 ft 0 in on centers. A 2- by 8-in S4S will safely support the load with a span of 4 ft 0 in. Accordingly, the stringers will be spaced 4 ft 0 in on centers. The load per linear foot of stringer will be 4 by 96 = 384 lb. For a 2- by 10-in S4S stringer the safe span, based on an allowable fiber stress of 1,800 lb/in^2, is 8.7 ft. A shore spacing of 8 ft 0 in will be used. A shore will be required at the end of each stringer. The ends of the centering strips at the edges of the floor bay will be supported by the beam forms.

  The area of floor will be 16 × 24 = 384 ft^2
The quantity of lumber will be:

Centering strips, 24 ÷ 3 ft = 8 + 1 = 9 pc, 2 × 8 in × 16 ft 0 in	= 192 fbm
Ledgers at beams, 2 pc, 1 × 6 in × 24 ft 0 in	= 24 fbm
Stringers, 3 pc, 2 × 10 in × 24 ft 0 in	= 120 fbm
Shores, 12 pc, 4 × 4 in × 10 ft 0 in	= 160 fbm
Scabs, 12 pc, 2 × 4 in × 2 ft 0 in	= 16 fbm
Sills, 12 pc, 2 × 8 in × 1 ft 6 in	= 24 fbm
Braces:	
6 pc, 1 × 6 in × 12 ft 0 in	= 36 fbm
4 pc, 1 × 6 in × 16 ft 0 in	= 32 fbm
Total quantity of lumber	= 604 fbm
Lumber per ft^2 of floor, 604 ÷ 384	= 1.6 fbm

TABLE 7-19
Concrete required for Meyer Steelform construction†

	Flange type (For 20-in widths)							Adjustable type (For 20-in widths)					
Depth of Steelform, in	Width of joist, in	Concrete, ft³/ft² of floor, with slab thickness over Steelforms of:			Concrete, ft³/lin ft of beam, using tapered ends of:		Depth of Steelform, in	Width of joist, in	Concrete, ft³/ft² of floor, with slab thickness over Steelforms of:			Concrete, ft³/lin ft of beam, using tapered ends of:	
		2 in	2½ in	3 in	2 in	4 in			2 in	2½ in	3 in	2 in	4 in
6	4	0.262	0.304	0.346	0.08	0.16	6	$3\frac{1}{2}$	0.262	0.304	0.345	—	0.14
	5	0.279	0.321	0.363	0.08	0.16		$4\frac{1}{2}$	0.279	0.321	0.362	—	0.13
	6	0.293	0.335	0.377	0.08	0.16		$5\frac{1}{2}$	0.295	0.337	0.378	—	0.13
8	4	0.298	0.340	0.382	0.12	0.24	8	$3\frac{1}{2}$	0.289	0.329	0.370	—	0.18
	5	0.320	0.362	0.404	0.12	0.24		$4\frac{1}{2}$	0.309	0.350	0.393	—	0.18
	6	0.339	0.381	0.423	0.11	0.22		$5\frac{1}{2}$	0.331	0.372	0.414	—	0.17
10	4	0.336	0.378	0.420	0.16	0.32	10	$3\frac{1}{2}$	0.312	0.353	0.395	—	0.23
	5	0.363	0.405	0.447	0.16	0.32		$4\frac{1}{2}$	0.340	0.381	0.424	—	0.22
	6	0.387	0.429	0.471	0.15	0.30		$5\frac{1}{2}$	0.367	0.408	0.450	—	0.21
12	4	0.377	0.419	0.461	0.21	0.42	12	$3\frac{1}{2}$	0.336	0.378	0.420	—	0.27
	5	0.408	0.450	0.492	0.20	0.40		$4\frac{1}{2}$	0.371	0.412	0.455	—	0.26
	6	0.437	0.479	0.521	0.20	0.40		$5\frac{1}{2}$	0.403	0.444	0.486	—	0.25
14	4	0.420	0.462	0.504	0.26	0.52	14	$3\frac{1}{2}$	0.361	0.403	0.444	—	0.31
	5	0.456	0.498	0.540	0.25	0.50		$4\frac{1}{2}$	0.402	0.443	0.485	—	0.30
	6	0.490	0.532	0.574	0.24	0.48		$5\frac{1}{2}$	0.438	0.480	0.522	—	0.29

For 30-in widths

Depth of Steelform, in	Width of joist, in	Concrete, ft³/ft² of floor, with slab thickness over Steelforms of:			Concrete, ft³/lin ft of beam, using tapered ends of:	
		$2\frac{1}{2}$ in	3 in	$3\frac{1}{2}$ in	4 in	5 in
6	5	0.293	0.334	0.374	—	0.13
	6	0.304	0.346	0.387	—	0.13
	7	0.315	0.357	0.398	—	0.13
8	5	0.322	0.364	0.405	—	0.18
	6	0.338	0.380	0.421	—	0.18
	7	0.351	0.392	0.434	—	0.17
10	5	0.353	0.395	0.436	—	0.24
	6	0.372	0.414	0.455	—	0.23
	7	0.389	0.430	0.472	—	0.23
12	5	0.386	0.427	0.469	—	0.30
	6	0.408	0.450	0.491	—	0.30
	7	0.430	0.470	0.512	—	0.29
14	5	0.420	0.460	0.493	—	0.37
	6	0.447	0.488	0.530	—	0.36
	7	0.470	0.511	0.553	—	0.35

For 30-in widths

Depth of Steelform, in	Width of joist, in	Concrete, ft³/ft² of floor, with slab thickness over Steelforms of:			Concrete, ft³/lin ft of beam, using tapered ends of:
		$2\frac{1}{2}$ in	3 in	$3\frac{1}{2}$ in	4 in
6	$5\frac{1}{2}$	0.300	0.341	0.383	0.10
	$6\frac{1}{2}$	0.311	0.353	0.395	0.10
	$7\frac{1}{2}$	0.322	0.364	0.406	0.09
8	$5\frac{1}{2}$	0.326	0.368	0.408	0.13
	$6\frac{1}{2}$	0.341	0.383	0.424	0.12
	$7\frac{1}{2}$	0.355	0.397	0.438	0.12
10	$5\frac{1}{2}$	0.352	0.394	0.434	0.16
	$6\frac{1}{2}$	0.371	0.413	0.454	0.15
	$7\frac{1}{2}$	0.389	0.430	0.472	0.15
12	$5\frac{1}{2}$	0.378	0.420	0.461	0.18
	$6\frac{1}{2}$	0.401	0.443	0.484	0.18
	$7\frac{1}{2}$	0.422	0.464	0.505	0.17
14	$5\frac{1}{2}$	0.404	0.446	0.486	0.21
	$6\frac{1}{2}$	0.430	0.472	0.514	0.21
	$7\frac{1}{2}$	0.456	0.497	0.539	0.20

† Amount of concrete given for tapered endforms is for one side of beam only.

The preceding quantity of lumber is based on an ideal bay, with no wastage because of end sawing. On most jobs this condition will not exist, and end wastage of centering strips and stringers will probably average about 10 percent. Adding this amount to the foregoing quantities will increase the total quantity of lumber by 10 percent of 312 = 31 fbm. The resulting quantity of lumber will be 635 fbm, giving 1.7 fbm/ft^2 of floor area.

If adjustable shores are used instead of wood shores, the quantity of lumber will be reduced by about 160 fbm. The elimination of scabs and sills will be offset by the use of more braces. The total quantity of lumber for the bay will be about 635 − 160 = 475 fbm. The quantity of lumber per square foot of floor area will be 475 ÷ 384 = 1.3 fbm. Thus the reduction in the quantity of lumber will be about 0.3 fbm/ft^2.

Table 7-20 gives the quantities of lumber in fbm per square foot of floor area of metal-pan and concrete-joist types of concrete floors, using wood shores and adjustable shores.

Labor Required to Build Forms and Install Metal Pans for Concrete Floors

The labor required to build the forms and install the pans for concrete floors may be divided into three operations. The first operation consists in cutting the centering strips, stringers, and shores to the correct lengths. The second operation consists in erecting the lumber in place. The third operation consists in installing the pans on the centering strips and securing them in place with nails driven through holes in the flanges. The labor required to remove the lumber and the pans should be included. If the lumber can be reused without additional cutting, only the

TABLE 7-20

Lumber required for floor for metal-pan and concrete-joist type of concrete floors,† fbm/ft^2

Height of ceiling, ft	4- × 4-in wood shores				Adjustable shores			
	Thickness of floor,‡ in				Thickness of floor, in			
	$8\frac{1}{2}$	$10\frac{1}{2}$	$12\frac{1}{2}$	$14\frac{1}{2}$	$8\frac{1}{2}$	$10\frac{1}{2}$	$12\frac{1}{2}$	$14\frac{1}{2}$
10	1.5	1.5	1.5	1.6	1.2	1.3	1.3	1.4
12	1.6	1.6	1.7	1.8	1.2	1.3	1.3	1.4
14	1.7	1.8	1.8	1.9	1.3	1.4	1.4	1.5
16	1.8	1.9	1.9	2.0	1.4	1.5	1.5	1.6
18	2.0	2.1	2.1	2.2	1.5	1.6	1.7	1.8

† The quantities are for construction where the ends of the centering strips are supported by beam forms.

‡ The thickness of the floor equals the depth of the pan plus a slab thickness of $2\frac{1}{2}$ in over the top of the pan.

TABLE 7-21
Labor-hours† required to build and remove forms and to install and remove metal pans for 100 ft² of floor

Labor	4- × 4-in wood shores				Adjustable shores			
	Thickness of floor, in				Thickness of floor, in			
	$8\frac{1}{2}$	$10\frac{1}{2}$	$12\frac{1}{2}$	$14\frac{1}{2}$	$8\frac{1}{2}$	$10\frac{1}{2}$	$12\frac{1}{2}$	$14\frac{1}{2}$
Cutting forms:								
Carpenter	1.2	1.2	1.25	1.25	1.1	1.1	1.2	1.2
Helper	0.7	0.7	0.8	0.08	0.6	0.6	0.7	0.7
Erecting forms:								
Carpenter	2.4	2.4	2.6	2.6	2.2	2.2	2.4	2.4
Helper	1.0	1.0	1.2	1.2	0.9	0.9	1.0	1.0
Removing forms:								
Helper	1.2	1.2	1.3	1.3	1.2	1.2	1.3	1.3
Installing pans:								
Mechanic	0.35	0.40	0.45	0.50	0.35	0.40	0.45	0.50
Helper‡	0.50	0.55	0.60	0.65	0.50	0.55	0.60	0.65
Removing pans:								
Helper	0.90	0.95	1.00	1.05	0.90	0.95	1.00	1.05

† The labor-hours are based on reasonably large simple areas, with skilled mechanics to install the pans. For complex jobs the labor-hours should be increased. If carpenters are used to install the pans, all labor-hours installing pans should be increased about 10 percent.

‡ If helpers are not used, add the time to that given for mechanics.

labor for the last two operations should be considered for additional uses. The lumber is usually cut with a power saw set up at the job.

If the pans are installed by carpenters instead of by experienced mechanics, the rates of installation will not be as rapid as those by mechanics. Metal-pan subcontractors use skilled mechanics, while contractors who rent the pans use carpenters to install them.

Table 7-21 gives the quantities of labor required to build forms and install metal pans for 100 ft² of floor.

Example 7-10. Estimate the cost of form lumber and metal pans in place and removed for a total floor area 39 ft 6 in wide by 96 ft 6 in long, of which 526 ft² will be beams. The floor thickness will be $10\frac{1}{2}$ in. The joints will run one way only. The ceiling height will be 10 ft 8 in. The structure will be a dormitory, with corresponding irregularities in the floor areas. Carpenters will install the pans. Adjustable shores will be used.

The gross area of the floor will be 39.5 × 96.5 = 3,812 ft². Deducting the area of the beams leaves a net floor area of 3,286 ft² for form purposes.

Assume that the lumber will cost $480.00 per 1,000 fbm at the job and that it will be used 8 times.

The metal pans will be rented, with the renter to pay the cost of freight to the job and back to the supplier.

The cost in place will be:

Materials:

Form lumber, 3,286 ft² × 1.3 fbm/ft² = 4,272 fbm @ $0.06 = $ 256.32
Nails, 4,272 fbm × 12 lb/1,000 fbm = 51.3 lb @ $0.95 = 48.74
Shores, 3,286 ft² ÷ 24 ft²/shore = 137 @ $2.25 = 308.25
Pan rental, 3,286 ft² @ $0.78 = 2,563.08
Freight on pans = 147.50
Form oil, 3,286 ft² ÷ 400 ft²/gal = 8.2 gal @ $7.50 = 61.50
Power saws, 4,272 fbm ÷ 600 fbm/h = 7.1 h @ $2.85 = 20.24

Labor:

Cutting and erecting forms:
Carpenters, 3,286 ft² × 3.3 h/100 ft² = 108.4 h @ $16.70 = 1,810.28
Helper, 3,286 ft² × 1.5 h/100 ft² = 49.3 h @ $12.10 = 596.53
Removing forms:
Helper, 3,286 ft² × 1.2 h/100 ft² = 39.4 h @ $12.10 = 476.74
Installing pans:
Carpenter, 3,286 ft² × 0.45 h/100 ft² = 14.8 h @ $16.70 = 247.16
Helpers, 3,286 ft² × 0.55 h/100 ft² = 18.1 h @ $12.10 = 219.01
Removing pans:
Helper, 3,286 ft² × 0.95 h/100 ft² = 31.2 h @ $12.10 = 377.52
Foreman, based on using 4 carpenters,
(108.4 + 14.8) ÷ 4 = 30.8 h @ $18.30 = 563.64
Total cost = $7,696.51
Cost per ft², $7,696.51 ÷ 3,286 = 2.34

For pans of types and sizes different from the ones used in this example, the estimator should make a similar determination of the costs.

Concrete Stairs

The construction of concrete stairways is complicated and presents a problem for the estimator. The costs per unit of volume or area varies a great deal, depending on the length of the tread, the width of the tread, the height of the riser, the shape of the supporting floor for the shores, whether the treads are square or rounded, and whether the ends of the treads and risers are open or closed with curbs. Some stairs are straight-run, while others have an intermediate landing. Some stairs are completely in the open, while others have a wall on one or both sides. All these conditions affect the cost of stairs.

The width of treads varies from 8 to 12 in, and the height of the risers varies from 6 to 8 in for different stairs.

Steel dowels should be set in the concrete floor and the beam supporting the landing prior to placing the concrete in the floor and landing. These dowels are tied to the reinforcing steel for the stairs as it is placed.

Lumber Required for Forms for Concrete Stairs

Figure 7-17 shows the details of one method of constructing forms for concrete stairs. The height from floor to floor is 10 ft 6 in. With a riser height of 7 in, 18 risers

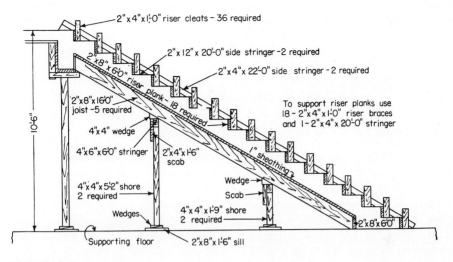

FIGURE 7-17
Wood forms for concrete stairs.

and 17 treads will be required. The treads will be 12 in wide, which gives a total horizontal length of 17 ft 0 in. The supporting floor is horizontal, which requires shores of variable lengths to support the forms.

Example 7-11. Estimate the cost of materials and labor for building forms for the stairs illustrated in Fig. 7-17.
The cost will be:

Sheathing, 16 × 6 + 20% waste = 115 fbm @ $0.48	=	$ 55.20
Joists, 5 pc, 2 × 8 in × 16 ft 0 in = 107 fbm @ $0.48	=	51.36
Side stringers:		
2 pc, 2 × 12 in × 20 ft 0 in = 80 fbm @ $0.52	=	41.60
2 pc, 2 × 4 in × 22 ft 0 in = 29 fbm @ $0.48	=	13.92
Riser planks, 18 pc, 2 × 8 in × 6 ft 0 in = 144 fbm @ $0.48 =		69.12
Riser cleats, 36 pc, 2 × 4 in × 1 ft 0 in = 24 fbm @ $0.48	=	11.52
Riser stringer, 1 pc, 2 × 4 in × 20 ft 0 in = 13 fbm @ $0.48 =		6.24
Riser braces, 18 pc, 2 × 4 in × 1 ft 0 in = 12 fbm @ $0.48	=	5.76
Stringers, 2 pc, 4 × 6 in × 6 ft 0 in = 24 fbm @ $0.55	=	13.20
Shores:		
2 pc, 4 × 4 in × 6 ft 0 in = 16 fbm @ $0.55	=	8.80
2 pc, 4 × 4 in × 2 ft 0 in = 6 fbm @ $0.55	=	3.30
Sills, 4 pc, 2 × 8 in × 1 ft 6 in = 8 fbm @ $0.48	=	3.84
Wedges, 5 fbm @ $0.48	=	2.40
Headers, 1 pc, 1 × 8 in × 6 ft 0 in = 4 fbm @ $0.48	=	1.92
Braces 3 pc, 1 × 6 in × 12 ft 0 in = 18 fbm @ $0.48	=	8.64
Nails, 600 fbm × 15 lb per 1,000 = 9.0 lb @ $0.95	=	8.55
Total cost	=	$305.37

The cost just given is for a single use. If the forms can be removed and used again on stairs of the same design, the cost per use will be reduced by approximately the number of uses. Frequently the forms for stairs are supported by shores resting on existing lower stairs. If this construction is used, the lengths of all shores will be increased to 9 ft 0 in for Fig. 7-17.

Labor Required to Build Forms for Concrete Stairs

Because of the irregularities of the forms for concrete stairs, it is difficult to estimate accurately the cost of building forms. It is necessary to cut the shores to length, attach the 4- by 6-in stringers, set the shores and stringers in place, and brace them securely. The 4- by 4-in wedges are nailed to the tops of the stringers and the joists before the sheathing is installed. The sheathing is cut and nailed to the joists. The side stringers, 2 by 12 in by 20 ft 0 in and 2 by 4 in by 22 ft 0 in, are cleated together, and the positions for the 2- by 8-in riser planks are marked on the stringers. The stringers and the riser planks are installed. Because of the length of the riser planks, 6 ft 0 in, a center support should be used to hold them against deflection. A 2- by 4-in riser stringer is placed at the center of the riser planks, and each riser plank is secured to it with a riser brace 2- by 4 in by 1 ft 0 in.

Two carpenters should be used to build the forms for the stairs. Each carpenter should fabricate and install about 25 fbm/h of lumber.

REINFORCING STEEL

Types and Sources of Reinforcing Steel

Reinforcing for concrete may consist of steel bars or welded-wire fabric, used separately or together. The cost of bars may be estimated by the pound, hundredweight, or ton, while the cost of welded-wire fabric may be estimated by the pound or square foot.

Usually bars are fabricated to the required lengths and shapes by commercial shops prior to delivery to a project. Such shops are equipped with machines that will perform the fabricating operations more economically than when fabricating is performed on the job. Upon request, these shops will furnish quotations covering the supplying and fabricating of all reinforcing for a given project. Estimators frequently request such quotations before preparing estimates.

Properties of Reinforcing Bars

Table 7-22 gives the sizes, areas, and weights of reinforcing bars.

TABLE 7-22
Sizes, areas, and weights of reinforcing bars

Bar no.	Size diam., in	Area, in^2	Weight, lb/ft
2	$\frac{1}{4}$	0.05	0.167
3	$\frac{3}{8}$	0.11	0.376
4	$\frac{1}{2}$	0.20	0.668
5	$\frac{5}{8}$	0.31	1.043
6	$\frac{3}{4}$	0.44	1.502
7	$\frac{7}{8}$	0.60	2.044
8	1.0	0.79	2.670
9	1.128	1.00	3.400
10	1.270	1.27	4.303
11	1.410	1.56	5.313
14	1.693	2.25	7.650
18	2.257	4.00	13.600

Estimating the Quantity of Reinforcing Steel

When the reinforcing steel consists of bars of different sizes and lengths, each size and length should be listed separately. A form such as the one used in Table 7-23 will simplify the listing and reduce the danger of errors. Each size and length should be assigned a number or a letter of the alphabet.

TABLE 7-23
Quantity of reinforcing steel

Bar mark	No. required	Bar size	Length	Weight, lb/ft	Total weight, lb
A	120	4	30 ft 0 in	0.668	2,405
B	56	4	20 ft 0 in	0.668	749
C	116	5	24 ft 0 in	1.043	2,900
D	42	6	12 ft 4 in	1.502	780
E	36	6	14 ft 8 in	1.502	794
F	28	6	19 ft 8 in	1.502	826
G	16	7	18 ft 6 in	2.044	604
H	72	7	22 ft 3 in	2.044	3,280
I	84	7	18 ft 8 in	2.044	3,200
J	24	8	24 ft 0 in	2.670	1,535
K	18	8	21 ft 6 in	2,670	1,037
				Total weight	18,110

Cost of Reinforcing Steel

The items which determine the cost of reinforcing steel delivered to a project are

1. The base cost of the bars at the fabricating shop
2. The cost of preparing shop drawings
3. The cost of shop handling, cutting, bending, etc.
4. The cost of selling
5. The cost of shop overhead and profit
6. The cost of transporting from the shop to the project
7. The cost of specialties, such as spacers, saddles, chairs, ties, etc.

It is customary to determine the weight of reinforcing steel based on the lengths and sizes of bars and the nominal weights given in Table 7-22, with no extra charge made for waste. Reinforcing bars are usually available in stock lengths of 40 and 60 ft.

Size Extras for Reinforcing Steel

If an estimator wishes to determine the approximate cost of reinforcing steel for a project, she or he should list each size bar separately and then determine the total weight by size. In addition to the base price cost of the reinforcing, an extra charge based on the sizes of the bars will be made. These extras are subject to change and should be verified before an estimate is prepared.

Quantity Extras for Reinforcing Steel

This extra cost varies with the total quantity of steel purchased.

Cost for Detailing and Listing Reinforcing Steel

Before fabricating reinforcing steel it is necessary for the fabricating shop to prepare drawings which show how the bars are to be fabricated. A charge is made for this service, based on the complexity of the drawings and the quantity of reinforcing.

Cost for Fabricating Reinforcing Steel

The cost for this operation varies with the sizes of the bars and the complexity of the operations.

Cost for Reinforcing Steel Delivered to a Project

If an estimator does not wish to consider the several extra costs, an estimate of the cost can usually be obtained from a supplier, who will quote a total cost for the entire lot, fabricated and delivered to the project.

Labor Placing Reinforcing-Steel Bars

The rates at which workers will place reinforcing-steel bars will vary with the following factors:

Sizes and lengths of bars

Shapes of the bars

Complexity of the structure

Distance and height the steel must be carried

Allowable tolerance in spacing bars

Extent of tying required

Skill of workers

Less time is required to place a ton of steel when the bars consist of large sizes and long lengths than when they are small sizes and short lengths.

Straight bars may be placed more rapidly than bars with bends and end hooks.

If the bars must be placed in complicated structures such as stairs, the rate of placing will be less than for simple structures such as walls, floors, etc.

Steel bars should be stored as near the structure as possible, preferably not more than 50 to 100 ft away, to reduce the time required to carry them. If steel must be carried to upper floors or parts of a structure, additional time will be required.

Rigid tolerances on the spacing of bars will reduce the rate of placing steel somewhat.

More time is required to tie bars at every intersection than when little or no tying is required.

Reinforcing steel may be placed by laborers or skilled steel setters. The last should place the steel at the fastest rate, with little or no supervision required.

Table 7-24 gives representative rates of placing reinforcing-steel bars. The rates are based on carrying the steel by hand not more than 100 ft from the stockpiles to the structure. The steel will be placed by either repairers or steel setters, but not by both.

TABLE 7-24
Rates of placing reinforcing-steel bars, h/ton

	Size of bars			
	$\frac{5}{8}$ in and less Length of bars, ft		$\frac{3}{4}$ in and over	
Class of worker	Over 15	Under 15	Over 15	Under 15
Bars not tied in place				
Laborers	14–16	16–18	11–13	13–15
Steel setters	11–13	13–15	7–9	9–11
Bars tied in place				
Laborers	16–18	18–20	12–14	14–16
Steel setters	12–14	14–16	8–10	10–12

Welded-Wire Fabric

For certain types of concrete projects, such as sidewalks, pavements, floors, canal linings, etc., it may be more economical to use welded-wire fabric for reinforcing instead of steel bars. This fabric is made from cold-drawn steel wire, electrically welded at the intersections of longitudinal and transverse wires, to form rectangles or squares. It is available in flat sheets or rolls, the latter frequently being 60 in wide and 150 ft long. It is usually priced by the square foot or roll, the price depending on the weight.

The quantity required will equal the total area to be reinforced, with 5 to 10 percent of the area added for side and end laps.

The fabric to be used is designated by specifying the style, such as 412-610. This style designates a rectangular fabric with longitudinal wires spaced 4 in apart and transverse wires spaced 12 in apart, using no. 6 gage longitudinal wires and no. 10 gage transverse wires.

Table 7-25 gives the properties of representative styles of welded-wire fabric. Many other styles are available.

Based on a width of 60 in and a length of 150 ft, the weight of a roll of fabric will vary from 75 to 1,000 lb or more.

Labor Placing Welded-Wire Fabric

Fabric is placed by unrolling it over the area to be reinforced, cutting it to the required lengths, lapping the edges and the ends, and tying at frequent spacings. On large regular areas a person should place it at a rate of 0.25 h/100 ft^2 while for irregular areas, requiring cutting and fitting, the rate of placing may be 0.5 h/100 ft^2.

TABLE 7-25
Properties of representative styles of welded-wire fabric

Style	Weight, lb per 100 ft²	Spacing of wire, in		Gage number		Sectional area of wires, in²/ft	
		Longitudinal	Transverse	Longitudinal	Transverse	Longitudinal	Transverse
44–1010	31	4	4	10	10	0.043	0.043
44–88	44	4	4	8	8	0.062	0.062
44–66	62	4	4	6	6	0.087	0.087
44–44	85	4	4	4	4	0.120	0.120
66–1010	21	6	6	10	10	0.029	0.029
66–88	30	6	6	8	8	0.041	0.041
66–66	42	6	6	6	6	0.058	0.058
66–44	58	6	6	4	4	0.080	0.080
66–22	78	6	6	2	2	0.108	0.108
48–1012	20	4	8	10	12	0.043	0.013
48–912	23	4	8	9	12	0.052	0.013
48–812	27	4	8	8	12	0.062	0.013
412–1012	19	4	12	10	12	0.043	0.009
412–812	25	4	12	8	12	0.062	0.009
412–610	36	4	12	6	10	0.087	0.014
412–49	49	4	12	4	9	0.120	0.017
412–48	51	4	12	4	8	0.120	0.021
416–812	25	4	16	8	12	0.062	0.007
416–610	35	4	16	6	10	0.087	0.011
416–49	48	4	16	4	9	0.120	0.013
416–28	64	4	16	2	8	0.162	0.015
612–66	32	6	12	6	6	0.058	0.029
612–44	44	6	12	4	4	0.080	0.040
612–22	59	6	12	2	2	0.108	0.054
612–14	61	6	12	1	4	0.126	0.040
612–03	72	6	12	0	3	0.148	0.047

CONCRETE

Cost of Concrete

The cost of concrete in a structure includes the cost of aggregate, cement, water, equipment, and of labor mixing, transporting, and placing the concrete. When ready-mixed concrete is used, some of the costs are transferred from the job to the central mixing plant. The cost of the several items just listed will vary with the size of the job, location, quality of the concrete, extent to which equipment is used instead of labor, and distribution of concrete within the job.

Quantities of Materials for Concrete

The estimator should determine the quantity of each class of concrete in the job. With this information she or he can determine the quantities and costs of aggregate, cement, and water for each class or for each structural element.

Concrete structures are designed for concretes having specified strengths, usually expressed in pounds per square inch in compression 28 days after it is placed in the structure. To produce a concrete with a specified strength, it is common practice to employ a commercial laboratory to design the mix. Such a laboratory specifies the weight or volume of fine aggregate, coarse aggregate, cement, and water to produce a batch or a cubic yard of concrete having the required strength.

Estimators seldom have the laboratory design information when estimating the cost of a project. Tables giving the approximate quantities of aggregate, cement, and water for different qualities of concrete are available. The information given in these tables is sufficiently accurate for estimating purposes.

Table 7-26 gives the approximate quantities of cement, water, and coarse and fine aggregates required to produce 1 yd^3 of concrete having the indicated 28-day compressive strengths. Note that the aggregate is saturated surface-dry when weighed. If the stockpiles of aggregates contain surface moisture, as they usually do, the quantity of water added should be decreased by the amount of water present on the surface of the aggregate. The weights of the aggregate should be increased to produce net weights that correspond to those given in the table.

Output of Concrete Mixers

The output of concrete mixers varies with the size of the batch, method of charging, method of discharging, and time the batch must be mixed.

The nominal size of a concrete mixer is indicated by a number representing the quantity of concrete per batch, measured in cubic feet. The batch may not always equal the indicated size. For example, it might be more practical to use 2 sacks of cement for a batch less than the maximum possible size instead of using 2.2 sacks to produce a batch equal to the size of the mixer.

TABLE 7-26
**Quantities of cement, water, and aggregates required for
1 yd³ of concrete having the indicated 28-day compressive
strength**

| Sack of cement | Water, gal | Weights of saturated surface-dry aggregate, lb | | 28-day compressive strength, lb/in² |
		Fine	Coarse	
		1-in coarse aggregate		
4.9	39.1	1,370	1,860	2,250
5.6	39.1	1,345	1,847	2,750
6.0	39.0	1,260	1,860	3,000
6.5	39.0	1,235	1,820	3,300
7.2	39.8	1,150	1,875	3,700
8.0	40.0	1,120	1,840	4,250
		2-in coarse aggregate		
4.5	36.0	1,350	1,980	2,250
5.1	35.6	1,275	1,980	2,750
5.5	35.7	1,265	1,980	3,000
6.0	36.0	1,200	1,980	3,300
6.7	36.8	1,140	2.010	3,700
7.4	36.8	1,110	2,000	4,250

When the full batch is discharged immediately into a hopper or a bucket instead of into several buggies or wheelbarrows, it will be possible to reduce the time per batch, thus increasing the number of batches per hour.

Table 7-27 gives the approximate output of construction types of concrete mixers for various sizes, based on a 1-min mixing time.

Labor Mixing and Placing Concrete

The labor required to mix and place concrete varies with the number of operations performed by labor; location of aggregate piles with respect to the mixer; location of cement storage; length of haul of concrete; condition of runways; the hauling equipment, buggies or wheelbarrows; and the distribution of the placing area. Factors which decrease the amount of labor, and thus the cost of labor, are reducing the length of haul of aggregate, storing cement near the mixer, locating the mixer near the center of placing concrete, constructing runways wide enough for easy travel and, if possible, level or sloping slightly from the mixer, using buggies instead of wheelbarrows, and distributing the placing over an area sufficiently large to eliminate congestion and interference. Also, the use of labor-saving pumping equipment will reduce the amount of labor required.

TABLE 7-27

Approximate output of construction types of concrete mixers based on a 1-min mixing time†

Size mixer	Batches per hour Range	Average	Approx. output, yd³/h
3½S	25–33	30	4.0
6S	25–33	30	6.5
11S	24–32	28	11.0
14S	20–30	24	12.0
16S	20–30	24	14.0
28S	18–26	22	23.0

† The higher number of batches per hour may be used if the job is organized to eliminate lost time in charging and discharging the mixer. For a poorly organized job, the lower number of batches should be used.

In estimating the labor-hours required for a given concrete pour, the estimator should allow for the time required for getting ready to start the pour, for cleaning out the mixer and the buggies, and for putting away tools and equipment after the pour is completed. This time, amounting to approximately 30 min, will be the same regardless of the length of pour. For this and other reasons, pours of less than 4 h should be avoided when possible.

Example 7-12. As an example illustrating the method of determining, the amount of labor required to mix and place 1 yd³ of concrete, assume that 80 yd³ of concrete is to be mixed and placed with an 11S mixer. Aggregate will be handled with wheelbarrows. Cement will be stacked near the mixer. The concrete will be hauled about 40 ft in buggies and deposited for a beam-and-slab floor. Two sacks of cement will be used per batch, which will be 11 ft³. A mixing time of 1 min is specified.

Assuming that the output of the mixer will be 11 yd³/h, the actual mixing time will be 7.3 h. Add 0.5 h for getting ready and cleaning up after the pour is completed. This gives a total time of 7.8 h.

The weight of sand per batch will be approximately $\frac{11}{27} \times 1,350 = 550$ lb, and the weight of gravel approximately $\frac{11}{27} \times 1,850 = 750$ lb. While the volume of a concrete wheelbarrow is about 3 ft³, the practical load for aggregate is about 2 ft³, or 220 lb. Use 3 wheelbarrows for sand and 4 for gravel. At least one extra laborer will be needed on each stockpile. This requires 5 workers handling gravel and four people handling sand. Concrete buggies of the type used on this job have a capacity of 5 to 6 ft³, but the load should be limited to about 3 ft². It will require 4 buggies to haul a batch.

The labor-hours for the job will be as follows:

Hauling gravel, 5 × 7.8 h	= 39.0 h
Hauling sand, 4 × 7.8 h	= 31.2 h
Handling cement, 1 person	= 7.8 h
Hauling concrete, 4 × 7.8 h	= 31.2 h

Helping empty buggies, 1 worker	=	7.8 h
Spreading and leveling concrete, 3 people	=	23.4 h
Moving runways and general utility, 2 people	=	15.6 h
Total laborers	=	156.0 h
Mixed operator	=	7.8 h
Foreman, 1	=	7.8 h
Labor per yd³, 156 ÷ 80	=	1.95 h
Mixer† operator per yd³	=	0.125 h
Foreman† per yd³	=	0.125 h

In mixing concrete with a 16S or a 28S mixer, it may be desirable to use a batching plant for the aggregate. This will require a clamshell for handling the aggregate. Eliminate the labor hauling gravel and sand, and add a clamshell operator plus a worker on the batching plant.

Example 7-13. A concrete bridge pier containing 210 yd³ of concrete is to be poured in one operation. A 28S concrete mixer will be used. Aggregate will be lifted to a two-compartment batching plant with a clamshell. The concrete will be transported directly from the mixer to the pier forms with a crane and a $1\frac{1}{2}$-yd³ bucket. Cement will be stored at the job in sacks about 10 ft from the mixer.

Using 2-in maximum size aggregate, a 2,500-lb concrete will require 5 sacks of cement per batch, assumed to be 1 yd³.

The mixer should mix 22 yd³/h of concrete. The actual pour should require 210 ÷ 22 = 9.5 h. Add 0.5 h to get ready and to clean up after the pour. This gives a total of 10 h.

The labor by classification will be:

Clamshell operator	= 10 h
Worker on batching plant	= 10 h
Laborers handling cement, 3	= 30 h
Mixer operator	= 10 h
Crane operator	= 10 h
Signaler on forms	= 10 h
Laborers spreading concrete, 3	= 30 h
Laborers on vibrator, 2	= 20 h
Utility workers, 2	= 20 h
Carpenter checking forms	= 10 h
Foreman	= 12 h

The labor required per yd³ of concrete will be:

Clamshell operator, 10 h @ 210 yd³	= 0.05 h
Mixer operator	= 0.05 h
Crane operator	= 0.05 h
Carpenter	= 0.05 h
Foreman, 12 h ÷ 210 yd³/h	= 0.06 h
Laborers‡ 120 h ÷ 210 yd³/h	= 0.57 h

† Additional time is allowed for the mixer operator to service the mixer and for the foreman, who must be at the job before the laborers arrive and after they leave.

‡ For many locations it may be necessary to classify some of these workers as other than laborers.

TABLE 7-28
Labor-hours per cubic yard required to mix and place concrete†

Size mixer	Method of handling	Common labor	Foreman	Mixer operator	Hoisting engineer	Crane operator	Carpenter
None	Hand	$3-3\frac{1}{2}$	0.33				
6S	Wheelbarrow	$2\frac{1}{4}$	0.17				
6S	Hoist spout	$1\frac{3}{4}$	0.17	—	0.17		
11S	Wheelbarrow	2	0.12	0.12	—	—	0.12
11S	Hoist spout	$1\frac{3}{4}$	0.12	0.12	0.12	—	0.12
14S	Wheelbarrow	2	0.10	0.10	—	—	0.10
14S	Hoist spout	$1\frac{3}{4}$	0.10	0.10	0.10	—	0.10
16S	Wheelbarrow	2	0.10	0.10	—	—	0.10
16S	Buggies	$1\frac{7}{8}$	0.10	0.10	—	—	0.10
16S	Clamshell, hoist and buggies	$1\frac{1}{2}$	0.10	0.10	0.10	0.10	0.10
28S	Clamshell, hoist and buggies	$1\frac{1}{2}$	0.08	0.08	0.08	0.08	0.08
	Mass concrete						
16S	Clamshell, crane and bucket	0.8	0.09	0.08	0.08	0.08	0.08
28S	Clamshell, crane and bucket	0.6	0.06	0.05	0.05	0.05	0.05

† The information given in the top portion of the table should be used for buildings and similar structures. The information given under "Mass concrete" should be used for large foundations, piers, dams, etc.

Table 7-28 gives approximate labor-hours by classification required to mix and place concrete for various sizes of mixers and methods.

Ready-Mixed Concrete

If ready-mixed concrete is available, it is frequently more economical and satisfactory to purchase it than to mix concrete at the job. This is especially true where working space around a job is limited and when small quantities of concrete are needed at various times during construction.

When ready-mixed concrete is used, the costs at the job will be reduced to handling and placing the concrete. If concrete can be delivered to any part of a structure, the cost of handling should be less than for job-mixed concrete.

Labor Placing Ready-Mixed Concrete

The labor required to place ready-mixed concrete will vary with the rate of delivery, type of structure, and location of the structure.

If the trucks can be driven to large foundations, constructed at or below the level of the natural ground, it may be possible to discharge the concrete directly into the forms by using a chute. Not more than five or six workers may be needed to spread and vibrate the concrete.

Concrete for a slab, constructed at or near ground level, may be discharged into a bucket, hoisted by a crane, and distributed over the slab area, with little handling necessary. A crew of six to nine should be able to place up to 15 yd^3/h of concrete.

Concrete for a floor slab, constructed above ground level, may be discharged into a bucket, hoisted with a crane or a tower unit, deposited into a floor hopper, and then hauled to its destination by power-driven or hand-pushed buggies.

Example 7-14. Estimate the labor required to place the concrete for a floor slab 100 ft long, 60 ft wide, whose average thickness is 6 in. The floor will be 16 ft above ground level. The concrete will be delivered to a 1-yd^3 bucket, hoisted with a crane, deposited into a 1-yd^3 floor hopper, hauled in hand-pushed buggies, deposited, and spread to the required thickness.

The volume of concrete will be $100 \times 60 \times 6/(12 \times 27) = 111$ yd^3. Concrete will be delivered at the rate of 15 yd^3/h. The crew should be about as follows:

1 crane operator
1 person on the ground handling the bucket
1 person emptying the bucket into the hopper
1 person filling the buggies
5 people pushing buggies
2 people helping empty the buggies
5 people spreading and screeding the concrete
1 carpenter on runways
1 person helping carpenter with runways
3 utility persons
1 foreman

Depending on the location of the project, the men might be classified as follows:

Laborers, 18
Crane operators, 1
Carpenter, 1
Foreman, 1
Time to place concrete, $111 \div 15$ $\quad = 7.4$ h
Add time to get ready and clean up $= \underline{0.6\ \text{h}}$
$\quad$ Total time $\qquad = 8.0$ h
Total labor-hours, $8 \times 18 = 144$
Labor-hours per yd^3, $144 \div 111$ $\quad = 1.3$
Foreman-hours per yd^3, $8 \div 111$ $\quad = 0.07$

Table 7-29 gives the representative labor-hours required to place ready-mixed concrete for various types of structures, based on delivering concrete at a rate of approximately 15 yd^3/h. For lower or higher rates of delivery, the labor-hours may be slightly higher or lower, respectively.

TABLE 7-29
Labor-hours required to place 1 yd^3 of ready-mixed concrete

Type of structure	Method of handling	Common labor	Foreman	Hoist or crane operator	Carpenter
Large foundation	Discharge directly from truck, using chute	0.5	0.07	—	0.07
Bridge pier	Crane and bucket	0.5	0.07	0.07	0.07
Slab at ground level	Crane and bucket	0.7	0.07	0.07	0.07
Slab above ground level	Crane or hoist, bucket, hopper, and buggies	1.3	0.07	0.07	0.07
Foundation wall	Crane, bucket, and tremies	0.7	0.07	0.07	0.07
Foundation wall	Hand buggies	1.0	0.07	—	0.07

Lightweight Concrete

When the strength of concrete is not a primary factor, and when a reduction in weight is desirable, lightweight aggregate, such as cinders, burned clay, vermiculite, pumice, or other materials, may be used instead of sand and gravel or crushed stone. Concrete made with this aggregate and portland cement may weigh as little as 20 lb/ft^3. However, most weights will be higher than this, with values usually running from about 40 to 100 lb/ft^3, depending on the aggregate and the amount of cement used.

The insulating properties of lightweight concrete are better than for conventional concrete.

Perlite Concrete Aggregate

While several lightweight aggregates are used, only Perlite will be discussed to illustrate the properties and costs of lightweight concrete. This aggregate is a volcanic lava rock that has been expanded by heat to produce a material weighing approximately 8 lb/ft^3. Concrete made from this aggregate may be used for subfloors, fireproofing steel columns and beams, roof decks, concrete blocks, etc.

The materials used in making concrete include portland cement, aggregate, water, and an air-entraining agent. Mixing may be accomplished in a plaster or a drum-type concrete mixer, or transit-mixed concrete may be purchased in some localities. Perlite is generally available in bags containing 4 ft^3, which weighs about 32 lb.

Table 7-30 gives representative properties of concrete made with Perlite aggregate. The dry weights will be approximately 70 percent of the wet weights.

TABLE 7-30
Representative properties of concrete made with perlite aggregate†

Mix proportions by volume				Materials required for 1 yd³				Com-pressive strength 28 days, lb/in²	Weight when placed, lb/ft³
Cement, Sacks	Perlite, ft³	Water, gal	Air-entraining agent, pt	Cement, sacks	Perlite, ft³	Water, gal	Air-entraining agent, pt		
1	4	9	1.00	6.75	27	61.0	6.75	440	50.5
1	5	11	1.25	5.40	27	59.5	6.75	270	45.5
1	6	12	1.50	4.50	27	54.0	6.75	180	40.5
1	7	14	1.75	3.85	27	54.0	6.75	130	38.0
1	8	16	2.00	3.38	27	54.0	6.75	95	36.5

† From Perlite Institute file 6a/Pe.

Cost of Perlite Concrete

When lightweight concretes are used for floor fills or roof decks, they are usually placed in thinner layers than when conventional concrete is used. The reduction in volume, without a corresponding reduction in the number of laborers, will result in a higher labor cost per cubic yard of concrete. The cost of labor required to mix and place this concrete for a floor fill or roof deck should be increased at least 50 percent over that required for slabs constructed with conventional concrete.

Example 7-15. Estimate the total cost and the cost per cubic yard for furnishing materials and mixing and placing 36 yd^3 of 1.6 Perlite concrete for a floor fill $2\frac{1}{2}$ in thick placed at ground level.

The cost will be:

Cement, 36 yd^3 × 4.5 sacks/yd^3 = 162 sacks @ \$4.90	= \$ 793.80
Perlite, 36 yd^3 × 27 = 972 ft^3 @ \$0.69	= 670.68
Water, 36 yd^3 × 60 = 2.2 k gal @ \$0.51	= 1.12
Air-entraining agent, 36 yd^3 × 6.75 pt/yd^3 = 243 pt @ \$0.29	= 70.47
Concrete mixer, 36 yd^3 × 0.15 h/yd^3 = 5.4 h @ \$4.96	= 26.78
Mixer operator, 5.4 h @ \$14.50	= 78.30
Laborers, 36 yd^3 × 3 h/yd^3 = 108 h @ \$12.10	= 1,306.80
Cement finisher, 36 yd^3 × 1 h/yd^3 = 36 h @ \$16.70	= 601.20
Foreman, 5.4 h @ \$18.30	= 98.82
Total cost	= \$3,647.97
Cost per yd^3, \$3,647.97 ÷ 36	= 101.33

TILT-UP CONCRETE WALLS

General Description

A method of building concrete walls that is economical and successful is to cast the walls on the concrete-floor slab, usually at ground level, and tilt them into position. For this method of construction, the walls are divided into several panels of convenient size. The concrete floor, on which the wall sections are to be cast, should be cleaned and coated with a bond breaker. Side forms, equal in height to the thickness of the wall, are placed around the panels to be poured. Stiffeners should be attached to the side forms to maintain true alignment. The reinforcing steel is placed in position, with steel dowels projecting through holes drilled in the side forms to permit the walls to be secured in position after they are tilted up. The concrete bond breaker is placed on the floor before the concrete is placed in the forms. After the concrete has cured properly, the forms are removed and the wall panels are tilted to final position as units of a wall.

Frames for window and door openings should be installed before the concrete is placed in the forms. Conduits can be placed in the walls easily.

A tight seal between the bottoms of the wall panels and the concrete floor or grade beam is obtained by spreadng a layer of cement grout on the floor under the wall prior to raising the panel.

The width of wall panels is limited to the shape and size of the building and the lifting capacity of the equipment used to tilt the wall up. The panels are tied into a solid continuous structure by joining two adjacent panels with weld plates or a reinforced-concrete column. The projecting reinforcing dowels are embedded in the column. Corners are secured in the same manner.

To reduce the danger of structural failure in the panels while they are being tilted into position, it is advisable to use *strongbacks,* or steel beams temporarily bolted to the panels. For this purpose, bolts should be embedded in the concrete. After the walls are erected, the strongbacks are removed and the bolts can be cut off flush with or slightly inside the concrete surface.

The advantages of this method of construction include low cost of forms, low cost of placing reinforcing, and low cost of placing concrete, since high hoisting into vertical forms is not necessary.

Considerable care should be exercised in securing a satisfactory parting membrane under the panel to be cast, or else the wall may partly bond to the concrete floor. Lifting will be difficult or even impossible if this should occur. A heavy unit of lifting equipment, such as a crane, will be needed to tilt up the panels.

If proper care is exercised, a very satisfactory job can be obtained.

Example 7-16. Estimate the total direct cost of furnishing materials, equipment, and labor to erect a concrete tilt-up wall in place. The total length of the wall will be 118 ft 9 in. The height will be 12 ft 0 in, and the thickness will be 9 in.

The wall will be cast in eight panels, each 13 ft 9 in wide. This width will permit the erection of 15- by 16-in connector and support columns located along the walls and at the corners of the building.

The reinforcing steel for the wall will consist of $\frac{3}{4}$-in-round horizontal bars, spaced 12 in on centers, and $\frac{3}{4}$-in-round vertical bars, spaced 8 in on centers. The horizontal bars will project 10 in outside the two vertical edges of the wall panels to tie the wall sections to the columns.

The concrete columns will be 15 by 16 in, with six 1-in-round reinforcing bars per column. Ties of $\frac{1}{4}$-in steel bars will be spaced 12 in apart around the column reinforcing bars. Figure 7-18 shows a section through a column.

Assume that lumber costs $440.00 per 1,000 fbm at the job.

The costs will be:

Wall forms, 4 sets required, assume 2 uses
    Sides:

8 pc, 2 × 10 in × 14 ft 0 in = 187 fbm @ $0.22	= $	41.14
8 pc, 2 × 10 in × 12 ft 0 in = 160 fbm @ $0.22	=	35.20

    Side stiffeners:

8 pc, 2 × 6 in × 16 ft 0 in = 128 fbm @ $0.22	=	28.16
8 pc, 2 × 6 in × 14 ft 0 in = 112 fbm @ $0.22	=	24.64
Nails, 8 lb @ $0.95	=	7.60
Bond breaker, (118.75 × 12.0) ÷ 350 ft²/gal @ $22.50	=	91.58

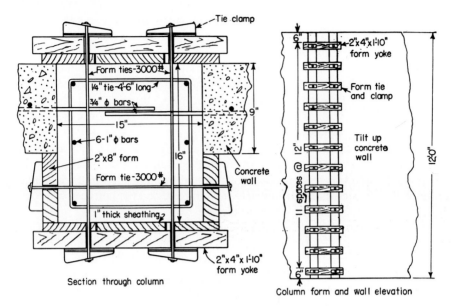

FIGURE 7-18

Details of column and form for a tilt-up concrete wall.

Column forms, 7 required, assume 2 uses
  Sheathing
    28 pc, 1 × 6 in × 12 ft 0 in = 168 fbm @ $0.22   =   36.96
    14 pc, 1 × 8 in × 12 ft 0 in = 112 fbm @ $0.22   =   24.64
    14 pc, 2 × 8 in × 12 ft 0 in = 224 fbm @ $0.22   =   49.28
  Yokes, 168 pc, 2 × 4 in × 1 ft 0 in = 112 fbm @ $0.22 =   24.64
  Form ties, 252 @ $0.38   =   95.76
  Tie clamps, 504 @ $0.037   =   18.65
Labor building forms
  Wall forms:
    Carpenters, 8 forms × 2 h each = 16 h @ $16.70   =   267.20
    Helpers, 8 forms × 1 h each = 8 h @ $12.10   =   96.80
  Column forms:
    Carpenters, 336 ft^2 × 6 h/100 ft^2 = 20 h @ $16.70 = 334.00
    Helpers, 336 ft^2 × 4 h/100 ft^2 = 13 h @ $12.10 = 157.30
    Foreman, 10 h @ $18.30   =   183.00
Reinforcing steel
  For walls:
    96 pc, $\frac{3}{4}$ in × 15 ft 4 in = 2,210 lb @ $0.24   =   530.40
    160 pc, $\frac{3}{4}$ in × 11 ft 10 in = 2,844 lb @ $0.24   =   682.56
    Tie wire, 12 lb @ $0.98   =   11.76
  For columns:
    42 pc, 1 in × 11 ft 10 in = 1,327 lb @ $0.24   =   318.48
    84 pc, $\frac{1}{4}$ in steel ties × 4 ft 6 in long = 63 lb @ $0.24 =   15.12
Labor placing reinforcing steel
  Steel setter, 3.2 tons × 12 h/ton = 38 h @ $15.80   =   600.40

Concrete, ready-mixed
    Walls, 8×13 ft 9 in $\times 12$ ft 0 in $\times 9$ in $= 36.7$ yd^3,
        36.7 yd^3 @ \$53.75   = 1,972.63
    Columns, 7×16 in $\times 15$ in $\times 12$ ft 0 in $= 5.2$ yd^3,
        5.2 yd^3 @ \$53.75   =   279.50
    Grout, 4 ft^3 @ \$3.60   =   14.40
Labor placing concrete
    Laborers, 42 yd^3 $\times$ 1.5 h/yd^3 $= 63$ h @ \$12.10   =   762.30
    Foreman, 8 h @ \$18.30   =   146.40
Tilting up the panels
    Crane, 8 panels $\times$ 1.5 h each $= 12$ h @ \$65.56   =   786.72
    Crane operator, 12 h @ \$17.25   =   207.00
    Crane oiler, 12 h @ \$11.25   =   135.00
    Laborers, 3×12 h $= 36$ h @ \$12.10   =   435.60
    Foreman, 12 h @ \$18.30   =   219.60
    Strongbacks, bolts, etc.   =   25.00
        Total cost   = \$8,659.42
        Cost per yd^3 of concrete, \$8,659.42 $\div$ 41.9   =   206.67
        Cost per ft^2 of wall surface, \$8,659.42 $\div$ 1,425   =   6.08

CONCRETE BRIDGE PIERS

Piers for highway and railway bridges are frequently built of concrete. If a solid-rock foundation is available at the site of the pier, it is customary to excavate into or down to the rock and to construct the pier on the rock. However, if the rock is not present or if it is so deep that excavating to it is too expensive, usually it will be necessary to drive piles to support the pier. The piles may be steel or concrete. They should extend into the bottom of the pier far enough to transmit the load safely from the pier to the piles.

The cost of the pier includes the cost of forms, reinforcing steel, and concrete. If irregular surface shapes, such as projections or depressions, are required, adequate provisions for the cost of these irregularities must be included in the estimate. Likewise, provision should be made for the cost of special surface finishes that may be required.

Forms for Piers

Forms for piers are made of wood or steel. In selecting the material for forms, consideration should be given to the number of times the forms will be used and to the kind of surface finish required for the pier. Additional costs for smooth-form linings may be less than the extra cost of labor necessary to produce the required surface finish after using rough forms of strip lumber.

Steel Forms

Steel forms may be economical if they can be used enough times to justify the higher initial cost. The salvage value of steel forms fabricated for special piers will usually be low. Forms are fabricated from steel plates and shapes, such as angles and I beams, into sections which will conform to the shape of the pier when the forms are assembled. The sections are lifted into position by cranes or other suitable hoisting equipment and fastened together to form a rigid structure. The cost of the form sections should be obtained from a fabricator. The cost of erecting will include the costs of labor and equipment at the job.

Wood Forms

If wood forms are used, they will include sheathing, studs, wales, and possibly braces. The opposite sides of the forms are held in position by form ties which pass completely through the forms and bear against the wales.

The sheathing may be S4S or D and M lumber, plywood, or a combination of planking with thin plywood lining. All elements of the forms should be designed to resist safely the pressure of the concrete for the temperature and the rate of pour.

For piers not more than 12 ft 0 in high, it is satisfactory to use 1-in-thick sheathing, 2- by 4-in studs, and double 2- by 4-in wales. For piers higher than 12 ft 0 in, it may be necessary to use 2- by 6-in or larger studs and double 2- by 6-in or larger wales. Since this height is approximate, the actual height for which an increase in the size of studs and wales is necessary should be determined for the particular job.

> **Example 7-17.** Figure 7-19 shows the dimensions and details of a reinforced-concrete pier for a highway bridge. Four identical piers are required for the project. Because a horizontal construction joint is permitted midway between the top and bottom of the pier, it is possible to use the side forms 2 times on each pier, or 8 times on the project. The end forms will be used one time on each pier, for a total of 4 times on the project.
>
> **Forms.** Concrete will be placed in the forms at a rate of 3 ft/h at a temperature of 70°F. Table 7-1 indicates a maximum pressure of about 536 lb/ft². Table 7-3 will be used to design the forms, based on a pressure of 550 lb/ft².
>
> > Sheathing, 1 × 6 in S4S, net width, $5\frac{1}{2}$ in
> > Studs, 2 × 4 in S4S, spaced 21 in on centers
> > Wales, double 2- × 4-in S4S, spaced 26 in on centers
> > Form ties, spaced 33 in on centers
>
> The area served by a form tie will be 26 × 33 in ÷ 144 = 5.96 ft². The stress on a tie will be 5.96 × 536 = 3,200 lb. If 3,000-lb ties are used, the maximum spacing should be reduced to (3,000/3,200)33 = 31 in. Use a spacing of 30 in on centers for the ties.
>
> If 2- by 6-in studs and double 2- by 6-in wales are used with 1-in sheathing, the quantity of lumber for studs and wales will be increased by about 17 percent. Unless the possible extra rigidity of the form panels provided by using the larger studs and wales is desired, it is more economical to use 2- by 4-in studs and wales, which will be used for this estimate.

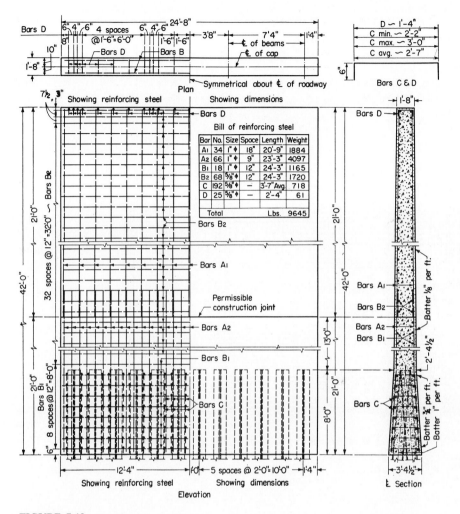

FIGURE 7-19
Details of forms for a concrete bridge pier.

The insides of the forms will be lined with $\frac{1}{4}$-in-thick waterproof plywood. Figure 7-20 shows the details of the form design.

Since the piers will have construction joints at the midelevations, the forms will be erected for the lower half of the piers and filled with concrete. After the concrete has set sufficiently, the forms will be removed and raised to the higher elevation. Because of the batter in the piers, it will not be possible to reuse the lower end forms at the higher elevation.

Cost of forms. The cost of form lumber per use will be based on the number of times the lumber can be used. Assume that the lumber for the side forms can be used an average of 8 times and that the lumber for the ends can be used an average of 4 times.

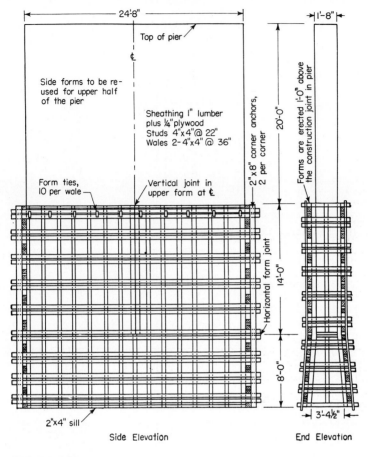

FIGURE 7-20
Details of forms for a concrete bridge pier.

The cost per use should be about as follows:

$$\text{Side lumber,} \frac{\$440.00 \text{ per } 1,000 \text{ fbm}}{8} = \$0.055 \text{ per fbm}$$

$$\text{End lumber,} \frac{\$440.00 \text{ per } 1,000 \text{ fbm}}{4} = \$0.11 \text{ per fbm}$$

Plywood for sides, \$0.36 per ft$^2 \div 8 = \$0.045$ per use
Plywood for ends, \$0.36 per ft$^2 \div 4 = \$0.090$ per use

The forms will be held in position by four guy wires attached near the tops of each side.

The costs will be:

Materials:

Side sheathing, $2 \times 22 \times 24.67 + 20\%$ waste
 = 1,300 fbm @ \$0.055   = \$ 71.50
End sheathing, 110 ft² + 20% waste = 132 fbm @ \$0.11 =   14.52
Side plywood, 1,038 + 10% waste = 1,142 ft² @ \$0.045 =   51.39
End plywood, 108 + 10% waste = 119 ft² @ \$0.090   =   10.71

Studs:

36 pc, 2 × 4 in × 8 ft 0 in = 192 fbm @ \$0.055   =   10.56
38 pc, 2 × 4 in × 14 ft 0 in = 355 fbm @ \$0.055   =   19.53

Sills:

8 pc, 2 × 4 in × 12 ft 0 in = 64 fbm @ \$0.055   =   3.52
8 pc, 2 × 4 in × 14 ft 0 in = 75 fbm @ \$0.055   =   4.13
1 pc, 2 × 4 in × 10 ft 0 in = 7 fbm @ \$0.11   =   0.77
1 pc, 2 × 4 in × 12 ft 0 in = 8 fbm @ \$0.11   =   0.88

Wales:

88 pc, 2 × 4 in × 14 ft 0 in = 821 fbm @ \$0.055   =   45.16
16 pc, 2 × 4 in × 6 ft 0 in = 64 fbm @ \$0.11   =   7.04
28 pc, 2 × 4 in × 4 ft 0 in = 75 fbm @ \$0.11   =   8.25
Wale supports, 176 pc, 1 × 4 in × 1 ft 0 in = 59 fbm @ \$0.11 =   6.49
Scabs for wales, 44 pc, 2 × 4 in × 3 ft 0 in = 88 fbm @ \$0.055   4.84

Corner anchors:

8 pc, 2 × 8 in × 14 ft 0 in = 149 fbm @ \$0.055   =   8.20
8 pc, 2 × 8 in × 8 ft 0 in = 86 fbm @ \$0.055   =   4.73
Form oil, 1,146 ft² ÷ 600 ft²/gal = 2 gal @ \$7.50   =   15.00
Form ties, 11 wales × 10 ties per wale = 110 @ \$0.375 =   41.25
Form tie clamps, 220 @ \$0.085   =   18.70
Nails, 3,475 fbm × 13 lb/1,000 fbm = 46 lb @ \$0.95 =   43.70
Bolts, 50 lb @ \$0.09 per use   =   4.50
Guy wire, 50 lb @ \$0.07 per use   =   3.50
  Total cost of form materials   = \$ 398.87

Labor:

Building forms:
 Carpenters, 1,146 ft² × 4 h/100 ft² = 46 h @ \$16.70 =   768.20
 Helper, 1,146 ft² × 1.5 h/100 ft² = 17 h @ \$12.10 =   205.70
Erecting and removing forms:
 Carpenters, 1,146 ft² × 4.5 h/100 ft² = 52 h @ \$16.70 =   868.40
 Helper, 1,146 ft² × 1.5 h/100 ft² = 17 h @ \$12.10 =   205.70
 Crane operator, 8 h @ \$17.25   =   138.00
 Crane oiler, 8 h @ \$11.25   =   90.00
 Foreman, based on using 4 carpenters = 24 h @ \$18.30 =   439.20
  Total labor cost   = \$2,715.20

Equipment:

Power saws, 3,475 fbm ÷ 400 fbm/h = 8.7 h @ \$2.75 = \$ 23.93
Crane, 8-ton truck-mounted diesel, 8 h @ \$61.87   =   494.96
  Total cost of equipment   = \$ 518.89

Total cost of forms for first use:

Materials	= $	398.87
Labor	=	2,715.20
Equipment	=	518.89
Total cost	=	$3,632.96
Cost per ft^2, $3,632.96 ÷ 1,146	=	3.17

Cost of reinforcing steel:

Steel reinforcing, 9,645 lb @ $0.24	=	$2,314.80
Tie wire, 24 lb @ $0.98	=	23.52

Labor placing reinforcing steel:

Steel setters, 4.8 tons × 12 h/ton = 57.6 h @ $15.80	=	910.08
Helpers, 4.8 tons × 4 h/ton = 19.2 h @ $9.45	=	181.44
Crane operator, 8 h @ $17.25	=	138.00
Crane oiler, 8 h @ $11.25	=	90.00
Crane, 8-ton truck-mounted diesel, 8 h @ $61.87	=	494.96
Total cost of reinforcing steel	=	$4,152.80
Cost per lb, $4,152.80 ÷ 9,645	=	0.43

Cost of concrete:

The concrete will be handled with a crane and bucket

The volume of concrete	=	84.3 yd^3
Add for overrun and waste	=	2.7 yd^3
Total volume	=	87.0 yd^3

Materials:

Cement, 87 yd^3 × 5.5 sacks/yd^3 + 2% waste = 488 sacks

488 sacks @ $4.90	=	$2,391.20

Sand, 87 yd^3 × 1,320 lb/yd^3 + 10% waste = 63 tons

63 tons @ $5.15	=	324.45

Gravel, 87 yd^3 × 2,060 lb/yd^3 + 5% waste = 94 tons

94 tons @ $5.90	=	554.60

Equipment:

Mixer, 87 yd^3 ÷ 14 yd^3/h = 6 h @ $14.50	=	87.00
Crane and bucket, 6 h @ $61.87	=	371.22
Internal vibrator, 6 h @ $0.76	=	4.56
Sundry equipment, 6 h @ $2.50	=	15.00

Labor mixing and placing concrete:

Assume 7 labor-hours for the placing

Mixer operator, 7 h @ $12.45	=	87.15
Crane operator, 7 h @ $17.25	=	120.75
Crane oiler, 7 h @ $11.25	=	78.25
Carpenter, 7 h @ $16.70	=	116.90
Laborers handling cement, sand, gravel, etc.		
10 × 7 h = 70 h @ $9.45	=	661.50
Foreman, 7 h @ $18.30	=	128.10
Utility laborers, 4, 28 h @ $8.10	=	226.80
Total cost of concrete	=	$5,167.48
Cost per yd^3 = $5,167.48 ÷ 87	=	59.40

The total direct cost of pier will be:

Forms	= $ 3,632.96
Reinforcing steel	= 4,152.80
Concrete	= 5,167.48
Total cost	= $12,953.74
Cost per yd^3, $12,953.74 ÷ 84.3 =	153.66

PROBLEMS

7-1. Estimate the direct cost of materials, equipment, and labor for wood forms for a concrete foundation wall for a building whose outside dimensions are 46 ft 6 in wide, 88 ft 0 in long, 9 ft 8 in high, and 9 in thick. The concrete will be placed in the forms at a rate of 3 ft/h at a temperature of 70°F. The concrete will be vibrated as it is placed.

Use $\frac{3}{4}$-in plywood for sheathing and 2- by 4-in lumber for studs and wales. Assume that the lumber can be used 5 times.

Use 3,000-lb form ties. Check or limit the allowable spacing of the ties for load safety.

Use the average national wage rates for carpenters and helpers, as given in this book.

The cost of materials will be

Lumber, per 1,000 fbm, $480.00
Nails, per lb, $0.98
Form ties, see book for costs
Form tie clamps, see book for costs

7-2. Estimate the total direct cost and the cost per square foot for the materials, equipment, and labor for wood forms for a retaining wall 126 ft 0 in long, 14 ft 6 in high, and 12 in thick.

Use 1- by 6-in D and M sheathing and 2- by 4-in lumber for studs and wales. Assume that the lumber will be used 4 times.

The forms will be filled at the rate of 4 ft/h at a temperature of 80°F.

Use the average national wage rates for carpenters and helpers, as given in this book.

The cost of materials will be:

Lumber, $480.00 per 1,000 fbm
Nails, $0.98 per lb
See this book for cost of form ties and clamps

7-3. In designing the forms for use in building a concrete wall, it is desirable to select lumber sizes that will require the smallest quantity of lumber. Two plans are under study, namely plan A and plan B. Determine which plan requires the least number of fbm per square foot of wall surface.

Item	Plan A	Plan B
Sheathing	$\frac{3}{4}$-in plywood	$\frac{3}{4}$-in plywood
Studs	2- × 4-in S4S	2- × 6-in S4S
Wales	2- × 4-in S4S	2- × 6-in S4S

Concrete will be placed in the forms at the rate of 6 ft/h at a temperature of 70°F.

7-4. Estimate the total direct cost and the cost per square foot of surface for materials, equipment, and labor for making, erecting, and removing wood forms for 32 concrete columns. The columns will be 18 by 18 in and 11 ft 8 in high. Adjustable steel clamps will be used with the forms.

Assume that the lumber will be used 4 times. Determine the cost for the first use and for subsequent uses which do not require remaking the forms. A 12-in-blade table saw will be used for sawing and ripping the lumber. Use the average national wage rates for carpenters and helpers.

The cost of materials will be:

Lumber $480.00 per kfbm
Nails, $0.98 per lb
Column clamps, $0.85 per use each

7-5. When concrete columns for a multistory building are designed, it is determined that loads from the floors require 16- by 16-in columns to support the second and third floors and 14- by 14-in columns to support the fourth and fifth floors. All columns will be 9 ft 8 in clear height. If these two column sizes are used, the area of the longitudinal reinforcing steel in the columns will be equal to 4 percent of the gross area of the columns.

If the columns that will support the fourth and fifth floors retain the 16- by 16-in size, the area of the reinforcing steel for these columns can be reduced to $1\frac{1}{2}$ percent of the area of the columns.

If the larger columns are specified for the fourth and fifth floors, the forms from the two lower floors can be reused, without modification or remaking. Only the cost of erecting and removing will be necessary. The amount of cost of concrete will be increased, whereas the quantity and cost of reinforcing steel will be reduced.

If the choice of size is based solely on the cost of labor making and removing forms, the cost of reinforcing steel, and the cost of concrete, should the larger size be retained for the higher floors?

The following costs will apply:

Carpenter and helper, use average wage rates in book
Reinforcing steel in place, $0.52 per lb
Concrete in place, $97.50 per yd^3

7-6. Estimate the total direct cost and the cost per square foot of surface area for furnishing, erecting, and removing forms for interior concrete beams. The beams will be 14 in wide and 16 in deep and have a total length of 196 ft. Wood shores, size 4 by 4 in S4S, will be used to support the beam forms. The distance from the lower floor to the bottom of the beam will be 10 ft 4 in.

Assume that the lumber will be used 4 times. Use the national average wage rates for carpenters and helpers.

The cost of materials will be:

Lumber, $480,00 per 1,000 fbm
Nails, $0.98 per lb

7-7. If the forms for the beams of Prob. 7-6 are reused for other beams of the same sizes, estimate the total direct cost and the cost per square foot of surface area.

7-8. Estimate the costs for Prob. 7-6, using adjustable shores instead of wood shores.

7-9. Estimate the total direct cost and the cost per square foot for furnishing, erecting, and removing forms for eight bays for concrete slab only for beam-and-slab type of concrete floors. Supporting stringers and shores will not be required for the forms.

Each bay will be 7 ft 0 in wide and 22 ft 4 in long, clear dimensions. The floor slab will be 6 in thick. Assume that the lumber will be used 4 times. Use the national average wage rates for carpenters and helpers.

Other costs will be:

Lumber, $480.00 per 1,000 fbm
Nails, $0.98 per lb
Electric saw, with blade, $0.57 per h

7-10. Estimate the total direct cost and the cost per square foot of floor area for furnishing, erecting, and removing wood forms for 1,642 ft^2 of slab only for beam-and-slab type of concrete floors, using adjustable shores. The ceiling height will be 11 ft 8 in. The slab will be 6 in thick. Supporting stringers will be required.

Assume that the lumber will be used 4 times. Material and labor costs will be the same as for Prob. 7-9.

7-11. Estimate the total direct cost and the cost per square foot for the forms for Prob. 7-10, using wood shores instead of adjustable shores.

7-12. Estimate the total direct cost and the cost per pound for furnishing and placing the reinforcing steel for the foundation wall of Prob. 7-1. The vertical reinforcing will consist of no. 6 bars 9 ft 6 in long, spaced 12 in apart around the wall, located at the center of the wall. The horizontal reinforcing will consist of no. 6 bars spaced not over 12 in apart, with the bottom bar placed 6 in above the bottom of the wall. The horizontal bars are available in lengths up to 40 ft. All splices must lap not less than 30 diameters of the bars. At each corner one bar from each row of horizontal reinforcing must be bent and must extend not less than 30 diameters of the bar beyond the bend for bond purposes. The reinforcing is to be securely wired in place by steel setters.

Material costs and labor costs per hour will be:

Reinforcing steel, $0.28 per lb
Tie wire, $0.95 per lb
Steel setters, $15.80 per h

7-13. Estimate the total direct cost and the cost per pound of reinforcing steel for a concrete floor slab resting on the ground. The floor will be 66 ft 8 in long, 52 ft 4 in wide, and 6 in thick. The reinforcing will be no. 4 bars spaced not more than 12 in apart each way in the slab. The bars laid next to and parallel to the edges of the slab will be placed 3 in from the edges of the slab. The ends of the bars will extend to within 2 in from the edges of the slab.

The bars are available in lengths not to exceed 40 ft 0 in. All splices must lap not less than 30 diameters of the bars. The reinforcing will be tied at each intersection by steel setters.

Material and labor costs per hour will be:

Reinforcing steel, $0.28 per lb
Tie wire, $0.95 per lb
Steel setters, $15.80

7-14. Estimate the total direct cost and the cost per 100 ft^2 for furnishing and placing 66-44 welded-wire fabric for a concrete slab 88 ft 0 in wide and 98 ft 6 in long located on the surface of the ground.

Material and labor costs per hour will be:

Welded-wire fabric, $17.85 per 100 ft^2
Tie wire, $0.95 per lb
Steel setters, $15.80 per h

7-15. Estimate the total direct cost and the cost per cubic yard of concrete for furnishing and placing a beam-and-slab type of concrete floor to be constructed above the ground. The slab will be 6 in thick and will have a total area of 3,460 ft^2. The exterior beams extending below the slab will have a cross section 14 in wide and 24 in from the top of the slab to the bottom of the beam. The total length of this beam will be 246 ft 0 in. The interior beams extending below the slab will have a cross section 14 in wide and 16 in deep below the bottom of the slab. The total length of these beams will be 216 ft 0 in.

Ready-mixed concrete will be delivered directly to the job and discharged into a tower bucket which will hoist it to a concrete floor hopper set at the elevation of the slab under construction. The rate of delivery will be 12 yd^3/h.

Power-driven buggies will haul the concrete from the floor hopper to the area of placement.

The hoisting and placing equipment will include:

1 heavy single-tube steel tower, 50 ft high @ $8.75 per h
1 tower hoist bucket 27-ft^3 capacity @ $2.15 per h
1 single-gate 27-ft^3 concrete-floor hopper @ $4.82 per h
1 double-drum 20-hp gasoline-engine-operated hoisting unit @ $28.50 per h
3 9-ft^3 power-driven concrete buggies @ $4.72 per h

The top of the slab will be hand-screeded to the specified thickness, but will not be finished.

Material and labor costs will be:

Concrete delivered to the job, $48.75 per yd^3
1 carpenter, $16.70 per h
1 carpenter helper, $12.10 per h
3 buggy operators, $14.50 per h each
2 screed workers, $9.45 per h each
1 hoisting engineer, $17.25 per h
6 laborers, $8.10 per h each
1 foreman, $18.30 per h

REFERENCES

1. *Formwork for Concrete, Publication SP-4*, American Concrete Institute, Detroit, 1971.
2. R. L. Peurifoy, *Formwork for Concrete Structures*, McGraw-Hill, New York, 1976.

CHAPTER

8

FLOOR FINISHES

Many types of finishes are applied to floors. Several of the more popular types will be discussed in this chapter.

CONCRETE-FLOOR FINISHES

In Chap. 7 the methods of estimating the cost of materials, equipment, and labor for concrete subfloors were discussed. The costs included all operations through the screeding of the rough concrete floors but did not include finishing the floors. These costs will be discussed in this chapter.

For most concrete floors, it is necessary to add other materials to the top of the subfloors to produce finished floors of the desired types. The types of finishes most commonly used are *monolithic topping* and *separate topping*, to either of which may be added coloring pigments or hardening compounds.

Monolithic Topping

One of the best and most economical finishes for concrete floors is monolithic topping. After the concrete for the floor has been screeded to the desired level and has set sufficiently, it is floated with a wood float or a power-floating machine until

211

all visible water disappears. Then the surface of the concrete is dusted with dry cement or a mixture of cement and sand, in approximately equal quantities. After the concrete or mixture is applied to the floor, it is lightly floated with a wood float and then finished with a steel trowel.

Materials Required for a Monolithic Topping

The thickness of the monolithic topping may vary from $\frac{1}{64}$ to $\frac{1}{8}$ in. For finishes less than $\frac{1}{16}$ in thick, only cement should be used, while for finishes thicker than $\frac{1}{16}$ in, fine sand may be added to the cement. Table 8-1 gives the quantities of materials required for finishing 100 ft^2 of concrete floor by using various proportions and thicknesses.

Labor Finishing Concrete Floors Using Monolithic Topping

In estimating the amount of labor required to finish a concrete floor, the estimator must consider several factors which may affect the total time required. After the concrete is placed and screeded, the finishers must wait until the concrete is ready for finishing. The length of wait may vary from as little as 1 h under favorable conditions to as much as 6 h under unfavorable conditions. The factors which affect the time are the amount of excess water in the concrete, the temperature of the concrete, and the temperature of the atmosphere. If the concrete is free of excess water and the weather is warm, finishing may be started soon after the concrete is placed. If the concrete contains excess water and the weather is cold and damp, finishing may be delayed several hours.

If concrete-placing operations are stopped at the normal end of a day, the concrete finishers may have to work well into the night to finish the job. This requires payment of overtime wages at prevailing rates.

Table 8-2 gives the labor-hours required to finish 100 ft^2 of monolithic concrete floor under various conditions.

TABLE 8-1
Quantities of materials for 100 ft^2 of monolithic topping

| Thickness, in | Proportion | | Cement, Sacks | Sand, ft^3 |
	Cement	Sand		
$\frac{1}{64}$	1	0	0.125	0
$\frac{1}{32}$	1	0	0.25	0
$\frac{1}{16}$	1	1	0.25	0.25
$\frac{1}{8}$	1	1	0.50	0.50

TABLE 8-2
Labor-hours required to finish 100 ft² of monolithic concrete floor

	Conditions		
Classification	Favorable	Average	Unfavorable
Cement finisher	0.75	1.25	1.75
Helper	0.75	1.25	1.75

Separate Concrete Topping

A separate concrete topping may be applied immediately after the subfloor is placed, or it may be applied several days or weeks later. The latter method is not considered as satisfactory as the first. The surface of the subfloor should be left rough to ensure a bond with the topping. The topping may vary from $\frac{3}{4}$ to 1 in thick. Concrete for the topping is usually mixed in the proportions 1 part cement, 1 part fine aggregate, and $1\frac{1}{2}$ to 2 parts coarse aggregate, up to $\frac{3}{8}$-in maximum size, with about 5 gal of water per sack of cement.

After the concrete topping has set sufficiently, it is floated with a wood float and finished with a steel trowel. It may be desirable to dust the surface with cement during the finishing operation.

Materials Required for a Separate Topping

The quantities of materials required for a separate topping will vary with the thickness of the topping and the proportions of the mix. Table 8-3 gives quantities of materials required for 100 ft² of floor with various thicknesses and proportions.

TABLE 8-3
Quantities of materials for 100 ft² of separate topping

Thickness, in	Proportions by volume			Cement, sacks	Sand, yd³	Gravel, yd³
	Cement	Sand	Gravel			
$\frac{1}{2}$	1	1	$1\frac{1}{2}$	1.75	0.07	0.10
	1	1	2	1.62	0.06	0.12
	1	$1\frac{1}{2}$	3	1.18	0.07	0.13
$\frac{3}{4}$	1	1	$1\frac{1}{2}$	2.62	0.10	0.15
	1	1	2	2.42	0.09	0.18
	1	$1\frac{1}{2}$	3	1.70	0.10	0.18
1	1	1	$1\frac{1}{2}$	3.50	0.13	0.20
	1	1	2	3.24	0.12	0.23
	1	$1\frac{1}{2}$	3	2.27	0.13	0.25

Labor Mixing, Placing, and Finishing Separate Topping

The concrete for a separate topping should be mixed with a small concrete mixer, such as a $3\frac{1}{2}$S or a 6S, if one is available, since 0.31 yd^3 will cover 100 ft^2 of area 1 in thick.

A finisher and a helper finishing 80 ft^2/h will require 0.25 yd^3/h of topping for a 1-in-thick layer. A crew of three finishers and three helpers should finish 240 ft^2/h under average conditions. This will require 0.75 yd^3/h of 1-in-thick topping. A $3\frac{1}{2}$S mixer with three laborers can mix and place the topping. If the topping is placed above the ground floor, a hoisting engineer will be needed on a part-time basis.

Table 8-4 gives approximate labor-hours required to mix, place, and finish 100 ft^2 of topping under average conditions.

Labor Finishing Concrete Floors with a Power Machine

Gasoline-engine- or electric-motor-driven power machines may be used to float and finish concrete floors (Fig. 8-1). One manufacturer, the Whiteman Manufacturing Company, makes two sizes, 46- and 35-in ring diameter. Each machine is supplied with three trowels, one set for floating and one for finishing the surface. The trowels are easily interchanged. The machines rotate at 75 to 100 r/min.

On an average job the 46-in-diameter machine should cover 3,000 to 4,000 ft^2/h one time over, including time for delays. The 35-in-diameter machine should cover 2,400 to 3,000 ft^2/h. For most jobs it is necessary to go over the surface 3 to 4 times during the floating and the same during the finishing operation.

TABLE 8-4
Labor-hours required to mix, place, and finish 100 ft^2 of separate topping

Thickness, in	Finishers per crew	Labor-hours†			
		Finisher	Helper	Laborer	Foreman
$\frac{1}{2}$	1	1.25	1.25	1.25	0
	2	1.25	1.25	1.25	0.63
	3	1.25	1.25	0.80	0.42
$\frac{3}{4}$	1	1.25	1.25	1.25	0
	2	1.25	1.25	1.25	0.63
	3	1.25	1.25	0.80	0.42
1	1	1.25	1.25	1.25	0
	2	1.25	1.25	1.25	0.63
	3	1.25	1.25	1.25	0.42
	4	1.25	1.25	1.00	0.31

† Add time for a hoisting engineer if one is required.

FIGURE 8-1
Power-operated trowel finishing a concrete slab.

In operating a power machine, a crew of two to three will be required—one operator and one or two finishers—to hand-finish inaccessible areas.

Table 8-5 gives approximate labor-hours required to finish 100 ft² of surface area, based on going over the surface 4 times during each operation

TABLE 8-5
Approximate labor-hours required to finish 100 ft² of surface area with power finishers

Operation	Machine operator	Finisher
46-in-diameter machine, 750 ft²/h		
Floating	0.133	0.266
Finishing	0.133	0.266
Total time	0.266	0.532
35-in-diameter machine, 600 ft²/h		
Floating	0.167	0.333
Finishing	0.167	0.333
Total time	0.334	0.666

Example 8-1. Estimate the cost of materials and labor required to hand-finish 5,000 ft^2 of floor area with a $\frac{1}{32}$-in-thick monolithic topping. Assume that the weather will be clear, with an average temperature of 70°F, which is an average condition. The placing of the concrete will start at 8:00 A.M. and will be completed at 4:00 P.M. One cement finisher will report at 8:00 A.M. to direct the screeding of the slab, and the rest of the finishers will report at 9:00 A.M. and will remain on the job until the finishing is completed. Assume that the finishing will be completed around 7:00 P.M. It will be necessary to pay overtime wages to the finishers and helpers, at $1\frac{1}{2}$ times the regular wage rates, for all time in excess of 8 h.

The number of finishers may be determined as follows:

$$\text{No. of labor-hours required, } \frac{5,000}{100} \times 1.25 = 62.5$$

Time required to finish job, 9:00 A.M. to 7:00 P.M. = 10 h
No. of finishers required, 62.5 h ÷ 10 h = 6.25

Use 6 finishers, which may cause a delay in time of completion until about 7:30 P.M.

The cost will be:

Cement, 5,000 ft^2 × 0.25 sacks per 100 ft^2 = 12.5 sacks @ $4.90 =	$	61.25
Cement finishers:		
Regular time, 6 × 8 h = 48 h @ $16.20	=	777.60
Overtime, 62.5 − 48 = 14.5 h @ $24.30	=	352.35
Helpers:		
Regular time, 6 × 8 h = 48 h @ $12.10	=	580.80
Overtime, 62.5 − 48 = 14.5 h @ $18.15	=	263.18
Add cost of foreman after 4:00 P.M. only		
Overtime rates = 3.5 h @ $27.45	=	96.08
Total cost	=	$2,131.26
Cost per ft^2, $2131.26 ÷ 5,000	=	0.43

TERRAZZO FLOORS

A terrazzo floor is obtained by applying a mixture of marble chips or granules, portland cement, and water laid on an existing concrete or wood floor. White cement is frequently used. The thickness of the terrazzo topping usually varies from $\frac{1}{2}$ to $\frac{3}{4}$ in. Several methods are used to install the topping on an existing floor.

Terrazzo Topping Bonded to a Concrete Floor

If terrazzo is to be place on and bonded to a concrete floor, the concrete surface should be cleaned and moistened. Then a layer of underbed not less than $1\frac{1}{4}$ in thick—made by mixing 1 part portland cement, 4 parts sand, by volume, and enough water to produce a stiff mortar—is spread uniformly over the concrete surface.

While the underbed is still plastic, brass or other metal strips are installed in the mortar to produce squares having the specified dimensions.

The terrazzo topping is made by dry-mixing about 200 lb of granulated marble and 100 lb of cement, or in other proportions as specified. Water is added to produce a reasonably plastic mix. This mix is placed on the underbed inside the metal strips, after which it is rolled to increase the density, then hand-troweled to bring the top surface flush with the tops of the metal strips.

After the surface has hardened sufficiently, it is rubbed with a machine-powered coarse carborundum stone. The surface is then covered with a thin layer of grout made with white cement. At the time of final cleaning, this coating of grout may be removed with a machine-powered fine-grain carborundum hone.

Terrazzo Placed on a Wood Floor

When terrazzo is placed on a wood floor, it is common practice to cover the floor with roofing felt and galvanized wire netting, such as no. 14 gage 2-in mesh, which is nailed to the floor. A concrete underbed not less than $1\frac{1}{4}$ in thick, as described earlier, is installed to a uniform depth over the floor. Metal strips and terrazzo topping are installed and finished, as described previously.

Labor Required to Place Terrazzo Floors

The labor rates required to place terrazzo floors will vary considerably with the sizes of the areas placed. The rates for large areas will be less than those for small areas. If terrazzo bases are required, additional labor should be allowed for placing and finishing the bases.

Table 8-6 gives representative labor-hours for the several operations required in placing terrazzo floors. Use the lower rates for large areas and the higher rates for small areas.

Example 8-2. Estimate the cost of 100 ft^2 of terrazzo floor placed on a concrete slab, with a $1\frac{1}{4}$-in concrete underbed, $\frac{3}{4}$-in-thick terrazzo topping, and brass strips to form 4-ft 0-in squares. Use white cement for the topping.

The quantities will be:
Underbed, $100 \times 1.25/12 = 10.5$ ft^3
Topping, $100 \times 0.75/12 = 6.25$ ft^3

The cost will be:

Sand, including waste, 12 ft^3 @ $0.47	= $	5.64
Gray cement, 3 sacks @ $4.90	=	14.70
Marble, 625 lb @ $0.18	=	112.50
White cement, including grout, 3.5 sacks @ $13.50	=	47.25
Brass strips, including waste, 60 lin ft @ $1.15	=	69.00
Labor cleaning concrete floor, 1 h @ $8.10	=	8.10
Mechanic placing underbed, 1.25 h @ $17.50	=	21.88
Helper placing underbed, 3.0 h @ $9.45	=	28.35

TABLE 8-6
Labor-hours required to place terrazzo floors

Operation	Labor-hours
Cleaning concrete floor, per 100 ft^2	0.75–1.25
Placing roofing felt on wood floors, per 100 ft^2	0.3–0.5
Placing netting on wood floor, per 100 ft^2	0.5–0.7
Mixing and placing 1$\frac{1}{4}$-in-thick underbed, per 100 ft^2:	
Mechanic	1.0–1.5
Helper	2.5–3.5
Mixing and placing $\frac{3}{4}$-in-thick terrazzo topping and metal strips, per 100 ft^2:†	
Mechanic	2.0–3.0
Helper	2.5–3.0
Finishing terrazzo topping, per 100 ft^2	7.0–8.0
Mixing and placing 100 lin ft of 6-in terrazzo cove base:	
Mechanic	14.0–18.0
Helper	14.0–18.0
Finishing 100 lin ft of 6-in terrazzo cove base	8.0–10.0

† These rates are for metal strips placed to form 5-ft 0-in squares. For other size squares, apply the following factors: 4 ft 0 in, 1.10; 3 ft 0 in, 1.20; 2 ft 0 in, 1.30.

Mechanic placing topping and strips 2.75 h @ $17.50 =		48.13
Helper placing topping and strips, 3.0 h @ $9.45	=	28.35
Helper finishing terrazzo topping, 7 h @ $9.45	=	66.15
Finisher and sundry equipment, 7 h @ $7.50	=	52.50
Total cost		= $502.55
Total cost ft^2, $502.55 ÷ 100	=	5.03

ASPHALT TILE

Asphalt tiles are available in sizes 3 by 3 in, 6 by 6 in, 9 by 9 in, 12 by 12 in, 6 by 12 in, 6 by 24 in, 9 by 18 in, 12 by 24 in, and 18 by 24 in and in thicknesses of $\frac{1}{8}$ and $\frac{3}{16}$ in. The cost of tile will vary with the thickness, color, and quality.

When the quantity of tile required for a given floor area is estimated, an allowance should be included for waste due to cutting at the edges of the floor and to irregularities in the locations of the walls around the floor. Wastage may amount to 2 to 15 percent of the net area of a floor.

Laying Asphalt Tile on a Concrete Floor

Before tile is laid, the floor should be cleaned, after which a primer may be applied at a rate of 200 to 300 ft^2/gal. An asphalt cement is then applied at a rate of approximately 200 ft^2/gal; then the tiles are laid and rolled with a smooth-wheel roller.

TABLE 8-7
Labor-hours required to lay 100 ft²
of asphalt tile

Type of floor	Size tile, in	Labor-hours
Concrete	12 × 12	1.5–2.0
	9 × 9	2.0–2.5
	6 × 6	2.8–3.3
Wood	12 × 12	1.8–2.2
	9 × 9	2.3–2.8
	6 × 6	2.0–3.5

Laying Asphalt Tile on a Wood Floor

Prior to the laying of the tile, the floor should be finished with a sanding machine or by some other approved method. A layer of felt is bonded to the floor with a linoleum paste, applied at a rate of approximately 150 ft²/gal, and rolled thoroughly. Then an asphalt cement or emulsion is applied at a rate of approximately 150 ft²/gal. The tiles are then laid and rolled.

Labor Laying Asphalt Tile

The labor required to lay asphalt tile will vary considerably with the size and shape of the floor covered. If specified color patterns are required, the amount of labor will be greater than that for single colors. Table 8-7 gives representative labor-hours for the operations required in laying asphalt tile.

> **Example 8-3.** Estimate the cost of materials and labor required to lay 100 ft² of asphalt tile $\frac{3}{16}$ in thick, using 12- × 12-in squares, on a concrete floor.
> The cost will be:

Asphalt primer, 0.5 gal @ $8.70	= $ 4.35
Asphalt cement, 0.5 gal @ $12.40	= 6.20
Tile, including waste, 105 ft² @ $0.57	= $59.85
Tile setter, 1.8 h @ $12.80	= 23.04
Total cost	= $93.44
Cost per ft², $93.44 ÷ 100	= 0.93

PROBLEMS

8-1. A concrete slab for the second-story floor of a building will be 58 ft 6 in wide, 136 ft 0 in long, and 6 in thick. The concrete will be delivered at a rate of 15 yd³/h, starting at 7:00 A.M., and continued until all concrete has been placed.

   The slab will be finished with a monolithic topping $\frac{1}{32}$ in thick, using hand finishers.

The finishers will report at 9:00 A.M. and will remain until the job is finished, estimated to be 3 h after the last concrete is placed. Overtime wages must be paid at $1\frac{1}{2}$ times the regular wage rates for all time in excess of 8 h. The cost of a foreman will be charged to finishing for only the time after the last concrete is placed in the slab.

Assume average working conditions.

Estimate the total cost and the cost per square foot for the materials and the labor for the topping. Assume that the unit costs for materials and labor will be the same as those used in Example 8-1.

8-2. Estimate the cost of finishing the floor in Prob. 8-1, using one or more, as needed, 35-in-diameter gasoline-engine-powered floor trowels. Use the costs from Prob. 8-1 in preparing your estimate. Power finishers cost $1.76/h.

8-3. Estimate the total direct cost and the cost per square foot for furnishing the materials and labor required for a 4,480-ft^2 terrazzo floor placed on a concrete slab, with a $1\frac{1}{4}$-in-thick underbed, $\frac{3}{4}$-in-thick terrazzo topping, and brass strips to form 4-ft 0-in squares. Use gray cement and marble chips for the topping.

The floor consists of large rooms and corridors at approximately ground level.

Use the unit costs for materials and labor from Example 8-2.

8-4. Estimate the total direct cost and the cost per square foot for furnishing and laying $\frac{3}{16}$-in-thick asphalt tile, size 9- by 9-in squares, on a concrete floor whose area is 2,360 ft^2.

Use the unit costs for materials and labor from Example 8-3.

CHAPTER 9

FLOOR
SYSTEMS

This chapter describes several types of floor systems which may be used as substitutes for concrete-beam-and-slab, concrete slab only, or pan-and-joist concrete floors. An examination of the cost developed for each of the floor systems and a comparison with the cost of an all-concrete floor will reveal that reductions in the cost frequently may be effected through the choice of the floor system.

Although these analyses are made for floor systems only, the results should demonstrate that similar analyses for other parts of structures may also permit the selection of methods and materials that will produce reductions in the costs.

STEEL-JOIST SYSTEM

General Description

Open-web steel joists (Fig. 9-1) are fabricated in the shop into the form of a Warren truss by an arc-welding process. The chord members consist of T sections, angles, or bars. The web is made of a single round bar, or angle, which is welded to the top and bottom chord members. A steel bearing seat is welded to each end of a joist to provide proper bearing area. The bearing seats are designed to permit the ends of the joist to rest on masonry walls or structural-steel beams.

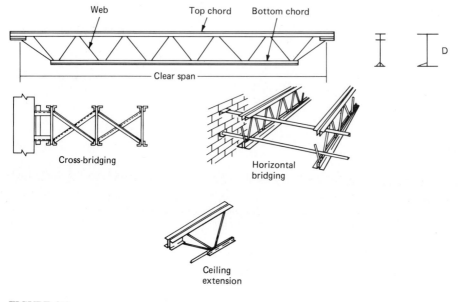

FIGURE 9-1
Steel joist.

Floor and Ceiling

A concrete slab 2 to 5 in thick is usually installed on corrugated-metal decking which rests on the tops of the joists. Suspended ceilings are supported by wire hangers that are attached to the bottom chords of the joists. If wood floors are to be installed, wood nailer strips may be attached to the top chords.

Bridging

Typical bridging for steel joists includes horizontal angle struts that are welded between adjacent joists. If cross bridging is required, angle struts are fabricated that are bolted or welded to the top chord of one joist and the bottom chord of an adjacent joist. Two struts are installed for each line of bridging to give a cross-bridging effect.

Cross bridging should be installed in accordance with the specifications of the designer or the manufacturer of the joists, which will be approximately as shown in Table 9.1. These are common bridging requirements for light joists. Longer spans can be obtained with heavier chord sizes with the above lines of bridging.

Metal Decking

To support the concrete floor or roof, a continuous layer of ribbed metal decking is installed over the joists, with ribs perpendicular to the joists. The decking is

TABLE 9-1
Cross bridging for steel joists

Lines of bridging	Span, ft	
	K series joists	LH series joists
Row near center	Up to 16	Up to 22
Rows at $\frac{1}{3}$ points	16–24	22–33
Rows at $\frac{1}{4}$ points	24–28	33–44

fastened to the top chords of the joists by welding or screw fasteners, spaced not over 15 in apart along the joists. Metal decking is fabricated under various trade names by several manufacturers. It is designated by shape, size, and gage. Table 9-2 provides a representative example of metal decking.

Floor decking should be lapped at least 2 in beyond the center of supporting joists at the end of sheets and should be securely fastened to the next sheet.

Ceiling Extensions

Ceiling extensions are provided at the same elevation as the bottom chord when a ceiling is to be attached. These extensions support the ceiling hanger wire along the bearing wall or supporting beam.

Joist End Supports

Each end of the joist is secured to either a load-bearing wall, or a structural beam and column framing system. If the joist is supported by a load-bearing masonry or concrete wall, the joist is welded to a steel bearing plate or clip angle that is anchored in the wall. For structural beam and column framing systems, the ends of the joist are welded directly to the beams and columns.

Size and Dimensions of Steel Joists

Steel joists are available in a great many sizes and lengths for varying loads and spans. Standard open-web joists, which are designed in accordance with the Steel

TABLE 9-2
Data for form decking

Joist spacing, in	Size of rib, in	Gage	Weight, lb/ft^2
Up to 30	$\frac{9}{16}$	28	0.86
30–36	$\frac{9}{16}$	26	1.01
36–60	1	26	1.06

Joist Institute standard specifications, are available for spans varying from 5 to 60 ft for K series joists. Spans up to 144 ft are available in the DLH series joists. Joists may be installed with any desired spacing from 18 to 72 in or more, provided that the maximum safe load is not exceeded. Table 9-3 provides representative samples of steel joists.

Cost of Steel Joists

The cost of steel joists varies so much with the number and sizes of the joists, accessories required, and location of the job that a table of costs is of little value to an estimator. Before preparing an estimate for a particular job, the estimator should submit the plans and specifications to a representative of a manufacturer for a quotation. Care should be taken to include in the quotation all necessary

TABLE 9-3
Representative dimensions of joists

Type	Depth, in	Span length, ft	Approximate weight, lb/lin ft
K series			
12K3	12	12–24	5.0
14K3	14	14–28	6.0
16K4	16	16–32	7.0
18K6	18	18–36	8.5
20K6	20	20–40	9.0
22K7	22	22–44	10.0
24K8	24	24–48	11.5
26K8	26	26–52	12.0
28K9	28	28–56	13.0
30K10	30	30–60	15.0
LH series			
18LH04	18	25–36	12.0
20LH06	20	25–40	15.0
24LH08	24	33–48	18.0
28LH09	28	41–56	21.0
32LH11	32	49–64	24.0
36LH12	36	57–72	25.0
44LH13	44	73–88	30.0
48LH15	48	81–96	36.0
DLH series			
52DLH13	52	89–104	34.0
56DLH14	56	97–112	39.0
60DLH14	60	105–120	40.0
64DLH16	64	113–128	46.0
68DLH16	68	121–136	49.0
72DLH17	72	129–144	56.0

accessories such as bridging and extensions. The quotation should specify whether the prices are f.o.b. the shop or the job. If the prices are f.o.b. the shop, the estimator must add the cost of transporting the materials to the job.

Labor Erecting Steel Joists

The labor cost of erecting steel joists and accessories is usually estimated by the ton. To arrive at a reasonable unit price per ton, it is necessary to determine the probable rate at which the joists will be erected. The rate will vary with the size and length of the joists, method of supporting them, type of end connections, type of bridging, height that they must be lifted, and complexity of the floor area.

In some locations, union regulations require that all labor erecting steel joists must be performed by union mechanics, while in other locations helpers are permitted to assist in the erection.

Table 9-4 gives the approximate labor-hours required to erect 1 ton of steel joists, including the installation of bridging and accessories. If helpers are not permitted to assist the ironworkers, the time shown for helpers should be added to that shown for the ironworkers.

Labor Installing Metal Decking

Sheets of metal decking are laid perpendicular to the joists and secured with welds or screw fasteners. Two skilled ironworkers working together should install 180 to 240 ft^2/h of decking, depending on the complexity of the floor area. This is equivalent to 7.5 to 11 labor-hours per 1,000 ft^2.

TABLE 9-4
Labor-hours required to erect 1 ton of steel joists

Length of span, ft	h/ton	
	Ironworker	Helper†
Irregular construction, small areas		
6–10	6.5	3.25
10–16	6.0	3.0
16–24	5.5	2.75
24–30	5.0	2.5
Regular construction, large areas		
16–24	4.5	2.25
24–32	4.0	2.00

† For each floor above the first floor, add 1.5 helpers-hours if the joists are carried up by hand.

Labor Placing Welded-Wire Fabric

Welded-wire fabric is frequently used to reinforce the concrete slab placed on steel joists. Table 7-25 gives the properties of representative samples of welded-wire fabric. Many sizes and weights are manufactured. An experienced worker should place approximately 400 ft²/h of welded-wire fabric on straight-run jobs. If cutting and fitting are necessary, the rate will be reduced. Since the fabric is furnished in rolls containing approximately 750 ft², it may be necessary to use mechanical equipment to hoist it to floors above the ground level.

Concrete for Slabs

The subject of mixing and placing concrete for the rough floor has been discussed in Chap. 7. Since the information given in Chap. 7 was developed primarily for floors having a greater thickness than is generally used with floors supported by steel joists, the amounts of labor required to haul, spread, and screed the concrete for joist-supported floors should be increased to provide for the additional time needed. An increase of approximately 25 percent for the operations affected should be adequate.

Example 9-1. Estimate the total direct cost of steel joists, metal decking, and a $2\frac{1}{2}$-in-thick lightweight concrete slab for a floor area 48 ft 0 in wide by 72 ft 0 in long. The floor will be divided into four bays, each 18 ft 0 in wide by 48 ft 0 in long. The floor is one story above the ground level.

The steel joist will be type 16K4, 18 ft 0 in long, spaced 24 in apart, with two rows of bridging per joist. The joist will be welded to the supporting steel members, by two 1-in-long welds at each end of each joist.

The joists will be covered with $\frac{9}{16}$-in 28-gage corrugated-metal decking weighing 0.86 lb/ft². The decking will be fastened to the joist with $\frac{1}{2}$-in welds spaced not over 15 in apart.

The concrete will be reinforced with style 6 × 6-10/10 welded-wire fabric. The top surface of the concrete will be screeded to the required thickness but will not be finished under this estimate.

The floor area will be 48 ft 0 in by 72 ft 0 in = 3,456 ft²
The volume of concrete will be:

3,456 ft² × (2.5 ÷ 12)27	= 26.7 yd³
Add for possible overrun =	2.0 yd³
Total volume	= 28.7 yd³

Joists:

Joists, 100 each @ $37.80	=	$3,780.00
Bridging, 24 in long, 384 struts @ $0.36	=	138.24
Welding electrodes, 5 lb @ $0.65	=	3.25
Freight on joists to job, 6.5 tons @ $15.50	=	100.75
Labor unloading 6.5 tons,		
6.5 tons × 1 h/ton = 6.5 h @ $9.45	=	61.43

Ironworkers erecting, 6.5 tons,
6.5 tons × 4.5 h/ton = 29.2 h @ $15.80 =   461.36
Helpers erecting, 6.5 tons,
6.5 tons × 2.25 h/ton = 14.6 h @ $12.40 =   181.04
Foreman, 8 h @ $18.30   =   146.40
Welding equipment, 8 h @ $3.75   =   30.00
Sundry equipment   =   60.00
   Total cost of joists   = $4,962.47

Metal decking:
  Metal decking, including 2 % waste and welding washers,
    3,456 × 1.02 = 3,525 ft^2 × 0.86 lb/ft^2 = 3,032 lb,
    3,032 lb @ $0.37   = $1,121.84
  Electrodes, 12 lb @ $0.57   =   6.84
  Ironworker, installing 3,525 ft^2,
    3,525 ft^2 × 9 h/1,000 ft^2 = 31.7 h @ $15.80   =   500.86
  Scaffolds   =   75.00
    Total cost of metal decking   = $1,704.54

Reinforcing steel:
  Welded-wire fabric, 3,456 ft^2 plus 5 % for waste,
    3,456 × 1.05 = 3,629 ft^2 @ $0.12   = $435.48
    Tie wire, 4 lb @ $0.98   =   3.92
  Steel setter placing fabric, 3,629 ft^2,
    3,629 ft^2 × 0.3 h/100 ft^2 = 10.9 h @ $15.80   = 172.22
    Total cost of reinforcing   = $611.62

Concrete slab:
Concrete, ready-mixed, 28.7 yd^3 @ $48.00   = $1,377.60
    Hoisting engineer, 28.7 yd^3,
    28.7 yd^3 ÷ 10 yd^3/h = 2.9 h @ $14.75   =   42.78
Buggy workers, 3 × 2.9 h = 8.7 h @ $9.45   =   82.22
Emptying buggies, 2 persons, 5.8 h @ $9.45   =   54.81
Spreading concrete, 4 workers, 11.6 h @ $9.45 =   109.62
Moving runways, 2 persons, 5.8 h @ $9.45   =   54.81
Foreman, 2.9 h @ $18.30   =   53.07
Hoisting equipment, 2.9 h @ $7.85   =   22.77
Runways   =   120.00
Sundry equipment   =   75.00
    Total cost of concrete   = $1,992.68

Summary of costs:
  Joists   = $4,962.47
  Metal decking   = 1,704.54
  Reinforcing steel   =   611.62
  Concrete   = 1,992.68
    Total cost   = $9,271.31
    Cost per ft^2 in place, $9,271.31 ÷ 3,456 =   2.68

COMBINED CORRUGATED-STEEL FORMS AND REINFORCEMENT FOR FLOOR SYSTEM

Description

A complete floor system which is suitable for floors supporting light to heavy loads is constructed with a deep corrugated-steel combined form and reinforcing unit and concrete. The steel sheets used in constructing this system are sold under the various trade names of manufacturers. The corrugated sheets are fabricated from high-strength steel, varying in thickness from 16 to 22 gage with lengths up to 40 ft. The deep corrugated-steel sheets serve as longitudinal reinforcing for positive moment. Embossments on the vertical ribs transfer positive shear from the concrete to the steel. Welded-wire fabric is installed in the slab for temperature and shrinkage crack control. Conventional reinforced-concrete design procedures are followed for simple and continuous spans.

Table 9-5 provides an illustrative example of data for corrugated sheets used for this type of floor system. The quantities of concrete, in cubic yards per square foot of floor area for various thickness of floor slab, will vary depending on the manufacturer. The thickness of slab shown in Table 9-5 is measured from the top of the slab to the bottom of the corrugation.

Installing Corrugated Sheets

The sheets are equally suited to concrete or steel-frame construction. They should be fastened to the supports by welds, screws, bolts, or other means. Conventional negative reinforcing steel is installed over the supports. Conduit for electric wires is laid on the sheets with openings made through them. Provisions for large openings, such as for stairs, should be prefabricated by the manufacturer.

To avoid excessive form stresses and deflection while supporting the wet concrete between permanent supporting members, one or two lines of temporary

TABLE 9-5
Data for corrugated sheets

Gage	Cover, in	Form depth, in	Slab depth, in	Volume of concrete, ft^3/ft^2
22	36	1.5	3.5	0.210
22	36	2.0	4.0	0.253
20	36	2.0	4.5	0.294
20	24	2.0	5.0	0.336
20	24	3.0	5.5	0.333
20	24	3.0	6.0	0.375

Source: Vulcraft Division of Nucor Corporation.

supports should be installed until the concrete has gained strength. The spacing of the temporary supports should conform to the recommendations of the manufacturer.

Labor Installing Corrugated Sheets

The labor required to install corrugated sheets will vary somewhat with the sizes and weights of the sheets, types of supports, and complexity of the floor. Based on a floor with 18-ft 0-in by 24-ft 0-in bays, with supporting steel beams spaced 8 ft 0 in apart, requiring 22-gage 1.25-in-deep corrugations, the rates of placing should be about as follows:

Operation	Labor-hours per 1,000 ft²
Sorting and arranging sheets	2.5
Laying and fastening decking	5.0
Installing welded-wire fabric	3.0
Installing negative reinforcing bars	8.0
Installing and removing temporary supports	11.0

Example 9-2. Estimate the total direct cost and the cost per square foot for materials and labor for a bay 18 ft 0 in wide by 24 ft 0 in long using corrugated forms and reinforcement and a concrete slab.

The sheets will be 22 gage with 1.5-in-deep corrugations, with 6 × 6-10/10 for shrinkage steel. The permanent supporting members will be steel beams, spaced 8 ft 0 in on centers, parallel to the 18-ft 0-in side of the bay.

The concrete will have a maximum depth of 3.5 in. Negative reinforcing, installed over each permanent beam, will consist of no. 4 bars, 4 ft 0 in long, spaced 6 in on centers.

Temporary intermediate supports for the corrugated sheets will consist of one 2- × 8-in wood stringer and adjustable shores, spaced 4 ft 6 in apart. The stringer will be installed parallel with the 18-ft 0-in sides of the bay.

The area of the bay will be 18 ft × 24 ft = 432 ft².

The cost will be:

Materials:
```
Corrugated sheets, f.o.b. at factory, 432 ft² @ $0.59 = $  254.88
Freight on sheets, 432 ft² @ $0.09              =      38.88
Welded-wire fabric, 432 ft² + 5% for waste,
    432 × 1.05 = 454 ft² @ $0.12               =      54.48
Negative reinforcing steel, 108 pc,
    108 pc × 4 ft 0 in × 0.668 = 288 lb @ $0.24  =      69.12
High chairs, tie wire, etc., 432 ft² @ $0.19     =      82.08
Concrete, ready-mixed, 432 ft²,
    432 ft² × 0.210 ft³/ft² = 90.7 ft³
    (90.7 ft³)(27 ft³/yd³) = 3.4 yd³ @ $48.00    =     163.20
```

Equipment:

Stringers, 5 uses, 112 fbm @ $0.096	=	10.75
Shores, 15 each @ $2.25	=	33.75
Hoisting equipment, 3 h @ $9.60	=	28.80
Welding equipment, 4 h @ $3.75	=	15.00
Sundry equipment	=	7.50

Labor:

Sorting and installing sheets, 432 ft^2,
432 ft^2 × 7.5 h/1,000 ft^2 = 3.2 h @ $15.80   =   50.56

Steel setter placing welded-wire fabric,
454 ft^2 × 3.0 h/1,000 ft^2 = 1.4 h @ $15.80   =   22.12

Ironworker placing negative reinforcing steel,
488 lb × 8.0 h/1,000 ft^2 = 3.9 h @ $15.80   =   61.62

Time to place concrete, 3.9 yd^3 ÷ 10 yd^3/h = 0.4 h

Laborers placing concrete, 6 × 0.4 h = 2.4 h,
2.4 h @ $9.45   =   22.68

Carpenter setting stringers, etc., 4.5 h @ $16.70	=	75.15
Foreman, 5.0 h @ $18.30	=	91.50
Hoisting engineer, for all materials, 4.1 h @ $14.75	=	60.48
Total cost	=	$1,142.55
Cost per ft^2, $1,142.55 ÷ 432	=	2.64

CHAPTER
10

MASONRY

Masonry Units

Masonry units which are commonly used for construction include brick, tile, concrete blocks, and stone, natural or artificial. They are bonded by a suitable mortar, with metal ties frequently added to increase the strength of the bond. Most of the units are available in several sizes, grades, and textures. Plans and specifications for a structure designate the kind of unit; the size, grade, texture, kind of mortar, and thickness of joint; and the quality of workmanship required.

Estimating the Cost of Masonry

In estimating the cost of a structure to be constructed entirely or partly of masonry units, the estimator should determine separately the quantity and cost of each kind of unit required. An appropriate allowance should be made for waste, resulting primarily from breakage. Determine the quantities and cost of materials for the mortar, including an allowance for waste. Estimate the cost of labor laying the masonry units. If construction equipment is used in mixing the mortar, in hoisting the masonry units, or for other purposes, the cost of such equipment should be added to the other costs.

TABLE 10-1
Quantity of lime per cubic foot of putty

Kind of lime	lb/ft^3
Hydrated lime	45
Pulverized quicklime	25
Lump quicklime	23

Mortar

Most mortar for masonry units is made by mixing portland cement, lime, and sand or by mixing a commercial masonry cement with sand. The quantity of each ingredient may vary to produce a mortar suitable for the particular job. Coloring is sometimes added. If a fine sand is used, the workability of the mortar will be much better than if coarse sand is used.

Lime for Mortar

The lime used in making masonry mortar is available as lump quicklime, pulverized quicklime, or hydrated lime. Quicklime or hydrated lime may be purchased in paper bags, containing 50 lb, or it may be purchased in bulk. Prices are frequently quoted by the ton.

Quicklime is approximately pure calcium oxide, while hydrated lime is calcium hydroxide. Before quicklime can be used in making mortar, it must be hydrated, or slaked, by mixing it with water and allowing it to season for several days. The putty will remain usable for several weeks after it is slaked. For quantities see Table 10-1. Table 10-2 gives the approximate quantities of materials required for 1 yd^3 of mortar for various mixing proportions.

BRICK MASONRY

Bricks may be classified by the material from which they are made, the method of molding them, the purpose for which they will be used, the size, etc. The costs, which vary considerably, are usually based on 1,000 units, either at the factory or delivered to the destination.

Sizes of Bricks

Bricks are manufactured in a great many sizes. The Common Brick Manufacturers Association adopted a standard size, with nominal dimensions of $2\frac{1}{4}$ by $3\frac{3}{4}$ by 8 in, corresponding to the thickness, width, and length, respectively. However, unequal degrees of burning will frequently cause variations in these dimensions for the finished products. Not all manufacturers have installed molds to produce standard

TABLE 10-2
Quantities of materials required for 1 yd³ of mortar

Masonry cement and sand

Proportions by volume		Quantity	
Cement	Sand	Cement, sacks	Sand, yd³
1	2	13.5	1
1	2.5	11.1	1
1	3	9.0	1
1	3.5	7.7	1
1	4	6.8	1

Pulverized quicklime, portland cement, and sand

Proportions by volume			Quantity		
Lime putty	Cement	Sand	Lime, lb	Cement, sacks	Sand, yd³
1	0	3	225	0	1
2	1	9	150	3	1
1	1	6	112.5	4.5	1
0.5	1	4.5	75	6	1
0.2	1	3	45	9	1
0.1	1	3	22.5	9	1

Hydrated lime, portland cement, and sand

Proportions by volume			Quantity		
Dry lime†	Cement	Sand	Lime, lb	Cement, sacks	Sand, yd³
1	0	3	450	0	1
2	1	9	300	3	1
1	1	6	225	4.5	1
0.5	1	4.5	150	6	1
0.2	1	3	90	9	1
0.1	1	3	45	9	1

† Hydrated lime is usually sold in paper bags weighing 50 lb and containing 1 ft³.

sizes. Although face bricks are made in sizes corresponding to the standard-size common bricks, other sizes are also available. It is good practice to specify the size of bricks desired, in estimating or purchasing them, instead of referring to them as "standard bricks."

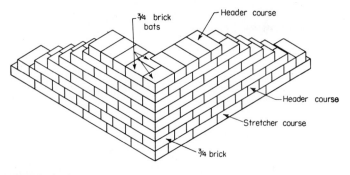

FIGURE 10-1
Brick wall laid with common bond.

In this book, the $2\frac{1}{4}$- by $3\frac{3}{4}$- by 8-in brick is used for all examples and estimates.

Thickness of Brick Walls

Brick walls are usually designated as 4, 8, 12, 16, and 20 in thick, although these dimensions are not entirely correct in all instances. The actual thickness will vary with the number of bricks of thickness and the thickness of the vertical mortar joints between rows of bricks. A three-brick-thick wall is referred to as a 12-in wall; but if $\frac{1}{2}$-in-thick mortar joints are used, the actual thickness will be $12\frac{1}{4}$ in. In a similar manner, a four-brick-thick wall will actually be $16\frac{1}{2}$ in thick.

Estimating the Quantity of Bricks

In estimating the cost of bricks for a structure, the estimator should determine the quantity of bricks for each type required. At least two methods of estimating the quantities of bricks are used. One method is to determine the total volume of the walls, usually in cubic feet, then to multiply this volume by the probable number of bricks per cubic foot. Another method is to determine the area of the wall, with a separate area for each different thickness, then multiply the respective area by the number of bricks per square foot of area, considering the thickness of the wall. The latter method should give more accurate results. For example, wall thicknesses may be 12 or 13 in without any variation in the number of bricks required, provided that there is no change in the thickness of the bed and end mortar joints. The area method of estimating quantities will be correct, whereas the volume will give different quantities.

Regardless of the method used, the estimator should adjust the volume or area for openings in the wall. Failure to make this adjustment introduces an error which is too large for accurate estimating.

The net exposed area of a standard-size brick laid flat is 8 by $2\frac{1}{4}$ in, which equals 18 in^2. The effective area is increased by the vertical and horizontal mortar joints. If each joint is $\frac{1}{2}$ in thick, the effective area will be $8\frac{1}{2}$ by $2\frac{3}{4}$ in, which equals $23\frac{3}{8}$ in^2. The number of bricks required per square foot of wall area for one-brick thickness will be 6.16. For an 8-in-thick wall, the number will be 12.32.

Table 10-3 gives the number of bricks per 100 ft^2 of wall area for various wall and mortar-joint thicknesses, using $2\frac{1}{4}$- by $3\frac{3}{4}$- by 8-in bricks. No allowance is made for waste.

Quantity of Mortar

The quantity of mortar required for brick masonry will vary with the thickness of the mortar joints and the extent to which all joints are filled with mortar. The inside vertical joints are not always filled with mortar, especially on secondary backup walls. It is difficult to accurately estimate the quantity of mortar that will be wasted in laying bricks; however, this will usually amount to 10 to 25 percent of the quantity required in laying the bricks.

Table 10-4 gives the quantities of mortar required for 1,000 standard-size bricks for various joint thicknesses.

Types of Joints for Brick Masonry

Several types of mortar joints are specified for brick masonry which will affect the rate of laying bricks. The more common types of joints are shown in Fig. 10-2.

Flush-cut joints are made by passing the trowel across the surface of the bricks and removing any excess mortar. This operation requires very little time.

TABLE 10-3
Number of standard bricks per 100 ft^2 of wall area

Nominal thickness of wall, in	Thickness of horizontal mortar joint, in				
	$\frac{1}{4}$	$\frac{3}{8}$	$\frac{1}{2}$	$\frac{5}{8}$	$\frac{3}{4}$
End joints $\frac{1}{4}$ in thick					
4	698	665	635	608	582
8	1,396	1,330	1,270	1,216	1,164
12	2,095	1,995	1,905	1,824	1,746
16	2,792	2,660	2,540	2,432	2,328
End joints $\frac{1}{2}$ in thick					
4	677	645	615	588	564
8	1,354	1,290	1,230	1,176	1,128
12	2,031	1,935	1,845	1,764	1,692
16	2,708	2,580	2,460	2,352	2,256
20	3,385	3,225	3,075	2,940	2,820

TABLE 10-4
Cubic yards of mortar required per 1,000 standard-size bricks for full joints, using common bond (no allowance included for waste)

Nominal thickness of wall, in	Thickness of horizontal mortar joint, in				
	$\frac{1}{4}$	$\frac{3}{8}$	$\frac{1}{2}$	$\frac{5}{8}$	$\frac{3}{4}$
Vertical joints $\frac{1}{4}$ in thick					
4	0.211	0.291	0.376	0.458	0.542
8	0.301	0.384	0.474	0.562	0.649
12	0.346	0.432	0.523	0.614	0.703
16	0.429	0.519	0.616	0.709	0.805
Vertical joints $\frac{1}{2}$ in thick					
4	0.262	0.348	0.433	0.518	0.627
8	0.365	0.456	0.546	0.637	0.751
12	0.414	0.506	0.600	0.693	0.809
16	0.433	0.520	0.621	0.715	0.832

Struck and weathered joints are made with a trowel after the mortar has gained some stiffness. This operation requires more time than cut joints.

Concave joints are made by tooling the mortar with a wood or nonstaining metal rod before the mortar gains final set.

Raked joints are made by removing the mortar to a depth of $\frac{1}{4}$ to $\frac{3}{8}$ in, by using a special tool.

Flush cut

Struck

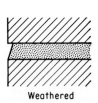
Weathered

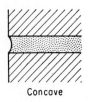
Concave

Convex

Raked

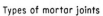
Types of mortar joints

FIGURE 10-2
Mortar joints for masonry.

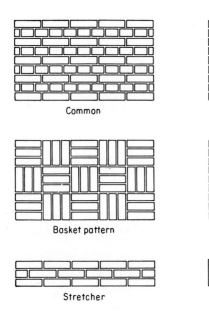

Common

English

Basket pattern

Flemish

Stretcher

Soldier

FIGURE 10-3
Bonds for brick masonry.

Bonds

Figure 10-3 illustrates the more commonly used bonds for brick walls. The amount of labor required to lay bricks will vary with the type of bond used.

Labor Laying Bricks

The labor required to lay bricks varies with a number of factors such as the quality of work, type of bricks, kind of mortar used, shape of the walls, kind of bond used, thickness of the wall, weather conditions, and inclination of the bricklayers.

If the walls must be plumbed accurately with straight courses and uniformly thick joints, the labor required will be greater than for a less rigid quality. If the joints must be tooled, more labor will be required than if the joints are simply cut flush with a trowel.

If the walls are irregular in shape with frequent openings, pilasters, or other changes in shape, the labor requirements will be greater than for long, straight walls.

If the walls contain various patterns requiring changes in the bond, the labor requirements will be greater than for a uniform bond. Less labor is required to lay bricks in a thick wall than in a thin one, since less time is spent per brick for plumbing, leveling, and moving scaffolds. The labor required to lay bricks in cold or wet weather will be greater than when the weather is mild and dry.

TABLE 10-5
Labor-hours required to lay 1,000 bricks by using lime-cement mortar and common bond†

Type of structure and quality of work	Type of joint	Hours per 1,000 bricks	
		Mason	Helper
Common brick			
Commerical buildings, warehouses, stores, shops,	Cut	7.5	7.5
walls 12 in or more thick	Struck	8.0	8.0
Public buildings, requiring first-grade workmanship,	Cut	9.0	8.0
walls 12 in or more thick, two or three stories high	Struck	9.5	8.5
Backing up face brick or stone partition, walls	Cut	7.5	7.5
8 in or more thick	Struck	8.0	8.0
Face brick			
Commerical buildings, stores, office buildings with	Cut	15.0	10
4-in face brick, ordinary workmanship	Struck	15.5	10
	Concave	16.0	10
Public buildings, schools, libraries, churches, etc.,	Cut	18.0	11
with 4-in face brick, first-class workmanship	Struck	18.5	11
	Concave	19.0	11

† For other than common bond, the labor-hours for brickmasons should be increased approximately 10 to 15 percent.

Table 10-5 gives the approximate labor-hours required to lay 1,000 bricks for various types of structures under different conditions. The indicated time includes mixing the mortar, hodding bricks and mortar, and laying the bricks.

Example 10-1. Estimate the total direct cost and the cost per 1,000 bricks for constructing the walls of a rectangular building 116 ft 0 in long, 68 ft 0 in wide, with brick walls 16 ft 0 in high. The walls will be 4 in thick, with the bricks laid in common bond. The total area of openings for doors and windows will be 484 ft^2. Face brick will be used for the wall. The mortar will be 1:6:6 hydrated lime, portland cement, and sand, respectively. Ordinary workmanship is specified, with $\frac{1}{2}$-in-thick struck joints.
The quantities are determined as follows:

The gross outside area of wall will be 2(116 + 68) × 16 = 5,888 ft^2
Deduct for four corners, 4 × 16   = − 64 ft^2
Deduct area of openings   = −484 ft^2
   Net area   = 5,340 ft^2

The quantity of bricks will be:
   Bricks, 5,340 ft^2 × 615 per 100 ft^2 = 32,841
   Add for waste, 2%   =   657
      Total quantity of bricks   = 33,498

The quantity of mortar will be:

Quantity of mortar, 33,498 bricks × 0.433 yd³ per 1,000 = 14.5 yd³
Add 10% for waste   =   1.5 yd³

Total quantity of mortar   = 16.0 yd³

The cost will be:

Bricks, 33,498 @ $195.00 per 1,000   = $ 6,532.11
Lime, 16 yd³ × 225 lb/yd³ = 1.8 tons @ $60.67   =   109.21
Cement, 16 yd³ × 4.5 sacks/yd³ = 72 sacks @ $4.90   =   352.80
Sand, including waste, 17 yd³ @ $8.55   =   145.35
Bricklayers, 33,498 bricks × 15.5 h per 1,000 = 519 h @ $16.50 = 8,563.50
Helpers, 33,498 bricks × 10.0 h per 1,000 = 335 h @ $9.45   = 3,165.75
Foreman, based on 6 bricklayers, 519 ÷ 6 = 87 h @ $18.30   = 1,592.10
Mixer operator, 87 h @ $8.10   =   704.70
Mortar mixer, 87 h @ $2.85   =   247.95
Scaffolds and sundry equipment   =   250.00

Total cost   = $21,663.47
Cost per 1,000 bricks in place, $21,663.47 ÷ 32.841   =   659.65

Face Brick with Common-Brick Backing

Buildings are frequently erected with a 4-in-thick veneer of face bricks, while the balance of the wall thickness is obtained with common bricks. If a common bond is used, it is customary to lay face bricks as headers every sixth course for bond purposes. Since a header course will require twice as many face bricks as a stretcher course, the total number of face bricks required must be increased over the number that would be required for a uniform 4-in thickness. This is equivalent to one extra stretcher course in six courses, amounting to an increase of $16\frac{2}{3}$ percent in the number of face bricks required. The number of common bricks may be reduced in an amount equal to the increase in the number of face bricks.

CONCRETE BLOCKS

Concrete blocks are made in various sizes from portland cement, sand and gravel, or cement and lightweight aggregate. Since the blocks are frequently made in local or nearby plants, with limited sizes available, it may not be possible to obtain all sizes and types in a given community. While blocks of any size and type may be obtained by purchasing them from a distant plant, the increased cost of freight may make their cost unreasonably high. The designer should check the source of supply before specifying given sizes for a project.

Quantities of Materials Required for Concrete Blocks

Table 10-6 gives the nominal and actual sizes, approximate weights, and quantities of mortar required for joints for the more popular sizes of blocks.

TABLE 10-6
Sizes, weights, and quantity of mortar for concrete blocks, using $\frac{3}{8}$-in joints†

Actual size: thickness, height, length, in	Approx. weight per block, lb		Quantities per 100 ft² of wall area	
	Standard	Light-weight	No. blocks	Mortar, yd³
$3\frac{5}{8} \times 4\frac{7}{8} \times 11\frac{5}{8}$	11–13	8–10	240	0.15
$5\frac{5}{8} \times 4\frac{7}{8} \times 11\frac{5}{8}$	17–19	12–14	240	0.16
$7\frac{5}{8} \times 4\frac{7}{8} \times 11\frac{5}{8}$	22–24	14–16	240	0.17
$3\frac{5}{8} \times 7\frac{5}{8} \times 11\frac{5}{8}$	17–19	12–14	150	0.11
$5\frac{5}{8} \times 7\frac{5}{8} \times 11\frac{5}{8}$	26–28	17–19	150	0.12
$7\frac{5}{8} \times 7\frac{5}{8} \times 11\frac{5}{8}$	33–35	21–23	150	0.13
$5\frac{5}{8} \times 3\frac{5}{8} \times 15\frac{5}{8}$	17–19	11–13	225	0.18
$7\frac{5}{8} \times 3\frac{5}{8} \times 15\frac{5}{8}$	22–24	14–16	225	0.19
$3\frac{5}{8} \times 7\frac{5}{8} \times 15\frac{5}{8}$	23–25	16–18	113	0.10
$5\frac{5}{8} \times 7\frac{5}{8} \times 15\frac{5}{8}$	35–37	24–27	113	0.11
$7\frac{5}{8} \times 7\frac{5}{8} \times 15\frac{5}{8}$	45–47	28–32	113	0.12

† These quantities do not include any allowance for waste. Add 2 to 5 percent for blocks and 20 to 50 percent for mortar.

The actual sizes of blocks usually will be $\frac{3}{8}$ in less than the nominal sizes. When $\frac{3}{8}$-in-thick mortar joints are used, the dimensions occupied by blocks will equal the nominal sizes.

The quantities given in Table 10-6 are based on 100 ft² of wall area, with no allowance for waste or breakage of the blocks. The quantities of mortar given do not include any allowance for waste, which frequently will amount to 20 to 50 percent of the net amount required.

Labor Laying Concrete Blocks

Concrete blocks are laid by masons. Joints are made by spreading mortar along the inside and outside horizontal and vertical edges. Joints may be cut smooth with a steel trowel, or they may be tooled as for brick. The joints are more resistant to the infiltration of moisture when they are tooled, because the tooling increases the density of the mortar.

Table 10-7 gives the labor-hours required to lay 1,000 concrete blocks of various sizes. The time given for masons includes laying the blocks and tooling the joints, if required. The rates provide for different classes of work. The time given for laborers includes supplying mortar and blocks for the masons. Add the cost of a hoisting engineer if required.

Example 10-2. Estimate the total direct cost and the cost per 1,000 blocks for furnishing and laying standard-weight concrete blocks, whose nominal sizes are 8 by 8 by 12 in, to back up the brick facing for a wall for a rectangular building 120 ft 0 in

TABLE 10-7
**Labor-hours required to handle
and lay 1,000 concrete blocks†**

Nominal-size block: thickness, height, length, in	Labor-hour per 1,000 blocks	
	Mason	Helper
Standard blocks		
4 × 5 × 12	33–38	33–38
6 × 5 × 12	38–44	38–44
8 × 5 × 12	44–50	44–50
4 × 8 × 12	38–44	38–44
6 × 8 × 12	44–50	44–50
8 × 8 × 12	50–55	50–55
6 × 4 × 16	38–44	38–44
8 × 4 × 16	44–50	44–50
4 × 8 × 16	38–44	38–44
6 × 8 × 16	44–50	44–50
8 × 8 × 16	52–57	52–57
Lightweight blocks		
4 × 5 × 12	30–35	30–35
6 × 5 × 12	35–40	35–40
8 × 5 × 12	40–45	40–45
4 × 8 × 12	35–40	35–40
6 × 8 × 12	40–45	40–45
8 × 8 × 12	45–50	45–50
6 × 4 × 16	35–40	35–40
8 × 4 × 16	40–45	40–45
4 × 8 × 16	35–40	35–40
6 × 8 × 16	40–45	40–45
8 × 8 × 16	47–52	47–52

† Add the cost of a hoisting engineer if required.

long, 80 ft 0 in wide, and 18 ft 0 in high. The wall will be 12 in thick, consisting of a 4-in-thick face of brick and 8-in-thick concrete blocks. The total area of the openings will be 516 ft². Mortar joints will be $\frac{3}{8}$ in thick, using 1:1:6 hydrated lime, portland cement, and sand, respectively.

The quantities are determined as follows:

Gross perimeter of building, 400 lin ft
Net perimeter of concrete block portion of building,
    400 ft 0 in − (8 × 4 in) = 397 ft 4 in
Gross area of concrete block portion of building,

397 ft 4 in × 18 ft 0 in	= 7,152 ft²
Deduct area of openings	= −516 ft²
Deduct area of four corners, 4 × 8 in × 18 ft	= −48 ft²
Net area of concrete portion	= 6,588 ft²

The cost will be:
  Materials:
    Concrete blocks, 6,588 ft² × 1.5 per ft² = 9,882,
      9,882 @ $0.93   = $9,190.26
      Add 2% for waste, 198 @ $0.93   =   184.14
    Mortar required,
      65.88 squares × 0.13 yd³ per square =   8.6 yd³
      Add 40% for waste, 0.4 × 8.6   =   3.4 yd³
        Total quantity   = 12.0 yd³
    Lime, 12.0 yd³ × 225 lb/yd³ = 1.4 tons @ $60.67   =   84.94
    Cement, 12.0 yd³ × 4.5 sacks/yd³ = 54 sacks @ $4.90   =   264.60
    Sand, including waste, 13 yd³ @ $8.55   =   111.15
  Labor:
    Masons, 9,882 blocks × 52 h per 1,000 = 514 h @ $16.20 =   8,326.80
    Helpers, 9,882 blocks × 52 h per 1,000 = 514 h @ $9.45 = 4,857.30
    Foreman, based on 6 masons, 514 ÷ 6 = 86 h @ $18.30   = 1,573.80
    Mixer operator, 86 h @ $12.40   = 1,066.40
  Equipment:
    Mortar mixer, 86 h @ $4.25   =   365.50
    Scaffolds and sundry equipment   =   125.00
      Total cost   = $26,149.89
      Cost per 1,000 blocks, $26,149.89 ÷ 9,882   =   2.65

STONE MASONRY

Several kinds of stone, both natural and artificial, are used in structures such as buildings, walls, piers, etc. Natural stones used for construction include sandstone, limestone, dolomite, slate, granite, marble, etc. Artificial limestone is available in many areas.

Each kind of stone and work should be estimated separately. The cost of stone in place may be estimated by the cubic yard, ton, cubic foot, square foot, or linear foot. Because of the various methods of pricing stonework, an estimator should be very careful to use the correct method in preparing the estimate.

Bonds for Stone Masonry

Figure 10-4 illustrates the more common bonds for stone masonry. Rubble masonry is formed of stones of irregular shapes which are laid in regular courses or at random with mortar joints. Ashlar masonry is formed of stones cut with rectangular faces. The stones may be laid in courses or at random with mortar joints.

Mortar for Stone Masonry

The mortar used for setting stones may be similar to that used for brick masonry. Sometimes special nonstaining white or stone-set cement may be specified instead

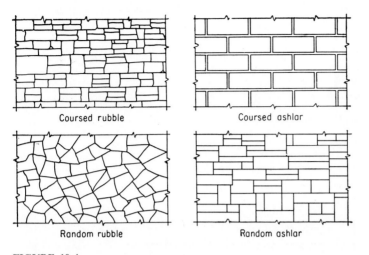

Coursed rubble

Coursed ashlar

Random rubble

Random ashlar

FIGURE 10-4
Bonds for stone masonry.

of gray portland cement. Hydrated lime is usually added to improve the working properties of the mortar.

The quantity of mortar required for joints will vary considerably with the type of bond, the thickness of the joints, and the size of stones used. Table 10-8 gives representative quantities of mortar required per cubic yard of stone.

Weights of Stone

Table 10-9 gives the ranges in weights of stones commonly used for masonry. The given weights are expressed in pounds per net cubic foot of volume.

TABLE 10-8
Quantities of mortar required per cubic yard of stone masonry

Type of bond	Quantity of mortar, ft^3
Coursed rubble	6.5–8.5
Random rubble	7.5–9.5
Cobblestone	6.5–9.5
Coursed ashlar, $\frac{1}{4}$-in joints	1.5–2.0
Random ashlar, $\frac{1}{4}$-in joints	2.0–2.5
Coursed ashlar, $\frac{1}{2}$-in joints	3.0–4.0
Random ashlar, $\frac{1}{2}$-in joints	4.0–5.0

TABLE 10-9
**Weights of building
stones**

Stone	Weight, lb/ft^3
Dolomite	155–175
Granite	165–175
Limestone	150–175
Marble	165–175
Sandstone	140–160
Slate	160–180

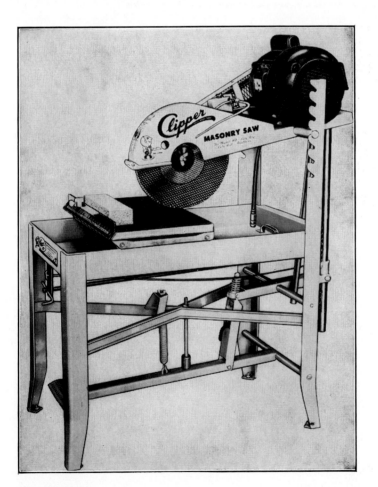

FIGURE 10-5
Masonry saw.

Cost of Stone

The cost of stone varies so much with the kind of stone, extent of cutting done at the quarry, and location of use that no estimate which requires accurate pricing should be made without obtaining current prices for the particular stone. The cost of freight to the destination must be added to the cost at the source to determine the cost at the job.

Stones suitable for rubble masonry may be priced by the ton for the specified kind of stone and sizes of pieces.

Stones may be purchased at a quarry in large rough-cut blocks, hauled to the job, and then cut to the desired sizes and shapes. The cost of such blocks may be based on the volume, with the largest dimensions used in determining the volume; or the cost may be based on the weight of the stone.

Stones used for ashlar masonry may be priced by the ton, cubic foot, or square foot of wall area for a specified thickness.

A cordovan limestone frequently used for random ashlar construction in Texas is available in such sizes as 4 in thick, 2 to 10 in or more in height, in steps of 2 in, and in random lengths. The top and bottom beds and the back sides will be sawed at the quarry, while the end faces may be sawed at the job, if desired. The exposed face is usually left rough. The cost of this stone will vary from $55.00 to $82.00 or more per ton delivered to a job within the delivery area.

This stone weighs about 140 lb/ft^3. Thus 1 ton will produce about 2,000 ÷ 140 = 14.3 ft^3. If the stone is 4 in thick, the area of wall covered per ton will be about 3 × 14.3 = 43 ft^2. With some loss from cutting and breakage, the net area per ton should be about 40 ft^2.

Consider a stone whose weight is 150 lb/ft^3 and whose cost is $82.00 per ton. If this stone is furnished with the top and bottom beds, and the back face sawed for a wall thickness of 4 in, in random lengths, then the cost per cubic foot and per square foot of wall area should be about as given in the accompanying table. The costs include an allowance of 10 percent for waste resulting from breakage and sawing end joints.

Item	Cost per ton	Cost, including 10% waste Per ft^3	Per ft^3
Stone f.o.b. job	$82.00	$6.77	$2.05
Cutting end joints	—	2.46	0.36
Total cost	$82.00	$9.23	$2.41

Labor Setting Stone Masonry

Table 10-10 gives representative values of the labor-hours required to handle and set stone masonry. The lower rates should be used for large straight walls, with

TABLE 10-10
Labor-hours required to handle and set stone masonry

	Labor-hours required			
	Per yd³		Per ft³	
Operation	Skilled	Helper	Skilled	Helper
Unloading stones from truck	0.2–0.4	0.8–1.6	0.01–0.02	0.03–0.06
Rough squaring	4.0–10.0	—	0.15–0.37	
Smoothing beds	6.0–12.0	—	0.22–0.45	
Setting stone by hand:				
Rubble	2.0–4.0	3.0–6.0	0.08–0.15	0.11–0.22
Rough squared	3.0–6.0	4.0–9.0	0.11–0.22	0.15–0.33
Ashlar, 4- to 6-in veneer	10.0–15.0	10.0–15.0	0.37–0.55	0.37–0.55
Setting stones by using a hand derrick:				
Heavy cut stone, ashlar,				
cornices, copings, etc.	2.0–3.0	12.0–18.0	0.08–0.11	0.45–0.67
Light cut stone, ashlar, sills,				
lintels, cornices, etc.	2.5–4.0	12.5–20.0	0.09–0.15	0.46–0.75
Setting stones by using a power crane:				
Heavy cut stone:				
Crane operator	1.0–2.0	—	0.04–0.08	
Stone setter	1.0–2.0	—	0.04–0.08	
Helpers	—	7.0–14.0	—	0.26–0.52
Medium cut stone:				
Crane operator	1.5–3.0	—	0.06–0.11	
Stone setter	1.5–3.0	—	0.06–0.11	
Helpers	—	9.0–18.0	—	0.33–0.67
Pointing cut stone:				
Heavy stone	0.5–0.8	0.2–0.4	0.02–0.03	0.01–0.02
Veneer, 4 to 6 in thick	1.0–1.5	0.2–0.8	0.04–0.06	0.02–0.03
Cleaning stone:				
Stone setter, 1.5 h/100 ft²				
Helper, 0.75 h/100 ft²				

plain surfaces, and the higher rates for walls with irregular surfaces, pilasters, closely spaced openings, etc. The rates for hoisting will vary with the size of stones, heights hoisted, and equipment used.

Example 10-3. Estimate the cost per 100 ft² for furnishing and setting stone in random ashlar bond for a building. The stone will be sawed on the top and bottom beds and on the front and back faces. It will be furnished in random lengths to be sawed to the desired lengths at the job. The thickness will be 4 in. The stone will be set with $\frac{1}{2}$-in mortar joints. The stone will be hoisted with a hand derrick.

The stone, which will weigh 150 lb/ft³, will be delivered to the job at a cost of $82.00 per ton. Allow 10 percent for waste.

The quantity of stone required for 100 ft^2 will be

$$\frac{100 \times 4 \times 1.1}{12} = 36.7 \text{ ft}^3$$

The weight of the stone will be $36.7 \times 150/2,000 = 2.76$ tons

The cost will be:

Stone, f.o.b. job, 2.76 tons @ $82.00	= $	226.32
Cutting end joints, 36.7 ft^3 @ $2.46	=	90.28
Mortar, 36.7 ÷ 27 = 1.36 yd^3 of stone,		
1.36 yd^3 × 4.5 ft^3/yd^3 = 6.1 ft^3 @ $2.41	=	14.70
Metal wall ties, 100 ft^2 ÷ 2 ft^2 per tie = 50 @ $0.86 =		43.00
Stone setters, 36.7 ft^3,		
36.7 ft^3 × 0.12 h/ft^3 = 4.4 h @ $17.60	=	77.44
Helpers, 36.7 ft^3,		
36.7 ft^3 × 0.6 h/ft^3 = 22 h @ $12.40	=	272.80
Stone setter pointing stone, 36.7 ft^3,		
36.7 ft^3 × 0.05 h/ft^3 = 1.8 h @ $17.60	=	31.68
Helper pointing stone, 36.7 ft^3,		
36.7 ft^3 × 0.03 h/ft^3 = 1.1 h @ $12.40	=	13.64
Stone setter cleaning stone, 100 ft^2,		
100 ft^2 × 1.5 h/100 ft^2 = 1.5 h @ $17.60	=	26.40
Helper cleaning stone, 100 ft^2,		
100 ft^2 × 0.75 h per 100 ft^2 = 0.75 h @ $12.40	=	9.30
Foreman, based on 2 stone setters, 3.9 h @ $18.30	=	71.37
Equipment, saws, derrick, scaffolds, etc.	=	250.00
Total cost	=	$1,126.93
Cost per ft^2, $1,126.93 ÷ 100	=	11.27

PROBLEMS

10-1. Estimate the total direct cost and the cost per 1,000 bricks for constructing a brick wall for a rectangular building. The outside dimensions will be 56 ft 6 in wide and 108 ft 0 in long, with walls 12 in thick and 15 ft 6 in high. The total area of the openings will be 324 ft^2. Common brick, laid in common bond, will be used, with $\frac{1}{2}$-in-thick cut mortar joints.

The mortar, which will be made of hydrated lime, portland cement, and sand, in the proportions 1:1:6, respectively, by volume, will be mixed in a $3\frac{1}{2}$-ft^3 mortar mixer.

Six bricklayers will be used on the job. Use the wage rates for the bricklayers and helpers (laborers) given in this book, with the foreman to be paid $1.75 per hour more than the bricklayers.

The material costs will be as follows:

Bricks, f.o.b. job, per 1,000, $180.00
Lime, per ton, $82.50
Cement, per sack, $4.90
Sand, per yd^3, $9.20
Scaffolds for the job, $425.00

10-2. Estimate the total cost and the cost per 1,000 bricks for Prob. 10-1 for bid purposes, using current local prices for materials and labor. Include the cost of a performance bond.

10-3. Estimate the total direct cost of constructing the wall of Prob. 10-1, using 8- by 8- by 12-in nominal-size standard-weight concrete blocks on the inside and a 4-in face-brick veneer on the outside. Use the same mortar as for Prob. 10-1. The blocks will be laid with $\frac{3}{8}$-in joints and the brick with $\frac{1}{2}$-in concave joints. The bricks are to be laid with first-class workmanship, in a common bond. Use one corrugated-metal wall tie for each 4 ft^2 of wall area.

Six bricklayers will be used on the wall. Use the same wage rates as for Prob. 10-1.

Estimate the costs of the blocks and bricks separately; then combine the costs for the total.

Material costs will be as follows:

Bricks, $230.00 per 1,000
Concrete blocks, $0.85 each
Lime, $60.67 per ton
Cement, $4.90 per sack
Sand, $8.55 per yd^3
Wall ties, $32.50 per 1,000
Scaffolds, $450.00 for job

10-4. Estimate the total cost of the wall for Prob. 10-3 for bid purposes, using current local prices for materials and labor. Include the cost of a performance bond.

10-5. Estimate the total direct cost and the cost per square foot for furnishing and laying cordovan limestone 4 in thick in random ashlar for the walls for a building whose net surface area will be 7,460 ft^2. The stones will be hoisted with a hand derrick.

Use the same prices for materials and labor as in Example 10-3.

Use $\frac{1}{4}$-in tooled mortar joints, with the mortar to be the same as for Prob. 10-1. Use one metal wall tie for each 4 ft^2 of wall area.

10-6. Estimate the total cost of the stone for Prob. 10-5 for bid purposes, including the cost of a bid bond. Use the same prices for materials and labor as for Prob. 10-5.

CHAPTER
11

CARPENTRY

This chapter will deal with the furnishing of materials, work, and equipment to construct houses, warehouses, shop buildings, roof trusses, etc., which are constructed primarily with lumber and timber.

The production rates given are based on using power equipment to fabricate the lumber when the use of such equipment is practical. Many of the tables for production give two rates, one for work requiring ordinary workmanship and the other for work requiring first-grade workmanship.

Lumber Sizes

The sizes of most lumber, with the exception of certain specialties such as moldings, are designated by the dimensions of the cross sections, using the nominal dimensions, which are the dimensions prior to finishing the lumber. After lumber is sawed lengthwise, it may be passed through one or more finishing operations, which will reduce its actual size to less than the identifying dimensions. Thus a 2- by 4-in plank will actually be $1\frac{1}{2}$ in thick and $3\frac{1}{2}$ in wide after it is surfaced on four sides. This lumber is designated as 2- by 4-in S4S.

If a 1- by 6-in plank is dressed and matched, with a tongue and groove, the actual dimensions will be $\frac{3}{4}$ in thick by $5\frac{1}{8}$-in net width. This lumber is designated as 1- by 6-in D and M and is often identified by the name *center match*.

Because the finishing and dressing operations will result in a reduction in the actual widths of the lumber, a quantity estimate must include an allowance for this side wastage or shrinkage. For a plank nominally 6 in wide, the net width will be $5\frac{1}{2}$ in. Thus the side shrinkage will be $\frac{1}{2}$ in. The side shrinkage for quantity purposes will be $(0.5 \times 100)/5.5 = 9.1$ percent. To provide enough lumber for a required area, the quantity must be increased 9.1 percent to replace side shrinkage only.

Lumber is available in lengths which are multiples of 2 ft, with the actual lengths not less than the designated lengths. If other than standard lengths are needed, they must be cut from standard lengths. This operation may result in end waste. For example, if a 1- by 6-in D and M plank 10 ft 6 in long is needed, it must be cut from a plank that is 12 ft 0 in long with a 1-ft 6-in end waste, unless the piece cut off can be used elsewhere. This will represent an end waste of $(1.5 \times 100)/10.5 = 14.3$ percent.

The combined side and end wastes for the cited examples will be

$$
\begin{aligned}
\text{Side waste} &= 9.1\,\% \\
\text{End waste} &= \underline{14.3\,\%} \\
\text{Total waste} &= 23.4\,\%
\end{aligned}
$$

Note that the side and end wastes are determined on the basis of the dimensions of lumber required instead of the nominal dimensions.

Table 11-1 lists the commonly used sizes of lumber, with the nominal and actual cross-sectional dimensions. Lumber that is less than 1 in thick is considered to be 1 in thick for quantity purposes. In the table the first dimension is the thickness and the second dimension is the width.

Grades of Lumber

Lumber is usually specified by the type (such as pine, fir, oak, etc.) size, and grade. Pine and fir are most commonly used for structural members, while oak, maple, redwood, etc., are used for floors, siding, and special purposes.

Pine and fir are graded according to quality as follows:

Select grades: A, B, C, and D
Common grades: 1, 2, 3, and 4

Select grades are used for trim, facings, moldings, etc., which may be finished in natural color or painted. Grade A is the best quality, and grade D is the poorest. Common grades are used for structural members such as sills, joist, studs, rafters, planks, etc. Grade 1 is the best quality, and grade 4 is the poorest. The lumber used for timber stuctures is generally designated as stress grade.

TABLE 11-1
Dimensions and properties of lumber

Nominal size, in	Standard dressed size, in	Area of section, in^2	Moment of inertia I	Section modulus S
		Boards		
1 × 3	$\frac{3}{4} \times 2\frac{1}{2}$	1.875	0.977	0.781
1 × 4	$\frac{3}{4} \times 3\frac{1}{2}$	2.625	2.680	1.531
1 × 6	$\frac{3}{4} \times 5\frac{1}{2}$	4.125	10.398	3.781
1 × 8	$\frac{3}{4} \times 7\frac{1}{4}$	5.438	23.817	6.570
1 × 10	$\frac{3}{4} \times 9\frac{1}{4}$	6.938	49.446	10.695
1 × 12	$\frac{3}{4} \times 11\frac{1}{4}$	8.438	88.989	15.820
		Dimension		
2 × 3	$1\frac{1}{2} \times 2\frac{1}{2}$	3.750	1.953	1.563
2 × 4	$1\frac{1}{2} \times 3\frac{1}{2}$	5.250	5.359	3.063
2 × 6	$1\frac{1}{2} \times 5\frac{1}{2}$	8.250	20.797	7.563
2 × 8	$1\frac{1}{2} \times 7\frac{1}{4}$	10.875	47.635	13.141
2 × 10	$1\frac{1}{2} \times 9\frac{1}{4}$	13.875	98.932	21.391
2 × 12	$1\frac{1}{2} \times 11\frac{1}{4}$	16.875	177.979	31.641
2 × 14	$1\frac{1}{2} \times 13\frac{1}{4}$	19.875	290.775	43.891
3 × 4	$2\frac{1}{2} \times 3\frac{1}{2}$	8.750	8.932	5.104
3 × 6	$2\frac{1}{2} \times 5\frac{1}{2}$	13.750	34.661	12.604
3 × 8	$2\frac{1}{2} \times 7\frac{1}{4}$	18.125	79.391	21.901
3 × 10	$2\frac{1}{2} \times 9\frac{1}{4}$	23.125	164.886	35.651
3 × 12	$2\frac{1}{2} \times 11\frac{1}{4}$	28.125	296.631	52.734
3 × 14	$2\frac{1}{2} \times 13\frac{1}{4}$	33.125	484.625	73.151
3 × 16	$2\frac{1}{2} \times 15\frac{1}{4}$	38.125	738.870	96.901
4 × 4	$3\frac{1}{2} \times 3\frac{1}{2}$	12.250	12.505	7.146
4 × 6	$3\frac{1}{2} \times 5\frac{1}{2}$	19.250	48.526	17.646
4 × 8	$3\frac{1}{2} \times 7\frac{1}{4}$	25.375	111.148	30.661
4 × 10	$3\frac{1}{2} \times 9\frac{1}{4}$	32.375	230.840	49.911
4 × 12	$3\frac{1}{2} \times 11\frac{1}{4}$	39.375	415.283	73.828
4 × 14	$3\frac{1}{2} \times 13\frac{1}{4}$	46.375	678.475	102.411
4 × 16	$3\frac{1}{2} \times 15\frac{1}{4}$	53.375	1,034.418	135.660
6 × 6	$5\frac{1}{2} \times 5\frac{1}{2}$	30.250	76.255	27.729
6 × 8	$5\frac{1}{2} \times 7\frac{1}{2}$	41.250	193.359	51.563
6 × 10	$5\frac{1}{2} \times 9\frac{1}{2}$	52.250	392.963	82.729
6 × 12	$5\frac{1}{2} \times 11\frac{1}{2}$	63.250	697.068	121.229
6 × 14	$5\frac{1}{2} \times 13\frac{1}{2}$	74.250	1,127.672	167.063
6 × 16	$5\frac{1}{2} \times 15\frac{1}{2}$	85.250	1,706.776	220.229
6 × 18	$5\frac{1}{2} \times 17\frac{1}{2}$	96.250	2,456.380	280.729

Dressed and matched†		Flooring		Shiplap	
S2S and CM		1 × 2	$\frac{3}{4} \times 1\frac{1}{8}$	1 × 4	$\frac{3}{4} \times 3\frac{1}{8}$
1 × 4	$\frac{3}{4} \times 3\frac{1}{8}$	1 × 3	$\frac{3}{4} \times 2\frac{1}{8}$	1 × 6	$\frac{3}{4} \times 5\frac{1}{8}$
1 × 6	$\frac{3}{4} \times 5\frac{1}{8}$	1 × 4	$\frac{3}{4} \times 3\frac{1}{8}$	1 × 8	$\frac{3}{4} \times 6\frac{7}{8}$
1 × 8	$\frac{3}{4} \times 6\frac{7}{8}$	1 × 6	$\frac{3}{4} \times 5\frac{1}{8}$	1 × 10	$\frac{3}{4} \times 8\frac{7}{8}$
1 × 10	$\frac{3}{4} \times 8\frac{7}{8}$			1 × 12	$\frac{3}{4} \times 10\frac{7}{8}$
1 × 12	$\frac{3}{4} \times 10\frac{7}{8}$				

† This lumber is also available in other thicknesses.

Cost of Lumber

Lumber is usually priced by the 1,000 fbm, with the exception of certain types used for trim, moldings, etc., which may be priced by the linear foot. A *board foot* is a piece of lumber whose nominal dimensions are 1 in thick, 12 in wide, and 1 ft long, or it may be 2 in thick, 6 in wide, and 1 ft long. The number of feet board measure in a plank 2 by 8 in by 18 ft 0 in long is $(2 \times 8 \times 18)/12 = 24$ fbm. This same operation should be applied to determine the number of feet board measure in any piece of lumber.

The cost of lumber varies with many factors, including the following:

1. Kind of lumber: pine, fir, oak, etc.
2. Grade of lumber: select or common
3. Size of pieces: thickness, width, and length
4. Extent of milling required: rough, surfaced, dressed and matched
5. Whether dried or green
6. Quantity purchased
7. Freight cost from mill to destination

The better grades of lumber cost more than the poorer grades. Select grades are more expensive than common grades. In the select grades, the costs decrease in the order A, B, C, and D, while in the common grades the costs decrease in the order 1, 2, 3, and 4.

The cost of lumber is higher for thick and wide planks than for thin and narrow ones. Lengths in excess of about 20 to 24 ft may require special orders. Prices quoted may be for lengths through 20 ft, with increasingly higher prices charged for greater lengths.

Rough lumber is the product of the sawing operation, with no further surfacing operation. After the lumber is sawed, it may be given a smooth surface on one or more sides and edges, in which case it is designated *surfaced one edge* S1E, *surfaced two sides*, S2S, or *surfaced on all edges and sides*, S4S. It may be surfaced on both sides with the edges matched with tongues and grooves, designated as *dressed and matched*, *D and M*. These operations reduce the thickness and width and increase the cost.

Freshly cut lumber contains considerable moisture which should be removed before the lumber is used, to prevent excessive shrinkage after it is placed in the structure. It usually is placed in a heated kiln to expel the excess moisture.

Nails and Spikes

The trade names, sizes, and weights of the more popular nails and spikes are given in Table 11-2.

Bolts

The bolts commonly used as connectors for timber structures are available in sizes varying from $\frac{1}{4}$ in to larger than 1 in and in lengths varying from 1 in to any desired length. The length does not include the head. Bolts with square heads and either square or hexagonal nuts are used. Two washers should be used with every bolt, one for the head and one for the nut. If the head and the nut bear against steel plates, it is satisfactory to use stamped-steel washers; but if there are no steel plates under the heads and nuts, cast-iron ogee washers should be used to increase the bearing area of the wood.

TABLE 11-2
Names, sizes, and numbers of nails and spikes per pound

								Roofing		
Size	Length, in	Common	Spikes	Casing	Finishing	Shingle	Barbed	No. 8	No. 9	No. 10
	$\frac{3}{4}$	—	—	—	—	—	714	205	252	290
	$\frac{7}{8}$	—	—	—	—	—	469	179	219	253
2d	1	876	—	1,010	1,351	—	411	158	193	224
3d	$1\frac{1}{4}$	568	—	635	807	429	251	128	156	183
4d	$1\frac{1}{2}$	316	—	473	584	274	176	108	131	154
5d	$1\frac{3}{4}$	271	—	406	500	235	151	93	113	133
6d	2	181	—	236	309	204	103			
7d	$2\frac{1}{4}$	161	—	210	238					
8d	$2\frac{1}{2}$	106	—	145	189					
9d	$2\frac{3}{4}$	96	—	132	172					
10d	3	69	41	94	121					
12d	$3\frac{1}{4}$	63	38	87	113					
16d	$3\frac{1}{2}$	49	30	71	90					
20d	4	31	23	52	62					
30d	$4\frac{1}{2}$	24	17	46						
40d	5	18	13	35						
50d	$5\frac{1}{2}$	14	10							
60d	6	11	9							
	7	—	7							
	8	—	4							
	9	—	$3\frac{1}{2}$							
	10	—	3							
	12	—	$2\frac{1}{2}$							

TABLE 11-3

Approximate pounds of nails required for carpentry (includes 10 percent waste)

Size and type lumber	No. nails per support	Size and kind of nails	Pounds per 1,000 fbm of lumber for nail spacings†					
			12 in	16 in	20 in	24 in	36 in	48 in
1 × 4	2	8d common	63	47	38	31	21	16
1 × 5	2	8d common	49	37	30	25	17	12
1 × 6	2	8d common	41	31	25	21	14	10
1 × 8	2	8d common	32	24	19	16	11	8
1 × 10	3	8d common	37	28	22	19	12	10
1 × 12	3	8d common	31	23	18	15	10	8
2 × 4	2	20d common	110	83	66	55	37	28
2 × 6	2	20d common	72	54	42	36	24	18
2 × 8	2	20d common	56	42	34	28	19	14
2 × 10	3	20d common	65	49	40	33	22	17
2 × 12	3	20d common	56	42	34	28	19	14
3 × 4	2	60d common	204	153	122	102	68	51
3 × 6	2	60d common	136	102	82	68	45	34
3 × 8	2	60d common	101	76	61	51	34	26
3 × 10	3	60d common	124	93	74	62	41	31
3 × 12	3	60d common	104	78	62	52	35	26
Framing, studs, etc.:								
2 × 4	—	16d common	15	11	9	7	5	4
2 × 6	—	16d common	11	8	6	5	4	3
2 × 8	—	16d common	11	8	7	6	4	3
Siding:								
Bevel, $\frac{1}{2}$ × 4	1	6d finish	12	9				
Bevel, $\frac{1}{2}$ × 6	1	6d finish	8	6				
Bevel, $\frac{1}{2}$ × 8	1	6d finish	6	$4\frac{1}{2}$				
Drop, 1 × 4	2	8d casing	51	38				
Drop, 1 × 6	2	8d casing	33	25				
Drop, 1 × 8	2	8d casing	27	20				
Flooring, D and M softwood:								
1 × 3	1	8d brads	51	38				
1 × 4	1	8d brads	37	28				
1 × 6	1	8d brads	25	19				

		lb/1,000 fbm						
Joists:								
2 × 6	—	16d common	17					
2 × 8	—	16d common	10					
2 × 10	—	16d common	8					
2 × 12	—	16d common	7					
Bridging, 1 × 4	4 ea.	8d common	83					
Rafters:								
2 × 4	—	16d common	18					
2 × 6	—	16d common	18					
2 × 8	—	16d common	15					
2 × 10	—	16d common	12					

TABLE 11-3
Approximate pounds of nails required for carpentry (includes 10 percent waste) (continued)

Size and type lumber	No. nails per support	Size and kind of nails	Pounds per 1,000 fbm of lumber for nail spacings†					
			12 in	16 in	20 in	24 in	36 in	48 in
		lb/100 ft²						
Flooring, D and M hardwood:								
$\frac{3}{8} \times 2$	—	3d finish	1.1					
$\frac{3}{16} \times 2\frac{1}{4}$	—	3d finish	0.9					
$\frac{13}{16} \times 2$	—	8d cut	6.0					
$\frac{13}{16} \times 2\frac{1}{4}$	—	8d cut	5.5					
$\frac{13}{16} \times 3\frac{1}{4}$	—	8d cut	4.0					
Plasterboard	—	3d barbed	2.5					
Wood shingles:								
$4\frac{1}{2}$-in exposed	2	3d shingle	4.5					
5-in exposed	2	3d shingle	4.1					
$5\frac{1}{2}$-in exposed	2	3d shingle	3.8					
Asphalt shingles	2 ea.	$\frac{7}{8}$ in barbed	4.2					

† If different sizes of common nails are used, make the following changes: 10d for 8d, increase weight 53 percent; 20d for 16d, increase weight 58 percent.

Timber Connectors

A factor which has contributed materially to the design and construction of timber structures was the introduction of modern timber connectors. These connectors are available through the Timber Engineering Company of Washington, D.C. They are classified under a general term as *Teco connectors*. They include split rings, toothed rings, shear plates, spike grids, and clamping plates. These connectors are illustrated in Fig. 11-1.

Split rings   Split rings, in sizes with $2\frac{1}{2}$- and 4-in diameters, are for use with wood-to-wood connections for medium and heavy loads. It is necessary to drill a hole for the bolt and a groove for the ring. A complete connector unit includes a split ring, bolt, and two steel washers.

Toothed rings   Toothed rings, with 2-, $2\frac{5}{8}$-, $3\frac{3}{8}$-, and 4-in diameters, are for use with wood-to-wood connections in thin members with light loads. A complete unit includes a toothed ring, bolt, and two steel washers.

Shear plates   Flanged shear plates of $2\frac{5}{8}$-in diameter pressed steel and 4-in-diameter malleable iron are for wood-to-wood or wood-to-steel connections. They are placed in precut grooves, or daps, in the timber members. A complete unit includes one or two plates, a bolt, and two steel washers.

Spike grids   Spike grids, of malleable cast iron, are designed for use with piles, poles, and flat timber members. The flat grid joins two flat timber

TABLE 11-4

Weights of bolts with square heads and hexagonal nuts in pounds per 100

Length under head, in	\frac{1}{4}	\frac{5}{16}	\frac{3}{8}	\frac{7}{16}	\frac{1}{2}	\frac{5}{8}	\frac{3}{4}	\frac{7}{8}	1	$1\frac{1}{8}$	$1\frac{1}{4}$
					Diameter of bolt, in						
1	2.7	5.0	7.2	11.2	14.9	28	43	—	—	—	—
$1\frac{1}{4}$	3.1	5.5	8.0	12.2	16.3	30	46	68	–	–	–
$1\frac{1}{2}$	3.4	6.1	8.8	13.3	17.7	32	49	73	103	144	190
$1\frac{3}{4}$	3.8	6.6	9.6	14.4	19.0	35	52	77	109	151	199
2	4.1	7.2	10.4	15.4	20.4	37	55	81	115	158	208
$2\frac{1}{4}$	4.5	7.7	11.1	16.5	21.8	39	58	85	120	165	216
$2\frac{1}{2}$	4.8	8.2	11.9	17.5	23.2	41	61	90	126	172	225
$2\frac{3}{4}$	5.2	8.8	12.7	18.6	24.6	43	64	94	131	179	234
3	5.5	9.3	13.5	19.7	26.0	45	68	98	137	187	242
$3\frac{1}{4}$	5.9	9.9	14.3	20.7	27.4	48	71	102	142	194	251
$3\frac{1}{2}$	6.2	10.4	15.1	21.8	28.8	50	74	107	148	201	260
$3\frac{3}{4}$	6.6	11.0	15.8	22.9	30.2	52	77	111	153	208	268
4	6.9	11.5	16.6	23.9	31.6	54	80	115	159	215	277
$4\frac{1}{4}$	7.3	12.0	17.4	25.0	33.0	56	83	119	165	222	286
$4\frac{1}{2}$	7.6	12.6	18.2	26.1	34.4	58	86	124	170	229	294
$4\frac{3}{4}$	8.0	13.1	19.0	27.1	35.7	61	89	128	176	236	303
5	8.3	13.7	19.8	28.2	37.1	63	93	132	181	243	312
$5\frac{1}{4}$	8.6	14.2	20.5	29.3	38.5	65	96	136	187	250	321
$5\frac{1}{2}$	9.0	14.8	21.3	30.3	39.9	67	99	141	192	257	329
$5\frac{3}{4}$	9.3	15.3	22.1	31.4	41.3	69	102	145	198	264	338
6	9.7	15.9	22.9	32.4	42.7	71	105	149	204	271	347
$6\frac{1}{4}$	10.0	16.4	23.7	33.5	44.1	74	108	153	209	278	355
$6\frac{1}{2}$	10.4	16.9	24.5	34.6	45.5	76	111	158	215	285	364
$6\frac{3}{4}$	10.7	17.5	25.2	35.6	46.9	78	114	162	220	292	373
7	11.1	18.0	26.0	36.7	48.3	80	118	166	226	299	381
$7\frac{1}{4}$	11.4	18.6	26.8	37.8	49.7	82	121	170	231	306	390
$7\frac{1}{2}$	11.8	19.1	27.6	38.8	51.1	84	124	175	237	313	399
$7\frac{3}{4}$	12.1	19.7	28.4	39.9	52.4	87	127	179	242	320	407
8	12.5	20.2	29.2	41.0	53.8	89	130	183	248	327	416
$8\frac{1}{2}$	—	21.3	30.7	43.1	56.6	93	136	192	259	341	434
9	—	22.4	32.3	45.2	59.4	98	143	200	270	356	451
$9\frac{1}{2}$	—	23.5	33.9	47.4	62.2	102	149	209	281	370	468
10	—	24.6	35.4	49.5	65.0	106	155	217	293	384	486
$10\frac{1}{2}$	—	—	37.0	51.6	67.8	111	161	226	304	398	503
11	—	—	38.6	53.7	70.5	115	168	234	315	412	520
$11\frac{1}{2}$	—	—	40.1	55.9	73.3	119	174	243	326	426	538
12	—	—	41.7	58.0	76.1	124	180	251	337	440	555
$12\frac{1}{2}$	—	—	—	60.1	78.9	128	186	260	348	454	573
13	—	—	—	62.3	81.7	132	193	268	359	468	590
$13\frac{1}{2}$	—	—	—	64.4	84.5	137	199	277	370	482	607
14	—	—	—	66.5	87.2	141	205	285	382	496	625
$14\frac{1}{2}$	—	—	—	—	90.0	145	211	294	393	510	642
15	—	—	—	—	92.8	150	218	302	404	525	660
$15\frac{1}{2}$	—	—	—	—	95.6	154	224	311	415	539	• 677
16	—	—	—	—	98.4	158	230	320	426	553	694
Per inch additional	1.4	2.2	3.1	4.3	5.6	8.7	12.5	17.0	22.3	28.2	34.8

TABLE 11-5
Data for cast-iron ogee washers

Diam. of bolt, in	Diam. of washer, in	Thickness of washer, in	Weight per 100, lb
$\frac{3}{8}$	$2\frac{3}{16}$	$\frac{1}{2}$	31.1
$\frac{1}{2}$	$2\frac{3}{8}$	$\frac{1}{2}$	37.5
$\frac{5}{8}$	$2\frac{3}{4}$	$\frac{5}{8}$	56.0
$\frac{3}{4}$	3	$\frac{11}{16}$	75.0
$\frac{7}{8}$	$3\frac{1}{2}$	$\frac{3}{4}$	106.0
1	$3\frac{3}{4}$	$\frac{7}{8}$	150.0
$1\frac{1}{4}$	$5\frac{3}{4}$	$1\frac{3}{16}$	450.0
$1\frac{1}{2}$	$6\frac{1}{8}$	$1\frac{1}{2}$	662.0

members; the single curve joins a curved pile and a flat surface. A complete unit includes a grid, a bolt, and two washers.

Clamping plates   Clamping plates are used as railroad-tie spacers by placing them between the ties and the guard timbers.

Installing Toothed Rings and Spike Grids

Toothed rings and spike grids are installed by using a high-strength threaded rod assembly in the bolthole. Pressure exerted by the assembly will force the connectors to penetrate the wood members. After the wood members are compressed sufficiently, the assembly is removed and replaced with a machine bolt of the proper size.

Table 11-6 gives the weights and dimensions of timber connectors.

Split Ring

Toothed Ring

Shear Plate

Shear Plate

Flat Grid

Single Curved Grid

Flat Clamping Plate

Flanged Clamping Plate

FIGURE 11-1
Timber connectors.

TABLE 11-6
Weights and dimensions of timber connectors

Timber connector	Size, in	Dimensions of metal, in		Shipping weight per 100, pound
		Depth	Thickness	
Split rings	$2\frac{1}{2}$	0.75	0.163	28
	4	1.00	0.193	70
Toothed rings	2	0.94	0.061	9
	$2\frac{5}{8}$	0.94	0.061	12
	$3\frac{3}{8}$	0.94	0.061	15
	4	0.94	0.061	18
Shear plates:				
Pressed steel	$2\frac{5}{8}$	0.375	0.169	35
Malleable iron	4	0.62	0.20	90
Spike grids:				
Flat	$4\frac{1}{8} \times 4\frac{1}{8}$	1.00	—	48
Single curve	$4\frac{1}{8} \times 4\frac{1}{8}$	1.38	—	70
Circular	$3\frac{1}{8}$ diam.	—	—	26
Clamping plates:				
Plain	$5\frac{1}{4} \times 5\frac{1}{4}$	—	0.077	59
Flanged	$5 \times 8\frac{1}{2}$	2.00	0.122	190

Fabricating Lumber

Fabricating lumber includes such operations as sawing, ripping, chamfering edges, boring holes, etc. The rates at which these operations can be performed will vary greatly with several factors, including the following:

1. Extent to which power equipment is used
2. Amount of fabrication required
3. Length of finished pieces
4. Size of pieces
5. Number of similar pieces fabricated with one machine setting
6. Care in sorting and storing lumber at the job

Lumber can be sawed 5 to 10 times as fast with a power saw as with a handsaw, provided that the operations are not too complicated.

It requires considerably more time to fabricate rafters and stair stringers than it does to fabricate studs and joists.

Since a certain amount of time is required to handle and measure lumber, more time is required to fabricate a given quantity of lumber consisting of short lengths and small sections than when it consists of long lengths and large sections.

FIGURE 11-2
Tools used to prepare lumber.

Time is required to set a machine for a given operation. If only a few pieces are fabricated, after which the machine must be reset, the total time per piece will be higher than when a great many similar pieces are fabricated. This may be illustrated by analyzing several operations.

Example 11-1. Assume that you are to cut to length, with square corners, 60 pieces of 2- by 12-in by 19-ft 6-in-long lumber for floor joists, using a stationary cutoff saw with a long table. The saw blade will be set to the correct position, and a stop will be set at one end of the table to eliminate measurements. One helper will supply stock lumber, a mechanic will saw the lumber, and another helper will remove and stack the fabricated pieces.

The time required should be about as follows:

Operation	Units	Unit time, min	Total time, min
Set saw and stop	1	10	10
Saw lumber	60	0.75	45
Total time			55

$$\text{Quantity of lumber} = \frac{60 \times 2 \times 12 \times 19.5}{12} = 2{,}340 \text{ fbm}$$

$$\text{Time per 1,000 fbm} = \frac{55}{60 \times 2.34} = 0.39 \text{ h}$$

If the same saw is used to fabricate 120 studs 2 by 4 in by 8 ft 0 in long, requiring a square cut at each end, the time should be about as follows:

Operation	Units	Unit time, min	Total time, min
Set saw and stop	1	10	10
Saw lumber	120	0.33	40
Total time			50

$$\text{Quantity of lumber} = \frac{120 \times 2 \times 4 \times 8}{12} = 640 \text{ fbm}$$

$$\text{Time per 1,000 fbm} = \frac{50}{60 \times 0.64} = 1.31 \text{ h}$$

If this saw is used to fabricate 100 pieces of cross bridging 1 by 4 in by 18 in long, requiring a bevel cut at each end, the time should be about as follows:

Operation	Units	Unit time, min	Total time, min
Set saw and stop	1	10	10
Saw lumber	100	0.2	20
Total time			30

$$\text{Quantity of lumber} = \frac{100 \times 1 \times 4 \times 1.5}{12} = 50 \text{ fbm}$$

$$\text{Time per 1,000 fbm} = \frac{30}{60 \times 0.05} = 10 \text{ h}$$

These examples illustrate the necessity of analyzing each operation to determine the probable time required. Simply assuming that work will be done at some given rate per unit of lumber, regardless of the size and length of lumber and the kind of operation, is not sufficiently accurate for estimating purposes.

ROUGH CARPENTRY HOUSES

House Framing

Houses may be constructed by using a balloon, braced, or platform frame. Figure 11-3 shows typical wall and partition sections for balloon and platform frames. For the balloon frame, the studs for the outside walls of a two-story building extend

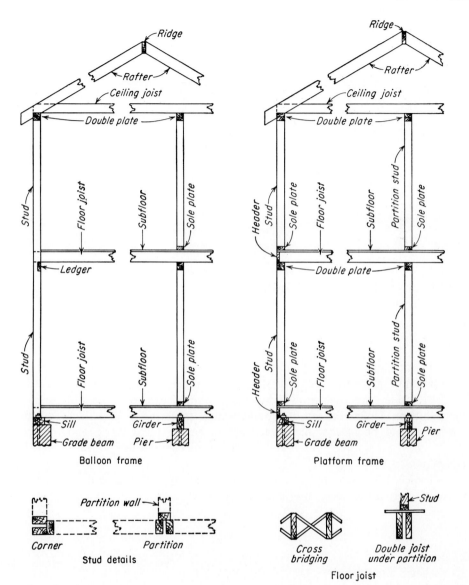

FIGURE 11-3
Typical wall section through a frame house.

TABLE 11-7
**Labor-hours required per
1,000 fbm of sills**

| Size sill, in | Labor-hours | |
	Carpenter	Helper
2 × 4	24–30	8–10
2 × 6	18–22	6–8
2 × 8	14–18	5–6
4 × 6	15–20	5–7
4 × 8	14–19	5–7
6 × 8	13–18	4–6

from the sill to the top plate on which the rafters rest. For the platform frame, the studs for the external walls are limited to one story in length. The lengths of studs for interior or partition walls for both types of frames are limited to one story.

Sills

Sills are attached to masonry-grade beams, usually with anchor bolts set in the masonry at 6- to 8-ft intervals. The sills, which may be 2- by 4-in, 2- by 6-in, 2- by 8-in, 4- by 6-in, or 6- by 8-in S4S lumber, are bored to permit the bolts to pass through them. Unless the top of a grade beam is smooth and level, a layer of stiff mortar should be placed under each sill to ensure full bearing and uniform elevation. At the corners and at end joints, the sills should be connected by half-lapped joints.

Table 11-7 gives the labor-hours required to fabricate and install sills, including boring the holes and tightening the nuts.

TABLE 11-8
**Labor-hours required per
1,000 fbm of floor girders**

| Size girder, in | Labor-hours | |
	Carpenter	Helper
4 × 6	15–20	5–7
6 × 8	13–18	4–6
8 × 10	11–16	4–6
3-2 × 8	14–18	5–6
3-2 × 10	12–15	4–5

TABLE 11-9
Labor-hours required per 1,000 fbm of joists

Size joists, in	Labor-hours	
	Carpenter	Helper
2 × 6	11–13	4–5
2 × 8	9–12	4–5
2 × 10	8–11	4–5
2 × 12	7–10	4–5
Cross bridging, per 100 sets	7–9	1–2

Floor Girders

Floor girders are installed under the first-floor joists to provide intermediate support for the joists. The girders may consist of single pieces such as 4- by 6-in, 6- by 8-in, etc., lumber, or they may be made by securely nailing together several pieces of lumber 2 in thick. They are supported by piers usually spaced 8 to 10 ft apart. End joints should be constructed with half-laps, located above piers.

Floor and Ceiling Joists

Floor and ceiling joists are generally spaced 12, 16, or 24 in on centers, using 2- by 6-in, 2- by 8-in, 2- by 10-in, or 2- by 12-in lumber, rough or S4S. When the unsupported lengths exceed 8 ft, one row of cross bridging should be installed at spacings not greater than 8 ft. The bridging may be made from 1- by 3-in or 1- by 4-in lumber, with two 8d nails at each end of each strut.

TABLE 11-10
Labor-hours required per 1,000 fbm for studs with one bottom and two top plates

Size studs, in	Stud spacing, in	Labor-hours	
		Carpenter	Helper
2 × 4	12	20–25	5–6
	16	22–26	5–6
	24	24–28	5–6
2 × 6	12	19–24	5–6
	16	21–25	5–6
	24	23–27	5–6

Studs

Studs for one- and two-story houses usually are 2- by 4-in lumber. Studs for walls, which must provide passages for plumbing pipes, should be 2- by 6-in lumber. The studs may be spaced 12, 16, 20, or 24 in on centers, with the 16-in spacing used for houses whose walls will be plastered, since plaster laths are furnished in 48-in lengths.

At all corners and where partition walls frame into other walls, it is common practice to install three studs, as illustrated in Fig. 11-4. At all openings for windows and doors, two studs should be used, with one stud on each side of the opening cut to support the header over the opening.

All the studs for a wall of a room are cut to length, properly spaced, and nailed to a bottom and top plate, on the subfloor if it is installed, then tilted into position. A second top plate may then be installed to join the sections or to tie partition walls to main walls.

Lateral rigidity for walls may be provided by plywood, diagonal steel tension straps, diagonal sheathing, or cut-in braces installed at each corner of the building.

Rafters

The labor required to frame and erect rafters will vary considerably with the type of roof. Double-pitch or gable roofs, with no dormers or gables, will require the least amount of labor, while hip roofs, with dormer or gables framing into the main roof, will require the greatest amount of labor.

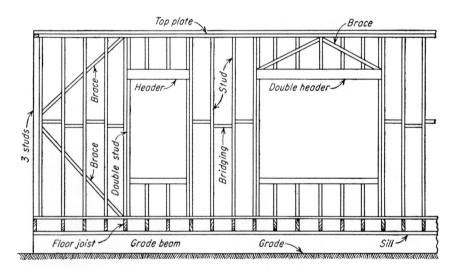

FIGURE 11-4
Typical framing detail for studs.

TABLE 11-11
Labor-hours required per
1,000 fbm for rafters

Size rafters, in	Labor-hours	
	Carpenter	Helper
Gable roofs—no dormers		
2 × 4	27–31	7–8
2 × 6	25–29	7–8
2 × 8	24–28	6–7
2 × 10	22–26	6–7
Hip roofs—no dormers		
2 × 4	30–34	8–9
2 × 6	29–33	8–9
2 × 8	28–32	7–8
2 × 10	26–30	7–8

Subfloors

Subfloors are usually constructed with tongue-and-groove plywood sheets. The thickness of the plywood will vary, depending on the required floor load and the spacing of the floor joists. A $\frac{5}{8}$- or $\frac{3}{4}$-in thickness is common. An 8d common nail is placed at 6-in spacing along the floor joists. It is a good practice to both nail and glue the plywood sheathing to the floor joist, to increase the strength of the subfloor and to reduce the tendency of a squeaking noise while someone is walking across the floor.

Subfloors may also be constructed with 1- by 6-in or 1- by 8-in planks, laid straight or diagonally. Two 8d common wire nails should be used at each joist. More labor will be required for diagonal than straight flooring. Also the lumber wastage will be greater for diagonal flooring.

TABLE 11-12
Labor-hours required per 1,000 ft² of
plywood subflooring

Size flooring, in	Joist spacing, in	Labor-hours	
		Carpenter	Helper
$\frac{5}{8}$	12	6.0–6.5	2.2–2.5
	16	5.5–6.0	2.2–2.5
	20	5.0–5.5	2.2–2.5
$\frac{3}{4}$	16	5.5–6.0	2.2–2.5
	20	5.0–5.5	2.2–2.5
	24	4.5–5.0	2.2–2.5

TABLE 11-13
Labor-hours required per 1,000 fbm for subfloors

Size flooring, in	Joist spacing, in	Labor-hours	
		Carpenter	Helper
Straight flooring			
1 × 6	12	11–12	4–5
	16	10–11	4–5
	24	9–10	4–5
1 × 8	12	10–11	4–5
	16	9–10	4–5
	24	8–9	4–5
Diagonal flooring			
1 × 6	12	13–14	5–6
	16	12–13	5–6
	24	11–12	5–6
1 × 8	12	12–13	5–6
	16	11–12	5–6
	24	10–11	5–6

Roof decking

Roof deckings are usually constructed of $\frac{1}{2}$- or $\frac{5}{8}$-in-thick sheets of plywood or waferboard. The thickness of the sheets will vary, depending on the required roof load and the spacing of the rafters. The decking is usually installed with a nail gun that is operated by compressed air.

Roof decks may also be constructed with lumber. The lumber may be 1- by 6-in or 1- by 8-in S4S, 1- by 6-in or 1- by 8-in center match, or 1 by 8-in shiplap,

TABLE 11-14
Labor-hours required per 1,000 ft² of plywood roof decking

Thickness, in	Rafter spacing, in	Labor-hours	
		Carpenter	Helper
$\frac{1}{2}$	12	6.5–8.0	2.5–3.5
	16	6.0–7.5	2.5–3.5
	20	5.5–7.0	2.5–3.5
$\frac{5}{8}$	16	6.0–7.5	2.5–3.5
	20	5.5–7.0	2.5–3.5
	24	5.0–6.5	2.5–3.5

TABLE 11-15
Labor-hours required per 1,000 fbm for roof decking

Size decking, in	Rafter spacing, in	Labor-hours	
		Carpenter	Helper
1 × 6	12	12–15	5–7
	16	11–14	5–7
	20	10–13	5–7
1 × 8	12	11–14	5–7
	16	10–13	5–7
	20	9–12	5–7

fastened with at least two 8d common nails per rafter. The labor will be slightly higher for a two-story than a one-story building. Also the labor will be higher for a steep roof than a flat roof.

Wall Sheathing

Sheathing for walls usually consists of $\frac{1}{2}$- or $\frac{5}{8}$-in-thick sheets of plywood, waferboard, insulating fiberboard, or gypsum board. Plywood sheets are installed horizontally across the vertical wood studs and fastened with 8d common nails.

Sheathing for exterior walls may also consist of 1- by 6-in or 1- by 8-in S4S planks, 1- by 6-in or 1- by 8-in shiplap, or 1- by 6-in or 1- by 8-in center match. The planks may be installed diagonally or horizontally. Diagonal sheathing will require more labor and will increase the wastage, but it will give the building much greater rigidity than horizontal sheathing. At least two 8d common nails should be used at each stud.

TABLE 11-16
Labor-hours required per 1,000 ft² of plywood wall sheathing

Thickness of sheathing, in	Stud spacing, in	Labor-hours	
		Carpenter	Helper
$\frac{1}{2}$	16	6.5–7.5	2.2–2.7
	24	6.0–7.0	2.2–2.7
$\frac{5}{8}$	16	7.0–8.0	2.2–2.7
	24	6.5–7.5	2.2–2.7

TABLE 11-17
Labor-hours required per 1,000 fbm for wall sheathing

Type and size sheathing, in	Labor-hours	
	Carpenter	Helper
Horizontal sheathing		
1 × 6 S4S	11–13	4–5
1 × 8 S4S	10–12	4–5
1 × 6 shiplap	12–14	4–5
1 × 8 shiplap	11–13	4–5
1 × 6 center match	12–14	4–5
1 × 8 center match	11–13	4–5
Diagonal sheathing		
1 × 6 S4S	16–18	4–5
1 × 8 S4S	15–17	4–5
1 × 6 shiplap	17–19	4–5
1 × 8 shiplap	16–18	4–5
1 × 6 center match	17–19	4–5
1 × 8 center match	16–18	4–5

Board and Batten Siding

Vertical boards, usually 1 by 10 in or 1 by 12 in, and batts sometimes are used for external siding on houses. The batts are installed to cover the joints between the boards.

Drop Siding

Rustic and drop siding, which is available in dressed-and-matched or shiplap patterns, in the nominal and face widths given in Table 11-19, is available in nominal thicknesses of $\frac{5}{8}$ and 1 in. Thicknesses less than 1 in are considered to be 1 in in determining the number of feet board measure of lumber.

TABLE 11-18
Labor-hours required per 1,000 fbm of boards for board and batten siding

Size boards, in	Labor-hours	
	Carpenter	Helper
1 × 10 S4S	16–18	4–5
1 × 12 S4S	15–17	4–5

TABLE 11-19

Labor-hours required per 1,000 fbm for drop siding

Pattern	Width, in		Labor-hours	
	Nominal	Face	Carpenter	Helper
Ordinary workmanship				
Dressed and matched	4	$3\frac{1}{4}$	22–24	5–6
	5	$4\frac{1}{4}$	21–23	5–6
	6	$5\frac{3}{16}$	20–22	4–5
	8	$7\frac{1}{8}$	18–20	4–5
Shiplap	4	$3\frac{1}{8}$	22–24	5–6
	5	$4\frac{1}{8}$	21–23	5–6
	6	$5\frac{1}{16}$	20–22	4–5
	8	$6\frac{7}{8}$	18–20	4–5
	10	$8\frac{7}{8}$	16–18	4–5
First-class workmanship				
Dressed and matched	4	$3\frac{1}{4}$	27–29	6–7
	5	$4\frac{1}{4}$	26–28	6–7
	6	$5\frac{3}{16}$	25–27	5–6
	8	$7\frac{1}{8}$	23–25	5–6
Shiplap	4	$3\frac{1}{8}$	27–29	6–7
	5	$4\frac{1}{8}$	26–28	6–7
	6	$5\frac{1}{16}$	25–27	5–6
	8	$6\frac{7}{8}$	23–25	5–6
	10	$8\frac{7}{8}$	21–23	5–6

TABLE 11-20

Labor-hours required per 1,000 fbm for bevel siding

Width, in		Labor-hours	
Nominal	Exposed to weather	Carpenter	Helper
Ordinary workmanship			
4	$2\frac{3}{4}$	28–30	5–6
5	$3\frac{3}{4}$	24–26	5–6
6	$4\frac{3}{4}$	21–23	4–5
8	$6\frac{3}{4}$	18–20	4–5
First-class workmanship			
4	$2\frac{3}{4}$	34–36	5–6
5	$3\frac{3}{4}$	30–32	5–6
6	$4\frac{3}{4}$	26–28	4–5
8	$6\frac{3}{4}$	23–25	4–5

Bevel Siding

Bevel siding is available in $\frac{1}{2}$- and $\frac{5}{8}$-in nominal thicknesses and in the widths given in Table 11-20. The number of feet board measures for thicknesses less than 1 in is the product of the nominal width in feet and the length in feet. The quantity of bevel siding required to cover a given wall area will vary with the actual width and the width exposed to weather.

Ceiling and Partition

Beaded and V-pattern ceiling and partition are available in the nominal and face widths given in Table 11-21 for all thicknesses.

Wood shingles

Wood shingles are available in three lengths—16, 18, and 24 in—and in random widths. For use on roofs having a pitch not less than $\frac{1}{4}$, which is a rise-to-run ratio of 1:2, the recommended exposure to weather is as follows:

Total length, in	Length exposed to weather, in
16	5
18	$5\frac{1}{2}$
24	$7\frac{1}{2}$

Shingles are sold by the square but packed in bundles which will cover $\frac{1}{4}$ square each when the standard exposure is used. About 10 percent should be added for waste. Prices will vary with the quality and length.

Shingles are nailed to shingle laths, sizes 1 by 2 in to 1 by 4 in, which are installed perpendicular to the roof rafters. The spacing of the lath should

TABLE 11-21
Labor-hours required per 1,000 fbm for ceiling and partition

Width, in		Labor-hours	
Nominal	Face	Carpenter	Helper
3	$2\frac{3}{8}$	26–28	5–6
4	$3\frac{1}{4}$	24–26	5–6
5	$4\frac{1}{4}$	22–24	4–5
6	$5\frac{3}{16}$	20–22	4–5

TABLE 11-22
Labor-hours required to lay 100 ft² of surface area for wood shingles

Kind of work	Labor classification	Length exposed to weather, in					
		4	4½	5	5½	6	7½
Plain hip and gable roofs	Carpenter	3.5	3.2	3.0	2.8	2.5	2.0
	Helper	0.9	0.8	0.8	0.7	0.6	0.5
Irregular hip and gable roofs	Carpenter	5.0	4.6	4.3	4.0	3.7	3.2
	Helper	1.0	0.9	0.8	0.8	0.7	0.6
Plain walls	Carpenter	5.0	4.6	4.3	4.0	3.7	3.2
	Helper	1.0	0.9	0.8	0.8	0.7	0.6

correspond to the length of shingles exposed to the weather. Each shingle should be fastened with two nails.

Fascia, Frieze, and Corner Boards

The quantity of lumber for these boards should be estimated by the 1,000 fbm of lumber. The labor required to place the boards will be greater for short than for long boards.

Wood Furring and Grounds

The labor required to place wood furring strips and grounds is usually estimated by the 100 lin ft.

Door Bucks

Door bucks may be made from 2- by 4-in or 2- by 6-in stock lumber or from 2- by 6-in lumber having the back grooved out by the mill, depending on the thickness of the wall in which they will be used.

TABLE 11-23
Labor-hours required to place 100 lin ft of fascia, frieze, and corner boards

Item	Carpenter	Helper
Fascia and frieze	4–5	0.8–1.0
Corner boards	3–4	0.6–0.8

TABLE 11-24
Labor-hours required to place 100 lin ft of wood furring strips and grounds

Kind of work	Carpenter	Helper
Furring strips on masonry walls	2.0–4.0	0.4–0.6
Furring strips on wood floors	2.5–4.5	0.4–0.6
Grounds on masonry walls	2.5–4.0	0.5–0.8
Grounds on frame walls	1.5–3.0	0.4–0.6

The total time required per opening will include the time to make, set, plumb, and brace a buck, which will be slightly higher for large than for small openings.

Finished Wood Floors

The lumber used for finished wood floors includes pine, maple, oak, beech, and birch. Not all these varieties are generally available in all locations.

Table 11-26 lists some of, but not necessarily all, the sizes of wood flooring produced. Not all the sizes listed may be available in all locations. This is especially true with respect to hardwood flooring.

The percentages of side waste listed in Table 11-26 are determined by dividing the reduction in width by the net width. For example, if the nominal width is 4 in and the face width is $3\frac{1}{4}$ in, the loss in width is 0.75 in. For these dimensions the table lists the percentage to be added for side waste to be $(0.75 \times 100)/3.25 = 23.1$. The reason for using this procedure is that when the quantity of lumber required is determined, the area of the floor is known. If the quantity of side waste listed in Table 11-26 is added to the area of the floor for the particular nominal width of flooring selected, this allowance will be sufficient for the side waste. An additional amount may be required for end waste, resulting from sawing to shorter lengths.

TABLE 11-25
Labor-hours required to make and install wood door bucks

Size opening	Carpenter	Helper
Up to 3 ft 0 in by 7 ft 0 in	2.0–2.5	0.5–0.6
Over 3 ft 0 in by 7 ft 0 in	2.5–3.0	0.6–0.7

TABLE 11-26
Dimensions of wood flooring

Thickness, in†		Width, in		
Nominal	Worked	Nominal	Face	Addition for side waste, %
Softwood				
$\frac{1}{2}$	$\frac{7}{16}$	3	$2\frac{1}{8}$	41.2
$\frac{5}{8}$	$\frac{9}{16}$	4	$3\frac{1}{8}$	28.0
1	$\frac{3}{4}$	5	$4\frac{1}{8}$	21.2
$1\frac{1}{4}$	1	6	$5\frac{1}{8}$	17.1
$1\frac{1}{2}$	$1\frac{1}{4}$	6	$5\frac{1}{8}$	17.1
Hardwood				
1	$\frac{25}{32}$	$2\frac{1}{4}$	$1\frac{1}{2}$	50.0
1	$\frac{25}{32}$	$2\frac{3}{4}$	2	37.5
1	$\frac{25}{32}$	3	$2\frac{1}{4}$	33.3
1	$\frac{25}{32}$	4	$3\frac{1}{4}$	23.0

† Generally all widths of softwood flooring are available in each listed thickness. In some locations not all listed nominal widths of hardwood may be available. Also hardwood may be available in other thicknesses.

TIMBER STRUCTURES

Fabricating Timber for Structures

Fabricating timber members for structures involves starting with commercial sizes of lumber and sawing them to the correct lengths and shapes, boring holes, dapping, chamfering, etc. Fabricating may be done with hand tools or with power-driven equipment.

The time required to fabricate timbers is illustrated by the following examples.

Example 11-2. Determine the probable time required to cut a 2- by 8-in by 12-ft 0-in-long plank to a specified length, requiring two square cuts with a handsaw. The approximate time should be as given in the accompanying table.

Operation	No. of units	Time per unit, min	Total time, min
Pick up and place on sawhorses	1	0.3	0.3
Measure for length	1	0.2	0.2
Mark for cuts	2	0.1	0.2
Saw the plank	2	1.0	2.0
Total time			2.7

Total time, assuming a 45-min hour, $2.7 \times \frac{60}{45} = 3.6$ min

Average time per cut, $3.6 \div 2 = 1.8$ min

If the plank is cut by an electric handsaw, the only reduction in time will be in the two sawing operations. The total time should be about as follows:

Operation	Time, min
Pick up and place on sawhorses	0.3
Measure for length	0.2
Mark for cuts	0.2
Saw the plank, 2 cuts × 0.25 min	0.5
Total time	1.2

Total time, assuming a 45-min hour, $1.2 \times \frac{60}{45} = 1.6$ min

Average time per cut, $1.6 \div 2 = 0.8$ min

TABLE 11-27

Approximate time required to perform various operations in fabricating timbers, min

Nominal size of timber, in	Operation				
	Sawing		Ripping power per lin ft	Boring 1 hole	
	Hand	Power		Hand	Power
2 × 4	1.3	0.6	0.2	1.4	0.8
2 × 6	1.5	0.7			
2 × 8	1.8	0.8			
2 × 12	2.1	0.9			
3 × 6	1.8	0.8	0.3	2.0	1.0
3 × 8	2.1	0.9			
3 × 12	2.5	1.1			
4 × 4	1.8	0.8	0.4	2.5	1.2
4 × 6	2.2	1.0			
4 × 8	2.6	1.2			
4 × 12	3.0	1.5			
6 × 6	2.5	1.1	0.6	3.0	1.5
6 × 8	3.0	1.5			
6 × 12	4.0	2.1			
6 × 18	5.5	3.0			
8 × 8	4.0	2.0	0.8	3.5	1.7
8 × 12	6.0	3.0			
8 × 18	9.0	4.5			
8 × 24	12.0	6.0			
10 × 10	6.0	3.0	1.0	4.0	2.0
10 × 12	8.0	4.0			
10 × 18	11.0	5.5			
12 × 12	10.0	5.0	1.2	4.5	2.5
12 × 18	14.0	7.0			
12 × 24	18.0	9.0			

The time to bore two 1-in-diameter holes in a 2-in-thick plank should be about as follows:

Operation	Time, min
Pick up and place on sawhorses	0.3
Mark location of holes	0.3
Bore, 2 holes × 0.75 min	1.5
Total time	2.1

Total time, assuming a 45-min hour, $2.1 \times \frac{60}{45} = 2.8$ min
Average time per hole, $2.8 \div 2 = 1.4$ min

Table 11-27 gives the approximate time required to fabricate timbers. The rates given include a reasonable amount of time for handling the timbers in addition to the actual operation. The time for operations will vary with the number of timbers fabricated to a given pattern, the extent to which jigs are used for measurements, and the distance from the stockpile to the working area.

Example 11-3. Use the information in Tables 11-27 and 11-28 to determine the probable time required to fabricate and assemble a joint for three 3- by 8-in members, using two 4-in split rings and one $\frac{3}{4}$-in bolt per assembly.

Boring and grooving, $4 \times 1 = 4$ min
Installing 2 rings, 2×1 = 2 min
Installing 1 bolt, 1×3 = 3 min
Total time = 9 min

TABLE 11-28
Approximate time required to install connectors and bolts, min

Connector	Size, in	Power grooving, per groove	Installing Connector	Bolt
Split rings	All	1	1	3
Shear plates	—	1	1	3
Toothed rings	2	—	5	3
	$2\frac{5}{8}$	—	5	3
	$3\frac{5}{8}$	—	6	3
	4	—	6	3
Spike grids	$3\frac{1}{8}$	—	5	3
	$4\frac{1}{8} \times 4\frac{1}{8}$	—	7	3
Clamping plates	All	—	4	3

MILL BUILDINGS

General Descriptions

Wooden structures of the mill-building type are frequently used for textile mills, factories, and warehouses. The timber members include columns, girders, beams, trusses, flooring, and roofing. The construction is generally open, with no thin sections, sharp projections, or concealed spaces. Although such a structure is not fireproof, it is classified as slow-burning. By treating the timbers with fire-resistant chemicals, it is possible to reduce the rate of burning to less than would be experienced with untreated timbers. Connections between members are made with nails, lag screws, bolts, timber connectors, and metal or wood column caps and bases.

Columns

Columns for mill buildings are usually square in sizes from 8 to 12 in. The sides are chamfered or rounded to reduce the fire hazard. The bottoms of the columns rest on and are attached to the footings through fabricated metal column bases and bolts or lag screws. Column caps may be all wood, all metal, or a combination of wood and metal with bolts and timber connectors to provide for the transfer of loads and stresses from upper columns, girders, floor beams, or trusses. For multistoried buildings it is common practice to provide a joint for each column at each floor level. Figure 11-5 shows typical details for column bases, Fig. 11-6 for floor-beam and column connections, and Fig. 11-7 for roof-beam and column connections.

Beams and Girders

If the floor loads are not extremely heavy and the spacings of the columns are not too great, it will be safe and economical to use beams only to support the floors and

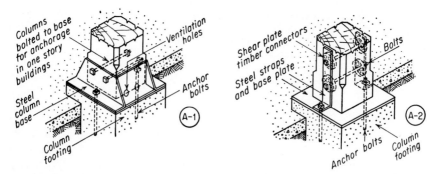

FIGURE 11-5
Typical details of bases for wood columns.

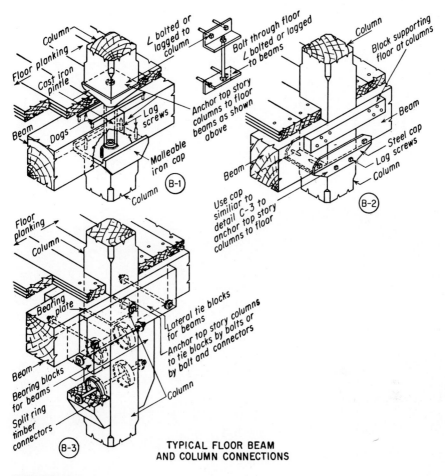

FIGURE 11-6
Typical details for floor-beam and column connections.

the roof. The beams are connected to and supported by the columns. If the floor loads are heavy and the column spacings are large, it will be necessary to use a beam-and-girder type of construction to support the floors and roof.

Floor

The floors may consist of tongue-and-groove planks, laid flat, or S4S planks laid on edge, designated as laminated floors. The tongue-and-groove planks are securely nailed to the floor beams. The laminated floor is constructed by laying the planks on edge, with the sides of adjacent planks brought in close contact by nails, driven horizontally, with spacings of 12 to 24 in along each plank laid. The laminated floor may or may not be fastened to the floor beam. End joints between the planks

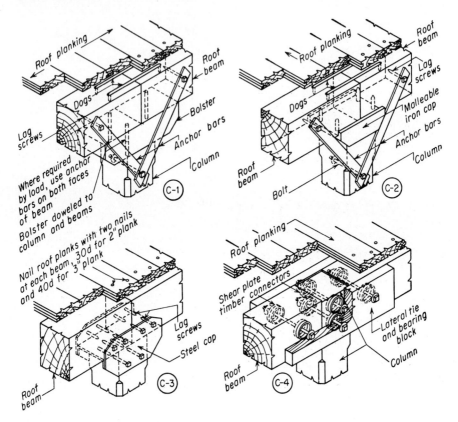

FIGURE 11-7
Typical details for roof-beam and column connections.

may be staggered over the floor beams, or they may be made between the floor beams, depending on the specifications for the structure.

Table 11-29 gives the quantities of lumber required for wood flooring, using various sizes of S4S planks and D and M flooring. Table 11-30 gives the weight of common nails required for 100 ft^2 of laminated-wood flooring or dressed and matched flooring.

Labor Fabricating and Erecting Columns

The labor operations required to fabricate and erect wood columns include sawing the two ends square, chamfering the four corners, drilling boltholes, and attaching bases and caps. The time will vary with the size and length of the columns and the kind of bases and caps used. For an 8- by 8-in by 12-ft 0-in-long column, where power equipment is used for fabricating, the time should be about as follows:

Operation		Time, min
End sawing, 2 × 2	=	4
Chamfering, 40 ft × 1.2	=	48
Drilling holes, 6 × 1.7	=	10
Attaching base	=	9
Attaching cap	=	9
Erecting	=	40
Total time	=	120

It will be necessary to provide approximately one laborer for two carpenters in fabricating and erecting columns, with the actual ratio depending on the extent to which hand hoisting above the first floor is necessary. Thus for the column just mentioned, the total labor required for fabricating and erecting at ground level would be

Carpenter   2.1 h
Helper   1.0 h

TABLE 11-29
Quantities of lumber required for 1 ft² of wood flooring

Nominal size plank, in	Dressed size plank, in	Side waste, %	Lumber, fbm/ft² of floor†
Laminated flooring			
2 × 4	$1\frac{1}{2} \times 3\frac{1}{2}$	33.3	5.33
2 × 6	$1\frac{1}{2} \times 5\frac{1}{2}$	33.3	8.00
2 × 8	$1\frac{1}{2} \times 7\frac{1}{4}$	33.3	10.67
3 × 6	$2\frac{1}{2} \times 5\frac{1}{2}$	20.0	7.20
3 × 8	$2\frac{1}{2} \times 7\frac{1}{4}$	20.0	9.60
4 × 6	$3\frac{1}{2} \times 5\frac{1}{2}$	14.3	6.85
4 × 8	$3\frac{1}{2} \times 7\frac{1}{4}$	14.3	9.15
Tongue-and-groove flooring			
2 × 4	$1\frac{1}{2} \times 3$	33.3	2.67
2 × 6	$1\frac{1}{2} \times 5$	20.0	2.40
2 × 8	$1\frac{1}{2} \times 6\frac{3}{4}$	18.5	2.37
2 × 10	$1\frac{1}{2} \times 8\frac{3}{4}$	14.3	2.29
2 × 12	$1\frac{1}{2} \times 10\frac{3}{4}$	11.6	2.23
3 × 4	$2\frac{1}{2} \times 3$	33.3	4.00
3 × 6	$2\frac{1}{2} \times 5$	20.0	3.60
3 × 8	$2\frac{1}{2} \times 6\frac{3}{4}$	18.5	3.55
3 × 10	$2\frac{1}{2} \times 8\frac{3}{4}$	14.3	3.43
3 × 12	$2\frac{1}{2} \times 10\frac{3}{4}$	11.6	3.25

† These quantities do not include any waste for end sawing lumber.

TABLE 11-30

Quantities of common nails required in lb per 100 ft^2 of wood flooring

Laminated flooring

Size nails	Spacing of nails, in		
	12	**18**	**24**
For 2-in nominal-thickness plank			
16d	15	10	8
20d	24	16	12
30d	31	21	16
40d	41	28	21
For 3-in nominal-thickness plank			
20d	15	10	8
30d	19	13	10
40d	26	18	13
60d	42	28	21
For 4-in nominal-thickness plank			
40d	19	13	10
50d	22	15	11
60d	30	20	15

D and M flooring

Nominal-size flooring	Size nails	Nails per support	Spacing of supports, ft				
			4	**6**	**8**	**10**	**12**
2 × 4	20d	2	13.0	8.5	6.3	5.0	4.2
2 × 6	20d	2	7.8	5.0	3.8	3.0	2.5
2 × 8	20d	2	5.7	3.7	2.8	2.2	1.9
2 × 10	20d	3	6.7	4.4	3.3	2.6	2.2
2 × 12	20d	3	5.4	3.5	2.7	2.1	1.8
3 × 4	40d	2	22.4	14.7	10.8	8.6	7.3
3 × 6	40d	2	13.4	8.6	6.6	5.2	4.3
3 × 8	40d	2	9.8	6.4	4.9	3.8	3.3
3 × 10	40d	3	11.6	7.6	5.7	4.5	3.8
3 × 12	40d	3	9.3	6.1	4.7	3.6	3.2

Table 11-31 gives the approximate time required for fabricating and erecting wood columns.

Labor Fabricating and Erecting Girders and Floor Beams

The labor operations required to fabricate and erect wood girders and floor beams include sawing the two ends, drilling boltholes and notching, if necessary, hoisting

TABLE 11-31
Approximate labor-hours required to fabricate and erect timbers, flooring, and roofing for a mill-type building

Member	Size	Labor-hours per 1,000 fbm lumber	
		Carpenter	Helper
Columns	8 in × 8 in × 12 ft 0 in	32	16
	10 in × 10 in × 12 ft 0 in	21	10
	12 in × 12 in × 12 ft 0 in	16	8
	12 in × 12 in × 16 ft 0 in	13	7
Girders	8 in × 12 in × 12 ft 0 in	8	8
	10 in × 12 in × 12 ft 0 in	8	8
	12 in × 16 in × 16 ft 0 in	6	6
	12 in × 24 in × 20 ft 0 in	4	4
Floor beams and purlins	4 in × 8 in × 12 ft 0 in	7	7
	4 in × 12 in × 18 ft 0 in	6	6
	6 in × 12 in × 16 ft 0 in	5	5
Laminated flooring	2 in × 4 in	8	5
	2 in × 6 in	7	5
	2 in × 8 in	6	5
	3 in × 8 in	5	4
	4 in × 8 in	4	4
	4 in × 10 in	4	4
D and M flooring and roofing	2 in × 6 in	8	4
	2 in × 8 in	7	4
	3 in × 8 in	5	3

into position, and attaching the connectors. If an accurate method of locating the boltholes is used, it will be possible to drill all holes with power equipment prior to erecting. Otherwise some drilling will be necessary after the members are hoisted into position. Such drilling is usually done with a portable electric drill. The time will vary with the size and length of the member, kind of connections used, and extent of hoisting necessary. For a 10- by 14-in by 12-ft 0-in-long beam, with power equipment used for fabricating, the time should be about as follows:

Operation	Time, min
End sawing, 2 × 9	18
Drilling holes, 4 × 2	8
Hoisting and connecting	40
Total time	66

It will be necessary to provide approximately one laborer for each carpenter in fabricating and erecting beams and girders above the ground floor. Table 11-31 gives the approximate time required to fabricate and erect timbers, flooring, and roofing for a mill-type building.

Example 11-4. Estimate the direct cost only for furnishing and erecting the structural members, flooring, and roofing in place for the mill building illustrated in Fig. 11-8. The structure will be erected on a concrete floor with the anchor bolts already in place. The column bases will be type A-2 of Fig. 11-5, and the column caps will be type B-2 of Fig. 11-6. All columns will be type B-2 of Fig. 11-6. All columns will be chamfered on four corners. The columns will be 10 by 10 in. The floor beams will be 10 by 14 in. The roof beams will be 10 by 10 in. The flooring will be 2 by 4 in laminated. The roofing will be 3- by 8-in D and M lumber, whose net width will be $6\frac{3}{4}$ in. All lumber will be structural-grade southern yellow pine.

The lumber will be fabricated with power equipment. The lumber will be hoisted into position by hand.

The cost will be:

Materials, f.o.b. job:
    Columns, 198 required,
        198, 10×10 in $\times$ 12 ft 0 in = 19,800 fbm @ \$0.59 = \$11,682.00
    Steel base plates, 99 required,
        99, $14 \times 14 \times \frac{1}{2}$ in = 2,751 lb @ \$0.98   =   2,696.12
    Steel straps, 198 required,
        198, $\frac{3}{8} \times 2\frac{1}{2} \times 10$ in = 526 lb @ \$0.95   =   499.70
    Shear plates, $2\frac{5}{8}$ in, 396 @ \$1.27   =   502.92
    Bolts, $\frac{1}{2} \times 13$ in, 198 @ \$1.95   =   386.10
    Steel caps, 198 @ \$15.93   =   3,154.14
    Lag screws, $\frac{1}{2} \times 4$ in, 1,560 @ \$0.89   =   1,388.40
    Supporting blocks, 792 pc,
        792 pc, 2×6 in $\times$ 3 ft 0 in = 2,376 fbm @ \$0.43   =   1,021.68
    Floor beams, 90 required,
        90, 10×14 in $\times$ 12 ft 0 in = 12,600 fbm @ \$0.59   =   7,434.00
    Roof beams, 90 required,
        90, 10×10 in $\times$ 12 ft 0 in = 9,000 fbm @ \$0.59   =   5,310.00
    Flooring, 9,600 ft^2 required,
        9,600 ft$^2 \times$ 5.33 fbm/ft^2 = 51,168 fbm @ \$0.47   = 24,048.96
    Nails, 30d $\times$ 18-in spacing,
        9,600 ft$^2 \times$ 21 lb/100 ft^2 = 2,016 lb @ \$0.98   =   1,975.68
    Roofing, 9,600 ft^2 required,
        9,600 ft$^2 \times$ 3.55 fbm per ft^2 = 34,080 fbm @ \$0.59 = 20,107.20
    Nails, 9,600 ft$^2 \times$ 3.8 lb per 100 ft^2 = 365 lb @ \$0.98 =   357.70
        Total cost of materials   = \$80,564.67

Labor fabricating and erecting:
    Columns:
        Carpenter, 19,800 fbm $\times$ 21 h/1,000 fbm = 416 h @ \$16.70 = \$ 6,947.20
        Helpers, 19,800 fbm $\times$ 10 h/1,000 fbm = 198 h @ \$12.40   =   2,455.20
    Floor beams:
        Carpenter, 12,600 fbm $\times$ 8 h/1,000 fbm = 101 h @ \$16.70   =   1,686.70
        Helpers, 12,600 fbm $\times$ 8 h/1,000 fbm = 101 h @ \$12.40   =   1,252.40

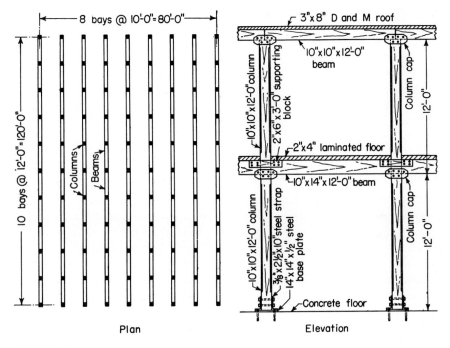

Plan   Elevation

FIGURE 11-8
Details of framing for a mill building.

Roof beams:
    Carpenter, 9,000 fbm × 8 h/1,000 fbm = 72 h @ $16.70   =   1,202.40
    Helpers, 9,000 fbm × 8 h/1,000 fbm = 72 h @ $12.40   =   892.80
Flooring:
    Carpenters, 51,168 fbm × 8 h/1,000 fbm = 410 h @ $16.70 =   6,847.00
    Helpers, 51,168 fbm × 5 h/1,000 fbm = 256 h @ $12.40   =   3,174.40
Roofing:
    Carpenters, 34,080 fbm × 5 h/1,000 fbm = 171 h @ $16.70 =   2,855.70
    Helpers, 34,080 fbm × 3 h/1,000 fbm = 102 h @ $12.40   =   1,264.80
Foreman, based on 6 carpenters,
    1,172.4 h ÷ 6 = 195.3 h @ $18.30   =   3,573.99
        Total cost of labor   = $32,152.59

Equipment, 195 h @ $4.75   =   926.25
Sundry tools and supplies   =   250.00
        Total cost of equipment   = $ 1,176.25

Summary of costs:
    Materials   = $ 80,564.67
    Labor   =   32,152.59
    Equipment   =   1,176.25
        Total cost   = $113,893.51
        Cost per ft^2 of building, $113,893.51 ÷ (2 × 80 × 120) =   5.93

ROOF TRUSSES

Wood roof trusses are used in buildings which require a clear floor area, without columns or other interior supporting structures. Trusses have been used for spans varying from 25 ft 0 in to 200 ft 0 in. Spacing of the trusses may vary from 12 ft 0 in to 20 ft 0 in, with a 16-ft 0-in spacing commonly used.

The introduction of metal timber connectors in the construction of trusses has increased the use of wood trusses.

Types of Roof Trusses

Several types are used: bowstring, Fink, Belgian, Pratt, and Warren among the open types; and the glued laminated arches among the solid-member types. The selection of the particular types of truss for a building depends on the span, the need to economize, and the desired appearance within the building.

Trusses are supported at the two ends, which may bear upon steel columns, wood columns, or pilasters. Pratt trusses are frequently supported by end columns which are made an integral part of the trusses. The roof is supported by joists or purlins. For roofing, 1-in-thick sheathing is generally used with joists and 2-in-thick sheathing with purlins.

Estimating the Cost of Wood Roof Trusses

The builder may completely fabricate all but the glued laminated trusses at the job; or may purchase them completely fabricated and partly assembled; or in some instances may purchase them completely fabricated and assembled, depending on the size of the trusses, length of haul to the job, and transportation clearances along the haul road. Trusses are generally assembled into complete units prior to erection, regardless of where they are fabricated.

If the trusses are to be prefabricated and assembled at a shop for shipment to the job, the estimator should obtain the cost from the fabricator. The cost of transportation to the job should be added to obtain the cost at the job. To these costs should be added the cost of erecting. If the trusses are to be fabricated at the job, the estimator must determine the costs of all materials required, of fabricating and erecting equipment, and of labor fabricating, assembling, and erecting. Small trusses may be erected by hand by using gin poles, but the larger trusses should be erected with power cranes.

Labor Fabricating and Assembling Wood Roof Trusses

Table 11-32 gives the approximate labor-hours required to perform various operations in fabricating members for a truss. In assembling the members into a

TABLE 11-32
Approximate time required to fabricate and assemble wood members for a 40-ft fink truss

Mark	Size, in	Pieces	Sawing			Drilling			Total time for carpenter, h	Time to assemble, h	
			No. cuts	Time per cut, min	Total time, h	No. grooves	Time per groove, min	Total time, h		Per piece	Total time
1	3 × 8	4	8	0.9	0.12	58	1.0	0.97	1.09	0.10	0.40
2	3 × 6	4	8	0.8	0.11	40	1.0	0.67	0.78	0.10	0.40
3	3 × 6	2	4	0.8	0.06	4	1.0	0.07	0.13	0.04	0.08
4	2 × 6	4	8	0.7	0.10	8	0.8	0.11	0.21	0.04	0.16
5	3 × 6	2	4	0.8	0.06	8	1.0	0.14	0.20	0.04	0.08
6	3 × 6	2	4	0.8	0.06	8	1.0	0.14	0.20	0.06	0.12
7	3 × 6	2	4	0.8	0.06	8	1.0	0.14	0.20	0.04	0.08
8	2 × 6	4	8	0.7	0.10	16	0.8	0.22	0.32	0.06	0.24
9	2 × 6	4	8	0.7	0.10	16	0.8	0.22	0.32	0.04	0.16
10	3 × 6	2	4	0.8	0.06	16	1.0	0.27	0.33	0.04	0.08
11	3 × 10	1	4	1.0	0.07	8	1.0	0.14	0.21	0.06	0.06
12	2 × 6	2	4	0.7	0.05	4	0.8	0.06	0.11	0.04	0.08
13	3 × 6	1	2	0.8	0.03	4	1.0	0.07	0.10	0.04	0.04
14	3 × 6	4	8	0.8	0.11	12	1.0	0.20	0.31	0.04	0.16
15	3 × 8	8	16	0.9	0.24	8	1.0	0.14	0.38	0.04	0.32
							Total time to fabricate =	4.89		Total time to fabricate =	2.46.

Number of carpenter hours to fabricate and assemble = 7.35
Quantity of lumber per truss = 502 fbm

Carpenter-hours per 1,000 fbm, $\dfrac{7.35 \times 1,000}{502} = 14.7$

Helper-hours per 1,000 fbm = 7.3

285

truss, considerable time will be saved if a jig or pattern is set up in a large working area, to assist in bringing the separate members into the correct position for the assembly. If all the members are prefabricated, the assembling operation will consist of placing the members into the correct positions, installing the metal timber connectors, if used, installing bolts and washers, and tightening the nuts. If the trusses are assembled at the job, this should be done as near the points of erection as possible, preferably on the floor of the building.

The labor-hours required to assemble a truss after the members are fabricated can be determined by estimating the time required to assemble each member of the truss. The sum of the separate times will give the time for the truss. Experienced estimators may determine the time on the basis of the quantity of lumber and an estimated rate of assembling, such as hours per 1,000 fbm of lumber.

With the truss illustrated in Fig. 11-9, the approximate time required to fabricate and assemble a truss under favorable conditions should be as given in Table 11-32. The rates are based on using two carpenters and one helper with power equipment for sawing, drilling holes, and grooving for the connectors.

Table 11-33 gives the approximate labor-hours required to fabricate and assemble wood trusses. The rates given are based on using power equipment for sawing and boring holes and on performing the operations under favorable conditions.

TABLE 11-33

Approximate labor-hours required per 1,000 fbm of lumber to fabricate and assemble wood roof trusses

Span, ft	FBM per truss	Labor-hours	
		Carpenter	Helper
Pitched truss			
20	120	17.8	8.9
30	220	16.8	8.4
40	500	14.7	7.3
50	800	13.9	7.0
60	1,150	12.7	6.3
80	2,060	11.1	5.5
Bowstring truss with glued laminated top chord†			
50	750	12.4	6.2
80	1,450	9.7	4.9
100	2,200	9.0	4.5
120	3,130	8.5	4.3
140	4,120	8.3	4.2
160	5,380	7.8	3.9

† For nailed top chord, reduce the labor approximately 10 percent.

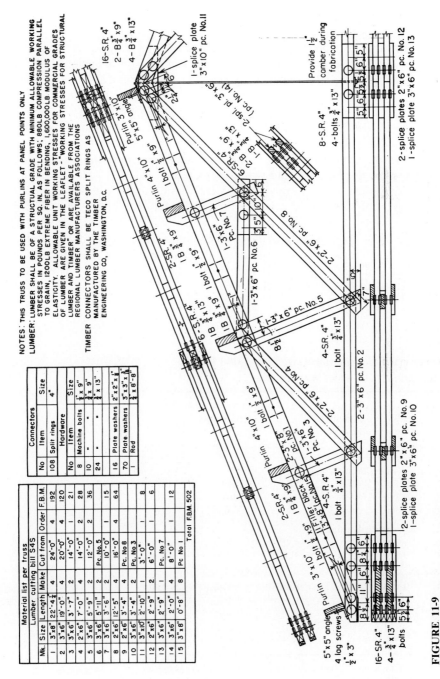

FIGURE 11-9

Details for a 40-ft 0-in Fink roof truss.

Hoisting Wood Roof Trusses

Wood roof trusses are usually hoisted into position with a gin pole, hand-operated derrick, or power crane, the equipment used being based on the weight, length, and number of trusses and the location of the job. From trusses whose weight does not exceed approximately 1,000 lb, a gin pole or a hand derrick may be used to lift the entire truss in a single operation. For trusses weighing approximately 1,000 to 2,500 lb, a gin pole or a hand derrick may be used to lift one end at a time. Trusses weighing more than 2,000 lb should be hoisted with a power crane, if one is available.

About $1\frac{1}{2}$ h will be required to hoist and erect a truss, by hand, if the entire truss is hoisted in one operation. If it is necessary to hoist each end separately, about 2 h will be required to hoist and erect a truss. The gang should include a foreman, four or five laborers, and two carpenters.

In hoisting trusses with power equipment about 1 to 2 h will be required to hoist and erect a truss, depending on the equipment, size of the trusses, and available working area.

Table 11-34 gives the approximate labor-hours required to hoist and erect wood trusses of various types and sizes. It is assumed that the trusses are prefabricated and assembled near the points where they will be hoisted. The trusses will be lifted 14 to 18 ft above the floor.

TABLE 11-34
Approximate labor-hours required per 1,000 fbm of lumber to hoist and erect wood trusses

Span	FBM per truss	Hoisting† time per truss, h	Labor-hours per 1,000 fbm of lumber		
			Foreman	Carpenter	Helper
Pitched truss—hand hoisting					
20	120	1.0	8.3	16.6	16.6
30	220	1.5	5.0	10.0	15.0
40	500	2.0	2.4	4.5	9.6
Pitched truss—power hoisting					
50	800	1.0	1.25	2.50	2.50
60	1,150	1.1	1.00	2.00	2.00
Bowstring truss					
80	1,450	1.3	0.90	1.80	1.80
100	2,200	1.2	0.55	1.10	1.10
120	3,130	1.4	0.45	0.90	0.90
140	4,120	1.7	0.41	0.82	0.82
160	5,380	2.0	0.37	0.74	0.74

† One power crane is used for the 50-, 60-, and 80-ft trusses and two cranes for the 100-, 120-, 140-, and 160-ft trusses.

Example 11-5

Pitched truss. Estimate the total direct cost of furnishing all materials, equipment, and labor required to fabricate, assemble, and erect the wood roof truss illustrated in Fig. 11-9. The project requires six trusses 16 ft 0 in apart. Lumber will be S4S structural-grade southern yellow pine. The lumber will be fabricated at the job, with power equipment. The trusses will be hoisted with a power crane.

The cost per truss will be:

Materials, f.o.b. job:

Lumber, 502 fbm @ $0.59	= $296.18
Split rings, 4 in, 108 @ $1.65	= 178.20
Machine bolts, $\frac{1}{2}$ × 9 in, 8 @ $0.72	= 5.76
Machine bolts, $\frac{3}{4}$ × 9 in, 10 @ $1.52	= 15.20
Machine bolts, $\frac{3}{4}$ × 13 in, 24 @ $1.97	= 47.28
Steel washers, 2 × 2 × $\frac{1}{8}$ in, 16 @ $0.43	= 6.88
Steel rod, $\frac{3}{4}$ in × 8 ft 8 in, 1 @ $12.51	= 12.51
Nails, 6 lb @ $0.98	= 5.88
Total cost of materials	= $567.89

Equipment:

Power crane, with operator, rented, $\frac{3}{4}$ h/truss @ $67.50	= $50.63
Power saws, 2 h/truss @ $2.85	= 5.70
Power drill, 2 h @ $1.85	= 3.70
Sundry equipment and tools	= 4.50
Total cost of equipment	= $64.53

Labor:

Fabricating and assembling:

Carpenters, 502 fbm, 502 fbm × 14.7 h/1,000 fbm = 7.4 h @ $16.70	= $123.58
Helpers, 502 fbm, 502 fbm × 7.3 h/1,000 fbm = 3.7 h @ $12.40	= 45.88
Foreman, based on 4 carpenters, 1.9 h @ $18.30	= 34.77

Erecting:

Carpenters, 4 × $\frac{3}{4}$ h/truss = 3 h @ $16.70	= 50.10
Helpers, 2 × $\frac{3}{4}$ h/truss = 1.5 h @ $12.40	= 18.60
Foreman, based on 4 carpenters, 0.8 h @ $18.30	= 14.64
Total cost of labor	= $287.57

Summary of costs:

Materials	= $567.89
Equipment	= 64.53
Labor	= 287.57
Total cost per truss	= $919.99

Example 11-6

Bowstring truss. Estimate the total direct cost of furnishing all materials, equipment, and labor required to fabricate, assemble, and erect the bowstring truss illustrated in

Fig. 11-10. The project requires seven trusses spaced 15 ft 6 in apart. The lumber will be S4S structural-grade southern yellow pine. The trusses will be fabricated and assembled at a shop and then transported to the job and hoisted into position with a truck-mounted power crane.

The cost per truss will be:

Materials, f.o.b. job:

Lumber, 748 fbm @ $0.59	= $441.32
Split rings, $2\frac{1}{2}$ in, 112 @ $1.10	= 123.20
Shear plates, 4 in, 16 @ $2.94	= 47.04
Bolts, $\frac{3}{4}$ × 13 in, 8 @ $2.04	= 16.32
Bolts, $\frac{1}{2}$ × 17 in, 16 @ $1.23	= 19.68
Bolts, $\frac{1}{2}$ × 14 in, 4 @ $1.06	= 4.24
Bolts, $\frac{1}{2}$ × 13 in, 26 @ $0.98	= 25.48
Bolts, $\frac{1}{2}$ × 11 in, 4 @ $0.90	= 3.60
Bolts, $\frac{1}{2}$ × 9 in, 18 @ $0.78	= 14.04
Washers, $2\frac{1}{2}$ in, 136 @ $0.14	= 19.04
Steel straps, 2 @ $57.00	= 114.00
Glue, 90 lb liquid mix per 1,000 ft² of glued area = 13 lb @ $1.70 =	22.10
Total cost of materials	= $850.06

Equipment:

Power crane, with operators, rented, 1.25 h/truss @ $72.50	= $ 90.63
Power saws, 5 h @ $1.85	= 9.25
Power planer, 3 h @ $1.95	= 5.85
Power drill, 4 h @ $1.60	= 6.40
Jigs and clamps	= 12.50
Sundry equipment and tools =	7.50
Total cost of equipment	= $132.13

Labor:

Fabricating and assembling:

Carpenters, 748 fbm required, 748 fbm × 12.4 h/1,000 fbm = 9.3 h @ $16.70 = $155.31	
Helpers, 748 fbm required, 748 fbm × 6.2 h/1,000 fbm = 4.7 h @ $12.40 =	58.28

Erecting:

Carpenters, 2 × 1.25 h/truss = 2.5 h @ $16.70	= 41.75
Helpers, 2 × 1.25 h/truss = 2.5 h @ $12.40	= 31.00
Foreman, fabricating and erecting, (9.3 + 2.5) ÷ 2 = 5.9 h @ $18.30	= 107.97
Total cost of labor	= $394.31

Summary of costs:

Materials	= $ 850.06
Equipment	= 132.13
Labor	= 394.31
Transporting to project =	85.00
Total cost per truss	= $1,461.50

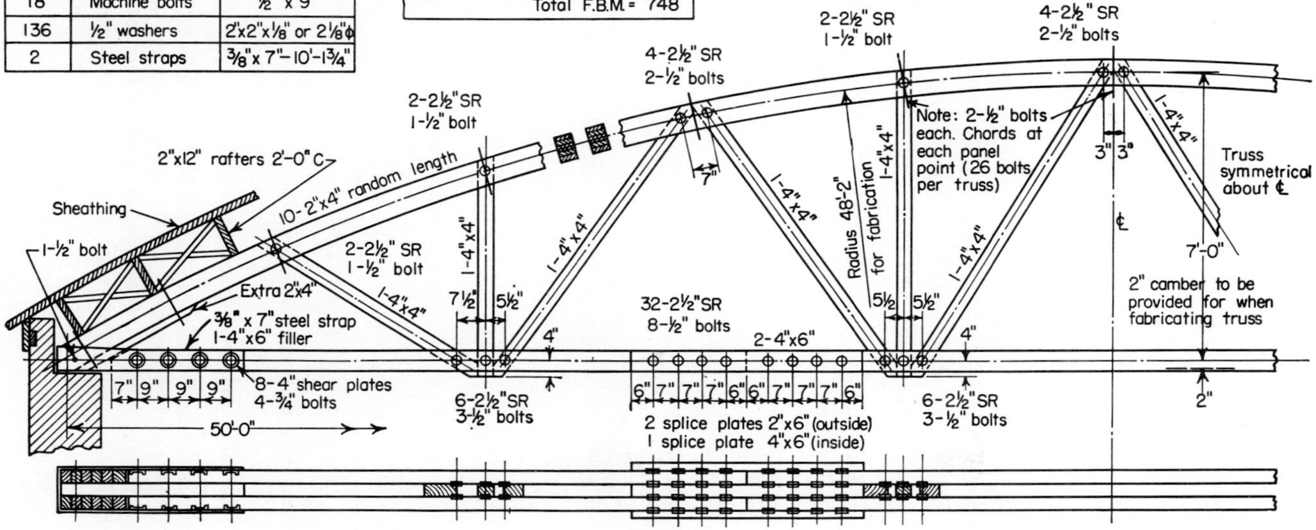

FIGURE 11-10
Details for a 50-ft 0-in bowstring roof truss.

NOTE: The conditions of the lumber, the quality of the glues and the procedure for the fabrication of the glued laminated top chord of this truss shall be as outlined in Part IX national design specifications for stress grade lumber and its fastenings, NLMA 1944

CAMBER: Camber of 2" shall be introduced into the top and bottom chord through fabrication

LUMBER: Lumber shall be structural grade with minimum allowable working stresses in pounds per square inch as follows:
1,000# compression parallel to grain
1,200# extreme fibre in bending
1,600,000# modulus of elasticity

CONNECTORS: Connectors shall be Teco split rings and shear plates as manufactured by the Timber Engineering Co., Washington, D.C.

LUMBER

Order	Size	Length	F.B.M.
	2" x 4"	Random	355
2	4" x 6"	18'-0"	72
4	4" x 6"	16'-0"	128
2	4" x 6"	10'-0"	40
2	4" x 4"	12'-0"	32
4	4" x 4"	18'-0"	96
2	2" x 6"	12'-0"	24

Total F.B.M = 748

HARDWARE

No.	Item	Size
112	Split rings	2½"
16	Shear plates	4"
8	Machine bolts	¾" x 13"
16	Machine bolts	½" x 17"
26	Machine bolts	½" x 13"
4	Machine bolts	½" x 11"
4	Machine bolts	½" x 14"
18	Machine bolts	½" x 9"
136	½" washers	2½"x2"x⅛ or 2⅛"⌀
2	Steel straps	⅜"x 7"-10'-1¾"

PROBLEMS

For all these problems use the national average wage rates for labor, as given in the text. Assume that S4S lumber will cost $480.00 per 1,000 fbm at the job and that common nails will cost $0.98 per pound, including an allowance for waste. Omit the cost of a foreman.

11-1. Estimate the total direct cost and the cost per 1,000 fbm for furnishing and installing 96 floor joists size 2 by 8 in by 17 ft 4 in, with the joists to rest on wood sills at each end of the joist and at the midpoint of the joist. Each joist will be fastened to the sills with twelve 16d common nails. Both ends of each joist must be sawed square.

11-2. Estimate the total direct cost and the cost per 1,000 fbm for furnishing and installing diagonal subflooring for a floor area of 2,240 ft². Use 1- by 6-in D and M lumber. The joists will be spaced 16 in apart.

11-3. Estimate the cost of Prob. 11-2 if the subflooring is $\frac{3}{4}$-in plywood.

11-4. The total length of the exterior and interior walls of a residence is 464 lin ft. Studs, size 2 by 4 in by 8 ft 0 in long will be spaced not over 16 in on centers for all walls and partitions. There will be 8 outside corners and 12 inside corners, each requiring three studs.

   The studs will rest on a 2- by 4-in plate and will be capped at the tops with two 2- by 4-in plates for the full length of the walls. Neglecting the door and window openings, prepare a bill of materials for the studs and the bottom and top plates, showing the number of pieces of each length required. Estimate the total direct cost and the cost per 1,000 fbm for furnishing and installing the studs and plates.

11-5. Estimate the total direct cost per 1,000 fbm of lumber for furnishing and installing 2- by 8-in rafters, spaced 24 in on centers, for a rectangular building whose outside dimensions are 28 ft 6 in wide and 74 ft 8 in long. The roof will be a gable type, with no dormers and with $\frac{1}{4}$ pitch. The rafters are to extend 18 in, measured horizontally, outside the outer surfaces of the walls. The rafters will be sawed as illustrated in Fig. 11-3.

11-6. Estimate the total direct cost per 1,000 fbm of lumber for furnishing and installing sheathing on the outside walls of a building. The walls to be covered will be 8 ft 6 in high and 324 ft 8 in long. Use $\frac{3}{4}$-in plywood. The studs will be spaced 16 in on centers.

11-7. Estimate the total direct cost of furnishing and installing 2,480 ft², net area, 1- by 6-in D and M siding, using first-grade workmanship. The siding will be installed horizontally on studs spaced 16 in on centers. The cost of materials will be:

   Lumber, $570.00 per 1,000 fbm   Nails, $1.05 per lb

11-8. Estimate the total direct cost and the cost per net 100 ft² of floor area for furnishing and laying 1- by 3-in D and M hardwood flooring on a wood subfloor. Use first-grade workmanship. The cost of materials will be:

   Flooring, $680.00 per 1,000 fbm   Nails, $1.10 per lb

11-9. Estimate the total direct cost and the cost per square for furnishing and laying wood shingles for a roof whose area will be 3,620 ft² consisting on irregular hips and gables. Use 18-in-long shingles, in random widths, with $5\frac{1}{2}$ in exposed to the weather. The roof, which is for a one-story home, will have a $\frac{1}{4}$ pitch. The cost of the materials will be:

   Shingles, $89.00 per square   Nails, $1.05 per lb

CHAPTER
12

INTERIOR FINISH, MILLWORK, AND WALLBOARD

Interior finish and millwork include such items as base, molding, chair rails, cornices, panel strips, window and door frames, windows and doors, casings, trim, wood cabinets, etc. The cost of these items will vary considerably with the kind and grade of materials and the grade of workmanship required.

In the tables giving the labor-hours required to install finish and millwork, the lower values are for ordinary workmanship and the higher values for first-grade workmanship. Because of the more rigid tolerances when first-grade workmanship is specified, the time required may be 25 to 50 percent greater than for ordinary workmanship.

Cost of Interior Finish and Millwork

The most dependable method of estimating the cost of interior finish and millwork is to list the quantity and cost of each item separately, together with the quantity and cost of the labor required to install each item. Finish material, such as molding, base, chair rail, and strip paneling, may be priced by the 100 lin ft, while window and door frames, sash, doors, cabinets, etc., usually are priced by the unit. The time

TABLE 12-1
Labor-hours required to install 100 lin ft of interior wood finish

Item	Carpenter	Helper
Picture molding	3.0–4.5	0.5–0.8
Chair rail	3.0–4.5	0.5–0.8
Base, one-member	3.0–4.5	0.5–0.8
Base, two-member	5.5–7.5	0.8–1.0
Base, three-member	7.0–9.0	0.9–1.1
Panel strips	3.5–4.5	0.5–0.8

required to install these items should be estimated by the 100 lin ft or by the unit, to conform with the method of pricing the materials.

Labor Required to Install Interior Finish

When molding, base, chair rails, etc., are installed in rooms with wood walls, they are nailed directly to the walls. However, if they are installed in rooms with plastered walls, they should be nailed to grounds set in and flush with the surface of the plaster.

Table 12-1 gives the labor-hours required to install various types of interior finish. Use the lower rates for ordinary workmanship and the higher rates for first-grade workmanship.

Wood Window Frames

Prefabricated window frames may be shipped from the mill unassembled, or they may be completely assembled with the exception of the interior casing. For the former condition, it is necessary to assemble the frames at the job prior to setting them in the openings.

Setting window frames involves carrying them to the openings, installing, plumbing, and bracing them in position. The time required will vary with the height to which they must be hoisted and the sizes of the frames, such as for single, double, or triple windows.

The interior trim for window frames usually is installed after the plaster walls are completed.

Table 12-2 gives the labor-hours required to perform various operations in installing wood window frames and the trim, for ordinary and first-grade workmanship.

Fitting and Hanging Wood Sash

Window sash most commonly used are of two types: double-hung, which are raised and lowered vertically, and casement, which open inward or outward. Double-

TABLE 12-2
Labor-hours per unit required to install a wood window frame

Operation	Carpenter	Helper
Assembling frame, single opening	1.0–1.5	0.2–0.3
Assembling frame, double opening	1.5–2.0	0.2–0.3
Assembling frame, triple opening	2.0–2.5	0.3–0.4
Setting frame, single opening	0.8–1.0	0.2–0.3
Setting frame, double opening	0.9–1.1	0.3–0.4
Setting frame, triple opening	1.0–1.2	0.4–0.5
Installing interior trim	1.0–2.0	0.2–0.4

hung sash are fitted with sash cord, or chain and weights, or counterbalances. Some fitting and finishing is required before a sash can be installed.

Table 12-3 gives the labor-hours required to perform the necessary operations in fitting and hanging wood sash windows. Use the lower rates for ordinary workmanship and the higher rates for first-grade workmanship.

Wood Door Frames and Jambs

The time required to set door frames and jambs will vary with the size of the opening, the type of frame or jamb set, and the grade of workmanship. The units, which are furnished by a mill, may require some fitting before they are set, plumbed, and fastened in position. The trim is installed after the walls are finished.

Table 12-4 gives the labor-hours required to set door frames, jambs, and trim. Use the lower rates for ordinary workmanship and the higher rates for first-grade workmanship.

Fitting and Hanging Wood Doors

Exterior doors are usually thicker and heavier than interior doors and require more time to fit and hang than interior doors. The operations include fitting the

TABLE 12-3
Labor-hours required to fit and hang a pair of wood sash windows

Operation	Carpenter	Helper
Fitting double-hung sash	0.5–0.8	0.1–0.2
Hanging double-hung sash	0.6–0.8	0.1–0.2
Fitting casement sash	0.6–1.0	0.1–0.2
Hanging casement sash	0.8–1.2	0.2–0.3

TABLE 12-4
Labor-hours required to set a wood door frame, jambs, and trim

Operation	Maximum size of opening, ft	Labor-hours	
		Carpenter	Helper
Set frame for exterior door	3 × 7	1.0–1.5	0.2–0.3
	6 × 7	1.5–2.0	0.3–0.4
	8 × 8	1.5–2.0	0.3–0.4
Set frame for interior door	3 × 7	0.5–1.0	0.1–0.2
	6 × 7	1.5–2.0	0.3–0.4
Install interior trim, two sides:			
Single casing	3 × 7	1.0–2.0	0.2–0.4
	6 × 7	1.5–2.5	0.3–0.5
Two-member trim	3 × 7	1.5–2.5	0.3–0.5
	6 × 7	2.0–3.0	0.4–0.6

door to the opening in the frame, routing the jamb and door for butts, mortising the door for the lock, installing the butts and lock, and hanging the door. Fitting, routing, and mortising may be done with hand tools or with power tools. Production rates are considerably higher when power tools are used.

Table 12-5 gives the labor-hours required to set the frame or jambs; install the stops, casing, butts; and lock, and fit and hang one door. The rates are based on using hand tools. If power tools are used to fit, route, and mortise, then reduce the

TABLE 12-5
Labor-hours required to set the frame or jambs, install stops and hardware, and fit and hang a door

Type of door and trim	Maximum size of opening, ft	Labor-hours	
		Carpenter	Helper
Exterior, single casing	3 × 7	3.0–6.0	0.6–1.0
	6 × 7	5.0–10.0	1.0–2.0
Exterior, two-member trim	3 × 7	3.5–6.5	0.7–1.3
	6 × 7	6.0–10.5	1.2–2.1
Interior, single casing	3 × 7	2.5–5.0	0.5–1.0
	6 × 7	4.5–9.0	0.9–1.8
Interior, two-member trim	3 × 7	3.0–6.0	0.6–1.2
	6 × 7	5.5–10.0	1.1–2.0
Sliding doors for closets, install			
track, two doors	5 × 8	6.0–8.0	1.2–1.6
Screen doors	3 × 7	1.0–1.5	0.2–0.3
	6 × 7	1.5–2.5	0.3–0.5

rates by about 10 percent. Use the lower rates for ordinary workmanship and the higher rates for first-grade workmanship.

WALLBOARDS

Wallboards include such items as gypsum boards, fiberboards, pressed wood, plywood, cement boards, etc., which are sold under a variety of trade names. They are used for walls, ceilings, and wainscotings in many buildings, especially houses. They may be installed on studs, joists, furring strips, plaster, or masonry, by using nails or cement to secure them in place.

The cost of wallboards in place will include the cost of materials and labor. The cost of materials can be obtained readily from the manufacturer or jobber. The cost of labor installing wallboards will vary considerably with the size, shape, and complexity of the area to be covered. Where areas are large and plain, the waste of materials will be low, and labor can install them rapidly; but where areas are small or irregular, with many openings, which require considerable cutting and fitting, the waste may be large and the cost of labor will be higher. The area to be covered should be carefully examined prior to preparing an estimate.

Gypsum Wallboards

Gypsum wallboards, which are made by molding gypsum between two sheets of paper, are used on walls and ceilings in many houses and other buildings. They are manufactured in sheets having the dimensions given in the accompanying table.

Thickness, in	Width, in	Length, ft	Approx. weight, lb/100 ft^2
$\frac{1}{4}$	48	6–12	270
$\frac{3}{8}$	48	6–12	400
$\frac{1}{2}$	48	6–12	530
$\frac{5}{8}$	48	6–12	660

Installing Gypsum Wallboard

Where gypsum wallboards are installed on wood studs and ceiling joists, the studs and joists should be spaced 12, 16, or 24 in on centers because these dimensions are divisible into the 48-in width. For walls, the boards are usually installed in vertical positions, with the length equal to the height of the wall, up to 12 ft. For ceilings, they may be installed perpendicular to or parallel with the joists. Some specifications require the use of nailing strips, such as 1- by 4-in planks, installed perpendicular to the ceiling joists on 12-in centers, to which the boards are fastened with nails.

TABLE 12-6
Quantities of nails required per 1,000 ft² of gypsum wallboard

| Thickness, in | Size nail | Quantity of nails, lb | | |
| | | Spacing of studs or joints, in | | |
		12	16	24
$\frac{1}{4}$	4d	5	4	3
$\frac{3}{8}$	4d	5	4	3
$\frac{1}{2}$	5d	8	6	4
$\frac{5}{8}$	6d	9	7	5

The boards are fastened with plasterboard nails, 4d, 5d, or 6d, with $\frac{5}{16}$- or $\frac{3}{8}$-in heads, depending on the thickness of the board. Recommended nail spacings are 8 in for wall and 7 in for ceiling installation. Table 12-6 gives recommended sizes of nails and the approximate quantities required per 1,000 ft² of wallboard.

The joints between adjacent boards are filled with a special cement, which is spread with a putty knife to cover a strip about $1\frac{1}{2}$ in wide on each side of the joint. After the cement is applied, the joint is covered with a perforated fiber tape about 2 in wide, and another layer of cement is applied and spread over the tape. After the cement dries, it is sanded to a uniformly smooth surface, which should eliminate all evidence of a joint, especially after the surface is painted. The nail heads are likewise covered with the cement and sanded flush with the surface of the wallboard.

TABLE 12-7
Labor-hours required to install gypsum wallboard and perforated tape

| Operation | Unit | Labor-hours | | |
| | | Spacing of studs or joists, in | | |
		12	16	24
Install wallboard:				
Large areas	100 ft²	1.7	1.5	1.2
Medium-size rooms	100 ft²	2.8	2.5	2.0
Small rooms	100 ft²	4.0	3.5	3.0
Install perforated tape and sand				
the surface	100 lin ft	4.0	4.0	4.0

Labor Installing Gypsum Wallboard

The labor required to install gypsum wallboard will vary considerably with the size and complexity of the area to be covered. For large wall or ceiling areas, which require little or no cutting and fitting of the boards, two carpenters should install a board 48 in wide and 8 to 9 ft long in about 10 to 15 min. This is equivalent to 1 to $1\frac{1}{2}$ labor-hours per 100 ft^2. However, where an area contains numerous openings, it is necessary to mark the boards and cut them to fit the openings. For such areas the labor required may be as high as 3 to 4 h per 100 ft^2. Cutting is done by scoring one side of the board with a curved knife, which is drawn along a straightedge. A slight bending force will break the board along the scored line.

Table 12-7 gives representative labor-hours required to install gypsum wallboard and perforated-tape joint system.

PROBLEMS

For these problems use the national average wage rates for labor, as given in this book. Assume that the cost of nails will be $1.05 per pound.

12-1. Estimate the total direct cost and the cost per unit for furnishing, assembling, and setting 24 double-opening wood window frames in wood walls, using first-grade workmanship. The sizes of the frames will be 2 ft 8 in wide and 5 ft 2 in high per opening. The cost of the frames per double opening will be $82.50.

12-2. Estimate the total direct cost per pair of sash for furnishing, fitting, and hanging wood sash for the window frames of Prob. 12-1, using first-grade workmanship. The cost of the glazed sash will be $32.75 per opening.

12-3. Estimate the direct cost per opening for furnishing and setting an interior wood door frame and installing single-casing trim on two sides for doors whose dimensions are 2 ft 8 in wide by 6 ft 8 in high. The frames will be furnished to the job with stops installed. The cost of materials per unit, including the casing, will be $22.50.

12-4. Estimate the direct cost per opening for furnishing and setting a wood frame, installing stops, and fitting and hanging an inside door, with two-member trim, using first-grade workmanship. The door size will be 2 ft 8 in wide by 6 ft 8 in high.

The cost of materials will be as follows:

Door frame, stop, and casing, $27.70
Door, good-grade birch, $62.50
Hardware, butts, and lock, $32.65

CHAPTER

13

LATHING AND PLASTERING

This chapter discusses methods of estimating the cost of materials and labor for placing lath and plaster, which are frequently used for finished surfaces for walls, ceilings, etc., in buildings.

Several types of materials are used for lath, as discussed under "Lathing." Various materials are used for plaster, as discussed under "Plastering."

Estimating the Cost of Lathing and Plastering

The cost of lathing and plastering may be estimated by the square yard, square foot, or, for some classes of work, the linear foot.

Within the lathing and plastering trades, the policy used to determine the area covered varies considerably. Some estimators do not deduct the areas of openings such as doors, windows, etc., when determining the area for cost purposes. Some estimators deduct one-half the areas of all openings larger than a specified dimension, such as 22 ft^2, but make no deductions for smaller areas. Still other estimators deduct the areas of all openings, regardless of size, to give the net area of the surface covered. When a cost per unit area is specified, the method of arriving at the total area used for payment purposes should be clearly stated.

When payment is made by the linear foot, it is also necessary to state clearly the basis on which the number of units will be determined.

Grades of Workmanship

Two grades of workmanship are commonly used in specifying the quality of work required, namely, ordinary and first grade. The cost of the work will be influenced considerably by the grade of workmanship specified.

Ordinary workmanship permits surfaces, grounds, corners, etc., to vary from true vertical or plane positions by as much as $\frac{1}{8}$ to $\frac{3}{16}$ in. First-grade workmanship requires that grounds, corners, etc., be in true alignment, both horizontally and vertically, and that areas of the finished surfaces shall not vary from a true plane by more than $\frac{1}{32}$ to $\frac{1}{16}$ in.

LATHING

Types of Lath

Plaster may be applied to masonry surfaces or to several types of lath, classified as follows:

1. Masonry
   a. Brick, stone, concrete, tile, and concrete and gypsum blocks
2. Metal lath
   a. Flat, self-furring, or ribbed lath
   b. Exposed or sheet metal
   c. Flat perforated sheet lath
   d. Wire lath, plain or stiffened
3. Gypsum and fiber lath and board

Metal Lath

Metal lath has substantially replaced wood lath as a base for plaster. Table 13-1 gives the properties of the more popular types of metal lath.

Flat expanded and rib metal lath are available in sheets 24 and 27 in wide and 96 in long, whose areas are 1.78 and 2.0 yd^2, respectively. The 24-in-wide sheets are packed in bundles of nine sheets, containing 16 yd^2, and the 27-in-wide sheets are packed in bundles of 10 sheets, containing 20 yd^2.

Applying Metal Lath on Wood Studs, Joists, and Furring Strips

Metal lath is fastened to wood supports with staples, large-head roofing nails, or common nails which are driven part way into the wood, then bent over the lath, spaced not more than 8 to 9 in apart. The sheets should be applied with the lengths perpendicular to the supports.

Side laps should be 1 in, with two strands of no. 18 gage soft wire used to tie them together at spacings of about 6 in. End laps should be at least $1\frac{1}{2}$ in, with wire

TABLE 13-1
Properties of metal lath

Type of lath	Weight, lb/yd²	Maximum spacing of supports, in	
		Vertical	Horizontal
Flat expanded metal lath	2.5	16	0
	3.4	16	16
$\frac{1}{8}$-in-rib metal lath	2.75	16	16
	3.4	24	19
$\frac{3}{8}$-in-rib metal lath	3.4	24	24
	4.0	24	24
$\frac{3}{4}$-in-rib metal lath	5.4	30	30
Sheet-metal lath	4.5	24	24
Stiffened wire lath	3.3	24	19

used to lace the adjacent sheets together at the lap. About 6 to 10 percent should be added for laps and waste on flat areas.

Applying Metal Lath on Steel Studs for Partition Walls

Where plastered partition walls, with total thicknesses of about $3\frac{1}{2}$ in, are to be constructed, metal lath is applied on both sides of lightweight $2\frac{1}{2}$-in metal channels, spaced 12 to 16 in on centers. These channels, which serve as studs, are anchored to the floor and ceiling, with floor and ceiling grounds being used for anchorage.

Metal lath is applied on each side of the channels, with wire at about 6-in spacings used to fasten the lath to the channels. The sides and ends of the lath are lapped 1 and $1\frac{1}{2}$ in, respectively, and fastened with wire ties spaced about 6 in apart.

If the channels are cut a full 9 ft 0 in long and spaced 16 in on centers for a wall 9 ft 0 in high and 100 ft 0 in long, whose total area is 100 yd², then $(100 \times 12)/16 = 75$ channels will be required. The total length of the channels will be $75 \times 9 = 675$ lin ft. Table 13-2 gives the quantities of materials required for 100 yd² of wall area, considering one side of the wall only.

Applying Metal Lath for Suspended Ceilings

Where metal lath is used in constructing suspended ceilings, it is necessary to support the lath below the ceiling. A commonly used method of supporting the lath is to install hanger wires in the concrete ceiling at spacings of 3 to 5 ft each way. These wires support steel runners, such as $1\frac{1}{2}$- or 2-in channels, spaced 3 to 5 ft apart. Cross furring, consisting of $\frac{3}{4}$- to 1-in steel channels, spaced 12 to 16 in on centers, is placed under and wired to the main runners to support lath (Table 13-3).

TABLE 13-2
Quantities of materials required for 100 yd² of metal-lath partition

Item	Stud spacing, in	Quantity
Channels	12	900 lin ft†
	16	675 lin ft†
Metal lath	—	106 yd²
Tie wire	—	10 lb

† These quantities include no allowance for waste.

TABLE 13-3
Sizes and maximum spacings of runners and furring for suspended ceilings

Size and kind of runners	Max. spacing of hanger wires, in
1½-in channel	48
2-in channel	60

Runners, max. spacing, in	Furring	
	Size and kind	Max. spacing, in
24	¾-in channel	24†
30	¾-in channel	24†
36	¾-in channel	24
42	¾-in channel	19
48	¾-in channel	16
54	1-in channel	19
60	1-in channel	12

† These spacings may be increased if a metal lath is used that will permit greater spacing.

TABLE 13-4
Quantities of materials required for 100 yd² of suspended metal-lath ceiling (includes waste)

Item	Quantity
Metal lath, $\frac{1}{8}$-in rib, 2.75 lb	108 yd²
Furring, $\frac{3}{4}$-in channels	710 lin ft
Runners, $1\frac{1}{2}$-in channels	240 lin ft
Hangers, 8-gage wires	60 each
Tie wire, 16 gage	12 lb

Table 13-4 gives the quantities of materials required for 100 yd² of suspended ceiling, using $\frac{3}{4}$-in-thick plaster, whose weight will be about 75 lb/yd². Hanger wires will be spaced 48 in each way. Use $1\frac{1}{2}$-in channels for main runners and $\frac{3}{4}$-in channels, spaced 16 in apart, for furring.

The weight on one hanger wire will be

Plaster, 1.78 yd² × 75 lb/yd²   = 134 lb
Lath, 1.78 yd² × 2.75 lb/yd²   =   5 lb
Furring, 12 lin ft × 0.3 lb/lin ft =   4 lb
Runners, 4 lin ft × 1.32 lb/lin ft =   5 lb
    Total weight   = 148 lb

Use 8-gage wire for hangers.

Labor Applying Metal Lath

The labor required to apply metal lath will vary considerably with the size and complexity of the areas covered, type of lath used, type of base to support the lath, and method of fastening. When an area contains openings, it usually is necessary to cut the lath to fit around the openings, which requires additional time.

Sufficient labor should be included to cover the installation of corner beads, moldings, bases, reinforcing for inside corners, etc., if such items are required.

Wood grounds are usually installed by carpenters and need not be considered here.

Table 13-5 gives the labor-hours required to apply metal lath under different conditions.

Example 13-1. Estimate the cost of applying metal lath, furring, and runners for the suspended ceiling in six rooms whose dimensions are 15 ft 0 in by 18 ft 0 in. The hanger wires are already in place.

TABLE 13-5
Labor-hours per unit required to apply metal lath

Class of work	Spacing of supports, in	
	12	16
On wood furring, per 100 yd²		
Walls and partitions	9–12	7–10
Flat ceilings	9–13	7–11
Beams and girders	22–32	18–25
Columns	22–30	20–28
Simple coves and cornices	20–25	15–20
Arch and groined ceilings	15–20	12–17
On metal furring, per 100 yd²		
Partitions, including furring, lath one side only	25–30	20–25
Flat suspended ceilings, including furring and runners	30–35	25–30
Beams and girders	25–32	20–25
Columns	25–32	20–25
Simple coves and cornices	25–30	20–25
Arch and groined ceilings	16–24	12–18
Accessories, per 100 lin ft		
Corner bead, molding, base	3–4	
Corner reinforcing	2–3	

The following material will be used:
  Runners, $1\frac{1}{2}$-in channels @ 48 in on centers
  Furring, $\frac{3}{4}$-in channels @ 16 in on centers
  Metal lath, $\frac{1}{8}$-in rib, 2.75 lb/yd²
  Tie wire, 16 gage

The quantities will be:
  Total area, 6 rooms × 15 × 18 ft = 1,620 ft²
    Total no. of yd², 1,620 ÷ 9 = 180 yd²
  Number of runners per room, (15 ft ÷ 4 ft) + 1 = 4.75, use 5
    Total length of runners, 5 per room × 6 rooms × 18 ft = 540 lin ft
  Number of furrings per room, [18 ft ÷ (16/12)] + 1 = 14.5, use 15
    Total length of furrings, 15 per room × 6 rooms × 15 ft = 1,350 lin ft

The cost will be:
  Runners, 540 lin ft @ $0.47 = $ 253.80
  Furring, 1,350 lin ft @ $0.32 = 432.00
  Metal lath, 180 yd² @ $3.80 = 684.00
  Tie wire, 180 yd² × 10 lb/100 yd² = 18 lb @ $1.65 = 29.70
  Lather, 180 yd² × 28 h/100 yd² = 50 h @ $16.20 = 810.00
  Scaffolding and tools = 15.00
    Total cost = $2,224.50
    Cost per ft², $2,224.50 ÷ 1,620 = 1.37

PLASTERING

Plaster is applied to masonry, wood lath, metal lath, and plasterboard inside a building to provide walls, ceilings, ornamental surfaces, etc. When it is applied on the outside of a building, it is called *stucco*. Several materials are used to make plaster and stucco.

The cost of plaster is usually estimated by the square yard, except for certain trim and novelty uses, whose cost may be estimated by the linear foot or square foot. There is no uniform method of determining the area of a surface to be plastered. Some contractors and estimators make no deductions for the areas of openings, while others deduct areas larger than specified sizes, and still others deduct the areas of all openings. An estimate should clearly state what method is used; otherwise, there may be a misunderstanding when the quantity is computed for payment purposes. Net areas are used in this book.

The labor required to apply plaster will vary considerably with the class of work specified. Two classes are usually recognized, ordinary workmanship and first-grade workmanship. Ordinary workmanship permits surface variations from a true plane by as much as $\frac{1}{8}$ to $\frac{3}{16}$ in, while first-grade workmanship permits variations not greater than $\frac{1}{32}$ to $\frac{1}{16}$ in. The labor required for first-grade workmanship may be 30 to 50 percent more than for ordinary workmanship.

Number of Coats Applied

Plaster may be applied in one, two, or three coats, the number depending on the base against which it is applied, the desired thickness, and the requirements of the specifications.

Scratch coat   When plaster is applied to wood and metal lath, it is common practice to apply three coats. The first coat is called the *scratch coat*. It should be thick enough to flow through all joints and openings in the lath and to form keys which will securely bond it to the lath. A thin covering of the lath is necessary in order that the surface may be scratched prior to final hardening to provide a good bond with the second coat. The total thickness should be $\frac{1}{8}$ to $\frac{1}{4}$ in, with about $\frac{1}{8}$-in coverage over the lath.

Brown coat   The second coat, called the *brown coat*, is applied to the scratch coat after the latter has dried. The brown coat contributes the greatest thickness to a plastered surface, usually being $\frac{3}{8}$ to $\frac{1}{2}$ in thick. Since this coat will determine the quality of workmanship for the job, it is necessary to finish it as carefully as the specifications require. Screeds, which span across the grounds, may be used to enable a plasterer to bring the surface to the desired smoothness. The surface is wood floated to provide a good bond with the finish coat.

Finish coat   This coat, which is also called the *white coat*, is the final coat. It is usually $\frac{1}{16}$ to $\frac{1}{8}$ in thick. After it is applied, it may be finished with a steel trowel to produce a hard smooth surface, or it may be finished with a sanded surface.

Total Thickness of Plaster

The total thickness of plaster may vary from $\frac{3}{8}$ to 1 in, depending on the base against which it is applied.

Where three coats are applied to metal lath, with the scratch coat $\frac{1}{4}$ in thick, the brown coat $\frac{3}{8}$ in thick, and the finish coat $\frac{1}{8}$ in thick, the total thickness will be $\frac{3}{4}$ in.

Where plaster is applied to brick, tile, concrete blocks, or plasterboard, it is common practice to omit the scratch coat, which will give a total thickness of $\frac{1}{2}$ to $\frac{5}{8}$ in.

Grounds having the required thickness should be used.

Materials Used for Plaster

The materials used for plaster include lime, gypsum plaster, cement, sand, or fiber, and sometimes special cements or plasters to produce desired effects. Plaster may be designated as lime, gypsum, or cement, depending on the chief cementing agent used. Fibers may be added to the scratch and brown coats to give them extra strength.

Lime is usually sold by the ton and may be obtained in bulk or in sacks. Hydrated lime is available in 50-lb sacks, which contain 1 ft³. Gypsum plaster is available in 100-lb sacks, which contain 1 ft³. Sand may be assumed to weigh 100 lb/ft³.

Table 13-6 gives the quantities of materials required for 1 yd³ of plaster. The mix proportions are by volume. The first number designates the quantity of cementing agent and the second number the quantity of sand.

Covering Capacity of Plaster

Neglecting waste, the quantity of plaster required to cover 100 yd² of area with a $\frac{1}{8}$-in-thick coat will be $(100 \times 9)/(8 \times 12) = 9.4$ ft³, or 0.35 yd³. The quantities for other thicknesses will be in the ratio of the thickness divided by $\frac{1}{8}$. Table 13-7 gives the quantities of plaster required to cover 100 yd² of area for various thicknesses.

Labor Applying Plaster

The labor required to apply plaster will vary considerably with the class of workmanship specified, as previously stated. Large flat surfaces will require less labor than small irregular surfaces. The number of helpers needed will vary with the speed at which the plaster can be applied and the height to which the plaster must be hoisted. If high scaffolds are required, additional labor will be needed.

Table 13-8 gives the labor-hours required to apply 100 yd² of plaster, based on net areas. Use the lower rates for ordinary workmanship and the higher rates for first-grade workmanship.

TABLE 13-6
Quantities of materials required for 1 yd³ of plaster (no waste included)

Kind of plaster	Proportions by volume	Cementing material, sacks†	Sand, yd³
Scratch or brown coat			
Hydrated lime‡	1:2	13.5	1
	1:3	9.0	1
Gypsum	1:2	13.5	1
	1:3	9.0	1
Finish coat			
Gypsum	1:0	27.0	0
Exterior stucco			
Hydrated lime	1:2	13.5	1
	1:3	9.0	1
Portland cement§	1:2	13.5	1
	1:3	9.0	1

† Sacks of hydrated lime and gypsum should contain 1 ft³ each, weighing 50 and 100 lb, respectively.

‡ If quicklime putty is used instead of hydrated lime, the following weight ratios should be used, compared to hydrated lime:

$$\text{For lump quicklime} = 0.65$$
$$\text{For pulverized quicklime} = 0.57$$

§ For portland cement stucco, add 8 lb or $\frac{1}{6}$ ft³ of hydrated lime for each sack of cement.

Example 13-2. Estimate the cost of applying a $\frac{3}{4}$-in-thick three-coat gypsum plaster on 100 yd³ of plain wall and ceiling, using ordinary workmanship. The use of fiber is required in the scratch coat. The thicknesses of the coats will be as follows: scratch coat, $\frac{1}{4}$ in; brown coat, $\frac{3}{8}$ in; finish coat, $\frac{1}{8}$ in. The plaster will be applied on metal lath. The scratch and brown coats will consist of gypsum and sand mixed in the ratio 1:3, respectively.

Assume that 10 percent of the sand delivered to the job for this use will be wasted.

TABLE 13-7
Quantities of plaster required to cover 100 yd² of area (no waste included)

	Thickness of plaster, in								
	$\frac{1}{16}$	$\frac{1}{8}$	$\frac{1}{4}$	$\frac{3}{8}$	$\frac{1}{2}$	$\frac{5}{8}$	$\frac{3}{4}$	$\frac{7}{8}$	1
Quantity, ft³	4.7	9.4	18.8	28.2	37.6	47.0	56.4	65.8	75.2
Quantity, yd³	0.18	0.35	0.70	1.05	1.40	1.74	2.10	2.44	2.80

TABLE 13-8
Labor-hours required to apply 100 yd² of plaster

Class of work	Plasterer	Helper
Scratch coat, three-coat work:		
Plain surface	5–6	5–6
Irregular surface	8–12	5–6
Brown coat, three-coat work:		
Plain surface	8–12	5-6
Irregular surface	10–15	5–6
Brown coat, two-coat work:		
Plain surface	8–12	5–6
Irregular surface	10–15	5–6
Finish coat	8–12	4–5
Sand finish coat	9–13	4–5

The cost will be:

Scratch coat:
Gypsum, $0.7 \times 9 = 6.3$ sacks @ \$3.69 = \$ 23.25
Sand, $0.7 \times 1.1 = 0.8$ yd³ @ \$8.55 = 6.84
Fiber, $0.7 \times 2 = 1.4$ lb @ \$2.30 = 3.22
Plasterer, 5 h @ \$16.75 = 83.75
Helper, 5 h @ \$12.40 = 62.00

Brown coat:
Gypsum, $1.05 \times 9 = 9.5$ sacks @ \$3.69 = 35.06
Sand, $1.05 \times 1.1 = 1.2$ yd³ @ \$8.55 = 10.26
Plasterer, 8 h @ \$16.75 = 134.00
Helper, 5 h @ \$12.40 = 62.00

Finish coat:
Gypsum, $0.35 \times 27 = 9.5$ sacks @ \$3.69 = 35.06
Plasterer, 8 h @ \$16.75 = 134.00
Helper, 4 h @ \$12.40 = 49.60

Scaffolds and sundry equipment = 7.50

Total cost = \$646.54
Cost per yd², $\$646.54 \div 100$ = 6.47

PROBLEMS

In solving the following problems, the hourly wage rates should be

Lathers   \$16.20
Plasterers   16.75
Helpers   12.40

13-1. Estimate the direct cost of furnishing and installing 100 yd² of flat metal lath weighing 3.4 lb/yd² on wood studs spaced 16 in on centers, using bent nails to secure the lath to the studs.

The cost of materials will be:

Metal lath, per yd² = $3.80
Tie wire, per lb   = 1.65
Nails, per lb   = 0.98

13-2. Estimate the direct cost of furnishing and installing 100 yd² of $\frac{3}{8}$-in-rib metal lath weighing 3.4 lb/yd² to two sides of $2\frac{1}{2}$-in metal studs, spaced 12 in on centers.

The cost of materials will be:

Metal lath, per yd² = $3.75
Tie wire, per lb   = 1.65

13-3. Estimate the total cost and the cost per square yard for furnishing and applying a $\frac{3}{4}$-in-thick three-coat gypsum plaster on 3,688 yd³ of irregular surface wall and ceiling, using first-grade workmanship. The plaster will be applied on metal lath.

The thicknesses of the coats will be: scratch coat, $\frac{1}{4}$ in; brown coat, $\frac{3}{8}$ in; and finish coat, $\frac{1}{8}$ in. For the scratch and brown coats, mix the gypsum and sand in the ratio 1:3, respectively, with no fiber required. The finish coat will be gypsum only.

The cost will be:

Gypsum, per sack   = $ 3.80
Sand, per yd³   =   9.10
Scaffolds and sundry equipment =   26.00

CHAPTER
14

PAINTING

Painting is the covering of surfaces of wood, plaster, masonry, metal, and other materials with a compound for protection or for the improvement of the appearance of the surface painted. Many kinds of paint are used, some of which will be described briefly.

Since various practices are used in determining the area of the surfaces to be painted, it is desirable to state what methods are used. Some estimators make no deductions for the areas of openings except those that are quite large, whereas other estimators deduct all areas that are not painted. The latter method seems to be more accurate. Practices vary in estimating the areas for trim, windows, doors, metal specialties, masonry, etc.

Most paints are liquids, which contain pigments such as titanium dioxide, aluminum, etc., suspended in linseed oil, tung oil, or an acrylic latex and alkyd resin emulsion. The vehicle carries the pigment during painting, and when it solidifies, it holds the pigment on the surface to provide the desired protection or appearance.

Paints may be applied in one, two, three, or more coats, with sufficient time allowed between successive coats to permit the prior coat to dry thoroughly. Paint may be applied with a brush, roller, or spray gun. The first coat, which is usually called the *prime coat*, should fill the pores of the surface, if such exist, and bond securely to the surface to serve as a base for the other coats.

The cost of painting includes materials, labor, and sometimes equipment, especially when the paint is applied with a spray gun. Because the cost of the paint is relatively small compared with the other costs, it is not good economy to purchase cheap paint.

Materials

It is beyond the scope of this book to discuss all the materials used for paints, but those most commonly used will be discussed briefly.

Ready-mixed paints These paints, which are mixed by the manufacturers, may be purchased with all the ingredients combined into a finished product.

Colored pigments Colored pigments are added to paints by the manufacturer or at the job to produce the desired color.

Linseed oil Linseed oil is frequently used as a vehicle for paints. The oil is processed from flax seed. When the oil dries, it forms a hard film on the surface to which it is applied. Boiling the oil before using it will speed its rate of drying and hardening. Raw oil may be used for inside painting, but a commercial drier should be added to increase the rate of drying.

Turpentine Turpentine, which is obtained by distilling the gum from pine trees, may be used to thin oil paints, especially the priming coat, to obtain better penetration into the wood.

Varnish Varnish, which is a solution of gums or resins in linseed oil, turpentine, alcohol, or other vehicles, is applied to produce hard transparent surfaces. Spar is a special varnish which is used on surfaces that may be exposed to water for long periods.

Shellac Shellac is a liquid consisting of a resinous secretion from several trees dissolved in alcohol. It may be applied to knots and other resinous areas of lumber prior to painting to prevent bleeding of the resinous substance through the paint.

Stains Stains, which are liquids of different tints, are applied to the surfaces of wood to produce desired color and texture effects. The vehicle may be oil or water.

Putty Putty, which is a mixture of powdered chalk or commercial whiting and raw linseed oil, is used to fill cracks and joints and to cover the heads of countersunk nails. It should be applied following the application of the priming coat of paint. If it is applied directly on wood, the wood will absorb most of the oil before the putty hardens, and the putty will not adhere to the wood.

Covering Capacity of Paints

The covering capacity of paint, which is generally expressed as the number of square feet of area covered per gallon for one coat, will vary with several factors, including the following:

1. The kind of surface painted, wood, masonry, metal, etc.
2. The porosity of the surface.
3. The extent to which paint is spread as it is applied.
4. The extent to which a thinner is added to the paint.
5. The temperature of the air. Thinner coats are possible during warm weather, resulting in greater coverage.

Table 14-1 gives representative values for the covering capacities of various paints, varnishes, and stains when applied to different surfaces.

Preparing a Surface for Painting

The operations required to apply a complete paint coverage will vary with the kind of surface to be painted, number of coats to be applied, and kind of paint used.

When new wood surfaces are being painted, it may be necessary to cover all resinous areas with shellac before applying the priming coat of paint. After this coat is applied, all joints, cracks, and nail holes may be filled with putty, which should be sanded smooth before the second coat of paint is applied.

TABLE 14-1
Surface area covered by 1 gal of paint, varnish, stain, etc.

Material	Surface area covered, ft²/gal
Applied on wood	
Prime-coat paint	450–550
Oil-base paint	450–550
Latex paint	450–550
Varnish, flat	500–600
Varnish, glossy	400–450
Shellac	600–700
Enamel paint	500–600
Aluminum paint	550–650
Oil stain on wood	500–700
Oil paint on floors	400–500
Stain on shingle roofs	125–225
Applied on plaster and stucco	
Size or sealer	600–700
Flat-finish paint	500–550
Calcimine on smooth plaster	225–250
Casein paint on smooth plaster	400–450
Casein paint on sanded plaster	400–450
Oil paint on stucco, first coat	150–160
Oil paint on stucco, second coat	350–375

It may be necessary to apply a filler to the surfaces of certain woods, such as oak, before they are painted. The surfaces may be sanded before and after applying the filler and sometimes following the application of each coat of paint or varnish, except the last coat.

Before paint is applied to plaster surfaces, it may be necessary to apply a sealer to close the pores and neutralize the alkali in the plaster.

The surfaces of new metal may be covered with a thin film of oil which must be removed with warm water and soap prior to applying the priming coat of paint.

Sometimes it is necessary to place masking tape, such as a strip of kraft paper with glue on one side, over areas adjacent to surfaces to be painted. This is especially true when paint is applied with a spray gun.

Labor Applying Paint

The labor required to apply paint may be expressed in hours per 100 ft^2 of surface area; per 100 lin ft for trim, cornices, posts, rails, etc.; or sometimes per opening for windows and doors. If the surfaces to be painted are properly prepared, and the paint is applied with first-grade workmanship, more time will be required than when an inferior job is permitted. As estimator must consider the kinds of surfaces to be painted and the requirements of the specifications before she or he can prepare an accurate estimate for a job.

The following factors should be considered in preparing an estimate for painting:

1. Treatment of the surface prior to painting, removing old paint, sanding, filling, etc.
2. Kind of surface, wood, plaster, masonry, etc.
3. Size of area, large flat, small, irregular.
4. Height of area above the floor or ground.
5. Kind of paint. Some flows more easily than others.
6. Temperature of air. Warm air thins paint and permits it to flow more easily.
7. Method of applying paint, with brushes or spray guns. Spraying may be 5 times as fast as brushing.

Table 14-2 gives representative labor-hours required to apply paint. The lower values should be used when the paint is applied under favorable conditions and the higher values when it is applied under unfavorable conditions or when first-grade workmanship is required.

Since the heavy pigment will settle to the bottom of a paint container, it may be necessary for a painter to spend 5 to 15 min/gal mixing paint before it can be used.

TABLE 14-2
Labor-hours required to apply a coat of paint and perform other operations

Operation	Unit	Hours per unit
Exterior work		
Oil paint on wall siding and trim	100 ft^2	0.60–1.00
Oil paint on wood-shingle siding	100 ft^2	0.65–0.90
Stain on wood-shingle roofs	100 ft^2	0.60–0.75
Oil paint on wood floors	100 ft^2	0.30–0.40
Painting doors, windows, and blinds	Each	0.50–0.75
Oil paint on stucco	100 ft^2	0.60–1.00
Oil paint on steel roofing and siding	100 ft^2	0.50–0.60
Oil paint on brick walls	100 ft^2	0.60–0.90
Interior work		
Size or seal plaster and wallboard	100 ft^2	0.40–0.50
Calcimine on plaster and wallboard	100 ft^2	0.30–0.40
Casein paint on plaster and wallboard	100 ft^2	0.20–0.30
Trim, mold, base, chair rail:		
Sanding	100 lin ft	0.80–1.00
Varnishing	100 lin ft	0.50–0.75
Enameling	100 lin ft	0.50–0.75
Flat painting	100 lin ft	0.50–0.75
Staining	100 lin ft	0.40–0.60
Painting doors and windows	Each	0.50–0.75
Floors:		
Sanding with power sander	100 ft^2	0.75–1.00
Filling and wiping	100 ft^2	0.50–0.60
Shellacking	100 ft^2	0.25–0.30
Varnishing	100 ft^2	0.40–0.50
Waxing	100 ft^2	0.40–0.50
Polishing with power polisher	100 ft^2	0.30–0.40
Structural steel		
Brush painting:		
Beams and girders, 150 to 250 ft^2/ton	Ton	0.80–1.00
Columns, 200 to 250 ft^2/ton	Ton	0.90–1.00
Roof trusses, 275 to 350 ft^2/ton	Ton	1.25–1.75
Bridge trusses, 200 to 250 ft^2/ton	Ton	1.00–1.25
Spray painting:		
Beams and girders	Ton	0.20–0.25
Columns	Ton	0.20–0.25
Roof trusses	Ton	0.25–0.35
Bridge trusses	Ton	0.20–0.30

Equipment Required for Painting

The equipment required for painting with brushes will include brushes, ladders, sawhorses, scaffolds, and foot boards. Inside latex paints may be applied with rollers. Where painting is done with spray guns, it will be necessary to provide one or more small air compressors, paint tanks, hose, and spray guns, in addition to the equipment listed for brush painting.

Cost of Painting

This example illustrates a method of determining the cost of furnishing materials and labor in painting a surface.

> **Example 14-1.** Estimate the cost of furnishing and applying paint for a three-coat application on 100 ft^2 of exterior new wood wall for a residence, using ordinary workmanship. The top of the wall will be 12 ft above the ground. The temperature of the air will be about 70°F. Ready-mixed oil paint will be used.
> The cost of the three coats will be:

Priming coat:
Paint, 100 ft^2 ÷ 500 ft^2/gal = 0.2 gal @ $12.50   = $ 2.50
Painter, 100 ft^2 @ 0.9 h × $16.10   = 14.49
Second coat:
Paint, 100 ft^2 ÷ 500 ft^2/gal = 0.2 gal @ $14.75   = 2.95
Painter, 100 ft^2 @ 0.8 h × $16.10   = 12.88
Third coat:
Paint, 100 ft^2 ÷ 550 ft^2/gal = 0.18 gal @ $14.75 = 2.66
Painter, 100 ft^2 @ 0.8 h × $16.10   = 12.88
Painter mixing and stirring paint, 0.2 h @ $16.10   = 3.22
Ladders, brushes, etc., 2.5 h @ $0.85   = 2.13
     Total cost   = $53.71
     Cost per ft^2, $53.71 ÷ 100   = 0.54

PROBLEMS

For these problems use the national average wage rates given in this book.

14-1. Estimate the total cost of furnishing and applying a prime coat and two finish coats of oil-base paint using brushes, on 3,380 ft^2 of exterior wood surface, for a one-story house. The cost of paint will be:

> Prime coat, $13.20 per gal
> Finish paint, $16.80 per gal

14-2. Estimate the total direct cost of furnishing and applying one sealing coat and two coats of flat wall paint on 3,680 ft^2 of smooth plaster walls and ceilings. The cost of materials will be:

> Sealer, $9.75 per gal
> Paint, $14.50 per gal

14-3. Estimate the total direct cost of sanding, filling, wiping, applying four coats of varnish, waxing, and polishing 3,640 ft² of oak flooring, using first-grade workmanship,

> Power sander, $2.10 per h
> Polisher, $0.67 per h
> Filler, $15.75 per gal
> Varnish, $18.50 per gal
> Wax, $13.50 per gal

CHAPTER
15

GLASS AND GLAZING

It is common practice to furnish wood sash with the glass already installed by the manufacturer. Steel and aluminum sash may be furnished with the glass installed for use in residences, but when such sash are furnished for industrial and commercial buildings, it is common practice to install the glass after the sash are set in the walls, usually just before completing a building, to reduce the danger of breakage.

The installation of glass is called *glazing*, and the individuals who install it are glaziers.

An estimate covering the cost of furnishing and installing glass should include a detailed list of the quantity for each size, kind, and grade of glass required and for each the cost of the glass, putty, and glaziers.

Glass

There are many kinds and grades of glass, not all of which will be covered in this book. Glass used for glazing purposes is called *window glass*, and it may be divided into sheet glass and plate glass.

Sheet glass The thickness of ordinary window glass is single-strength and double-strength.

TABLE 15-1
Thicknesses and weights of clear sheet glass

Designation	Thickness, in		Average weight	
	Min.	Max.	oz/ft^2	lb/ft^2
Single-strength	0.085	0.100	18.5	1.16
Double-strength	0.115	0.133	24.5	1.53
$\frac{3}{16}$ in	0.182	0.205	39.0	2.44
$\frac{7}{32}$ in	0.205	0.230	45.5	2.85
$\frac{1}{4}$ in	0.240	0.255	52.0	3.25
$\frac{3}{8}$ in	0.312	0.437	78.0	4.87
$\frac{1}{2}$ in	0.438	0.556	104.0	6.50

Plate glass   Plate glass used for glazing purposes is available in two grades, second silvering and glazing quality.

Figured sheet glass   Many kinds of figured sheet glass are available, with various patterns or figures rolled or otherwise produced on them.

The thicknesses and weights of clear sheet glass are given in Table 15-1.

Window glass is designated by size and grade, such as 20- by 36-in B double-strength. The first number represents the width and the second number the height of the sheet.

Glass is sold by the square foot in sheets whose dimensions vary in steps of 2 in. Fractional dimensions are figured for the next standard dimensions. Thus a sheet whose actual size is $17\frac{5}{8}$ by $14\frac{5}{8}$ in is classified for cost purposes as 18 by 16 in.

Costs and certain operations related to glass are frequently based on the united inches of a sheet of glass, which is the sum of the width and height of a sheet.

Glass for Steel and Aluminum Sash

Steel and aluminum sash are manufactured in standard units which require glass having designated sizes. Each pane of glass in a sash is called a *light*. Thus a sash that requires 3 lights for width and 4 lights for height is called a 12-light sash.

Table 15-2 lists representative net prices for various sizes of flat glass used for steel and aluminum sash effective in October 1987. All dimensions are given according to standard pricing procedures, even though the actual sizes of the lights may be smaller than the dimensions listed. For example, a sash may require lights whose actual dimensions are $12\frac{1}{2}$ by $20\frac{3}{4}$ in. Table 15-2 lists this glass as size 14 by 22 in, because this is the smallest standard size from which it can be cut.

The net cost of glass is obtained by applying applicable discounts to a schedule of semipermanent list prices. The list price may remain constant for several years, but the discounts may change frequently and may vary with the quality and quantity of glass purchased and the location of the purchase.

TABLE 15-2

Representative costs of flat window glass per light

Glass size, in	Lights per box	Cost per light	
		Single-strength	Double-strength
8 × 12	75	$0.57	$0.80
8 × 16	56	0.77	1.07
9 × 12	67	0.17	1.07
9 × 14	57	0.84	1.18
9 × 16	50	0.95	1.34
10 × 22	33	1.32	1.85
10 × 24	33	1.43	2.00
12 × 16	38	1.14	1.61
12 × 18	33	1.30	1.82
12 × 24	25	1.73	2.42
14 × 18	29	1.52	2.12
14 × 24	21	2.04	2.86
16 × 18	25	1.73	2.42
16 × 24	19	2.25	3.16
16 × 32	14	3.14	4.80

Putty Required to Glaze Steel and Aluminum Sash

Putty and glazing compounds are used to glaze glass in steel and aluminum sash. The compound must be used with aluminum sash and is frequently used with steel sash. Putty will cost about $1.50 per pound, while the compound will cost about $2.00 per pound.

Before a glass is installed in a metal sash, a quantity of putty or glazing compound should be spread uniformly against the metal of the rabbet to hold the glass away from the metal. After the glass is installed, additional putty is applied to complete the glazing. A pound of putty or compound will glaze about 5 lin ft of rabbet for steel and aluminum sash and about 8 lin ft for wood sash.

TABLE 15-3

Representative costs for cutting a box of glass of 50 ft^2 to odd sizes or irregular shapes

Operation	Cost per box
Cut to one odd size	$ 5.25
Cut to two odd sizes	9.80
Cut circles or ovals	17.60

TABLE 15-4
Labor-hours required to glaze 100 window lights

Size of glass, in	Putty, lb	Labor-hours	Size of glass, in	Putty, lb	Labor-hours
Set window or plate glass using wood stops					
12 × 14	—	18	30 × 36	—	46
14 × 20	—	24	36 × 42	—	54
20 × 28	—	36	40 × 48	—	62
Set glass in wood sash using putty					
12 × 14	54	12	30 × 36	138	25
14 × 20	72	15	36 × 42	162	32
20 × 28	100	20	40 × 48	184	40
Set glass in steel or aluminum sash					
24 × 10	114	18	18 × 14	107	17
30 × 10	134	24	24 × 14	127	20
38 × 10	160	29	30 × 14	147	24
			38 × 14	173	30
9 × 12	70	16			
16 × 12	94	18	18 × 16	114	18
18 × 12	100	19	20 × 16	120	19
24 × 12	120	20	22 × 16	127	20
30 × 12	140	23	32 × 16	160	29
38 × 12	167	28	38 × 16	180	32
42 × 12	180	32	46 × 16	204	42

Labor Required to Install Glass

Table 15-4 gives the labor-hours required to glaze windows during warm weather when the glazing is done from inside the building.

If glazing is done during cold weather, the putty or glazing compound will be more difficult to work, which will require about 25 percent more labor time than when glazing is done during warm weather. If glazing is done from the outside of a building, it will require about 20 percent more time than when it is done from inside a building.

Example 15-1. Estimate the direct cost of double-strength glass, glazing compound, and labor required to glaze 240 lights in aluminum sash. The actual sizes of the lights will be $15\frac{1}{2}$ by $31\frac{1}{2}$ in. This will require lights whose standard sizes are 16 by 32 in, which will require two cuttings per light.

The costs will be
No. boxes of glass, 240 ÷ 14 = 17.2
Add for possible breakage = 0.8
Total no. boxes = 18.0

Glass, 240 lights @ $4.80   = $1,152.00
Cutting glass to size, 18 boxes @ $9.80   =   176.40
Glazing compound, 240 × 160/100 = 384 lb @ $2.00 =   768.00
Glazier, 240 × 29/100 = 69.6 h @ $17.25   =   1,200.60
     Total cost   = $3,297.00

PROBLEMS

For all these problems, use the representative wage rates appearing in Table 2-1.

15-1. Estimate the direct cost for furnishing and installing double-strength glass in 160 aluminum sash. Each sash will require two sheets of glass size $15\frac{1}{2}$ by $31\frac{3}{4}$ in. Use glazing compound to set the glass. The installation will be done from outside the building in good weather.

15-2. Estimate the cost of furnishing and installing the glass of Prob. 15-1 when the glazing is done from the inside of the building during cold weather.

15-3. Estimate the cost of furnishing and installing 56 plate-glass lights 36 by 42 in in wood frames using wood stops. The glass, which is $\frac{1}{4}$ in thick, will cost $4.75 per ft^2.

15-4. Estimate the total cost of furnishing and installing 280 single-strength $11\frac{1}{2}$- by $23\frac{1}{4}$-in lights in aluminum sash, using glazing compound. The glazing will be done from inside the building during the summer. Use the material prices given in this book.

15-5. Estimate the total direct cost for furnishing and installing double-strength glass in 96 aluminum sash. Each sash will require two sheets of glass $15\frac{1}{2}$ by $23\frac{1}{2}$ in. Use glazing compound to set the glass. The installation will be done from outside the building during the summer.

15-6. Estimate the cost of furnishing and installing the glass of Prob. 15-5 when glazing is done inside the building during cold weather.

15-7. Estimate the cost of furnishing and installing 76 plate-glass lights in 30- by 36-in aluminum sash. The glass, which is $\frac{1}{4}$-in thick, will cost $4.35 per ft^2.

15-8. Estimate the total direct cost of furnishing and installing 156 double-strength $11\frac{1}{2}$- by $17\frac{3}{4}$-in glass lights in aluminum sash, using glazing compound. The glazing will be done from inside the building during the summer.

15-9. Estimate the cost of furnishing and installing 196 double-strength $15\frac{1}{4}$- by $17\frac{3}{8}$-in lights in steel sash, using glazing compound. The glazing will be done from outside the building during the summer.

15-10. Estimate the cost of furnishing and installing 216 size $15\frac{1}{4}$- by $22\frac{3}{4}$-in lights in aluminum sash, using glazing compound. The glazing will be done during the winter from inside the building.

CHAPTER
16

ROOFING AND FLASHING

Roofing Materials

Roofing refers to the furnishing of materials and labor to install coverings for roofs of buildings. Several kinds of materials are used, including but not limited to the following:

1. Shingles
   a. Wood
   b. Asphalt
2. Slate
3. Built-up
4. Clay tile
5. Metal
   a. Copper
   b. Aluminum
   c. Galvanized steel
   d. Tin

Area of a Roof

Several methods are used to measure the area of a roof for estimating purposes. Some estimators make no deduction for the areas of openings containing less than

100 ft^2 but deduct one-half of the areas of openings containing 100 to 500 ft^2. A method which may be used for measuring the area of slate roofs is to make no deductions for skylights, chimneys, etc., whose dimensions are less than 4 ft square, to deduct one-half the areas whose dimensions are between 4 and 8 ft square, and to deduct all the areas whose dimensions are larger than 8 ft square. An estimator should determine what method will be used before preparing an estimate.

When you are determining the area of a pitched roof, measure along the full length of the roof; and for the width, measure along the slope of the roof. Be sure that every dimension includes the maximum length or width to be covered, including eaves, overhang, etc.

The unit of area most commonly used for roofing is the *square*, which is 100 ft^2.

Materials used to cover ridges and valleys should be measured by the linear foot, with the width specified.

Roof Pitch

The pitch of a roof indicates its steepness. The terms frequently used and their values are as follows:

Pitch	Rise, in/ft
$\frac{1}{8}$	3
$\frac{1}{4}$	6
$\frac{1}{3}$	8
$\frac{1}{2}$	12
$\frac{2}{3}$	16

Roofing Felt

When certain materials such as asphalt, slate, and tile are used for roofing, the specifications may require the application of roofing felt to the entire roof prior to applying the covering material. This material is sold in rolls containing 108, 216, and 432 ft^2, which will cover 100, 200, and 400 ft^2, respectively, with a 2-in lap. Weights of felt used are 15 and 30 lb per square. Galvanized nails about $\frac{3}{4}$ in long, with $\frac{7}{16}$-in heads, spaced about 6 in apart along the edges, are used to hold the felt in place. About $\frac{3}{4}$ lb of $\frac{3}{4}$-in nails will be required per square. Asphalt felt, weighing 15 lb per square, will cost about \$2.50 per square, and 30-lb felt about \$5.00 per square.

The labor required to lay felt should be about 0.5 h per square.

WOOD SHINGLES

Wood shingles are discussed in Chap. 11 under "Carpentry."

ASPHALT SHINGLES

Asphalt shingles are manufactured in several styles, colors, and sizes, including organic asphalt, fiberglass asphalt, hexagons, etc.

The shingles are fastened to the wood decking with galvanized roofing nails, 1 to $1\frac{1}{2}$ in long, starting at the lower edge of the roof. Asphalt starting strips should be laid along the eaves of a roof prior to laying the first row of shingles, or the first row may be doubled.

The net area of a roof should be increased about 10 percent for gable roofs, 15 percent for hip roofs, and 20 percent for roofs with valleys and dormers, to provide for waste.

Table 16-1 gives the net quantities of asphalt shingles of various styles required to cover one square.

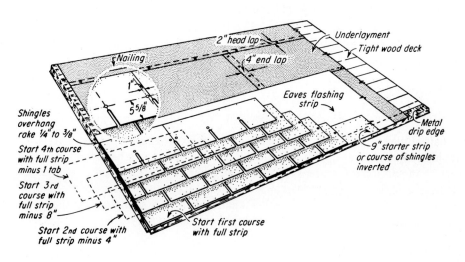

FIGURE 16-1
Three-tab strip asphalt shingles.

TABLE 16-1
Quantities of asphalt shingles and roofing nails per square

Style of shingle	Size, in	No. per square	Length exposed, in	No. nails per shingle	Nails per square, lb
Asphalt strip, 3 tab, 235 lb	12 × 36	80	5	4	1.0
Fiberglass asphalt, 340 lb	12 × 36	80	5	4	1.0
Hexagon strip, 170 lb	12 × 36	86	5	6	1.5

TABLE 16-2
Representative prices per square for asphalt shingles

Style of shingle	Size, in	Weight per square, lb	Cost per square
Asphalt strip, 3 tab	12 × 36	235	$21.95
Hexagon strip	12 × 36	170	23.95
Fiberglass asphalt	12 × 36	340	48.50
Hip and ridge, 100 lin ft	9 × 12		24.95

Cost of Asphalt Shingles

The cost of asphalt shingles will vary with the style, materials, and weight per square for the shingles selected. The quoted price per square is based on furnishing enough shingles to cover 100 ft^2 of roof in the manner indicated by the manufacturer, with no allowance for waste included. Table 16-2 gives representative prices for asphalt shingles.

Labor Required to Lay Asphalt Shingles

Asphalt shingles may be laid by carpenters or experienced roofers. The latter should lay them more rapidly. The rate of laying shingles will be lower for simple areas than for irregular areas. Hips, valleys, gables, dormers, etc., require the cutting of shingles for correct fit, which will reduce the production rates. Table 16-3 gives the labor-hours required to lay a square of asphalt shingles. Use the lower rates for gable roofs and the higher rates for roofs with valleys, hips, dormers, etc.

TABLE 16-3
Labor-hours† required to lay a square of asphalt shingles, using carpenters

Style of shingle	Carpenter	Helper
Individual	3.5–5.5	1.0–1.2
Strip, 3 tab	2.0–2.6	1.0–1.2
Hexagonal strip	2.0–2.6	1.0–1.2
Double coverage	2.5–3.5	1.0–1.2

† If experienced shinglers are used, reduce the hours by 15 to 20 percent.

SLATE ROOFING

Slate for roofing is made by splitting slate blocks into pieces having the desired thickness, length, and width. Although slate is available in various thicknesses up to about 2 in, the most common thickness used is $\frac{3}{16}$ in. Sizes vary from 10 by 6 in to 24 by 16 in, with the lengths varying in steps of 2 in and the widths varying in steps of 1 in.

Colors and Grades of Slate

The trade names of roofing slate, based on color, are as follows.

Black	Mottled purple and green
Blue black	Green
Gray	Purple variegated
Blue gray	Red
Purple	

The trade grades of slate are as follows:

No. 1 clear	No. 1 ribbon
Medium clear	No. 2 ribbon

Quantities of Slate

Slate is priced by the square, with sufficient pieces furnished to cover a square with a 3-in head lap over the second course under the given course. Thus for slates 16 in long, the length exposed to weather will be $(16 - 3)/2 = 6.5$ in. This same calculation will apply to any length for which the head lap is 3 in. If the slates are 8 in wide, the area covered will be $8 \times 6.5 = 52$ in^2. The number of slates required to cover a square will be $(100 \times 144)/52 = 277$. For other sizes the number required to cover a square can be determined in the same manner. An area should be increased 10 to 25 percent to allow for waste.

Weight of Slate Roofing

Slate weighs 170 to 180 lb/ft^3. The weights per square on a roof with a standard 3-in head lap are given in Table 16-4.

Nails for Slate Roofing

The weight of nails required for slate roofing will depend on the number, length, gauge, and type. The number required will be twice the number of slates laid. Table 16-5 gives the number of copper nails per pound.

TABLE 16-4
Weights of slate roofing
per square for a 3-in lap

Thickness, in	Weight per square, lb
$\frac{3}{16}$	700
$\frac{1}{4}$	900
$\frac{3}{8}$	1,400
$\frac{1}{2}$	1,800
$\frac{3}{4}$	2,700
1	4,000

Laying Slate

Roofing slate should be laid on asphalt felt, weighing 30 lb per square. The first course of slate should be doubled, with the lower end laid on a wood strip to give the slate the proper cant for the succeeding courses. Joints should be staggered. Each slate should be fastened with two nails driven through prepunched holes.

Hips and valleys will require edge mitering of the adjacent slates. The lengths of slates laid in courses along ridges must be reduced.

TABLE 16-5
Number of copper nails per pound

Length, in	Copper-wire slating nails		Copper-wire common nails	
	Gage	No. per lb	Gage	No. per lb
1	12	310	15	575
$1\frac{1}{4}$	10	144	14	400
	11	196		
	12	216		
$1\frac{1}{2}$	10	124	12	260
	11	160		
	12	192		
$1\frac{3}{4}$	10	112	11	140
			12	156
2	10	100	10	114
			11	128
$2\frac{1}{2}$	—	—	10	84
3	—	—	9	57

TABLE 16-6
**Labor-hours required to apply
a square of felt and roofing
slate**

Size slate,† in	Slater	Helper
12 × 8	5.0–7.0	2.5–3.5
16 × 8	4.0–6.0	2.0–3.0
16 × 12	3.5–5.2	1.8–2.7
18 × 12	3.2–5.0	1.6–2.5
20 × 10	2.9–4.5	1.4–2.2
20 × 12	2.8–4.3	1.4–2.1
22 × 12	2.5–3.9	1.2–2.0
24 × 12	2.3–3.6	1.2–1.8
24 × 16	2.0–3.2	1.0–1.6

† The first number is the length, and the
second number is the width of a slate.

Labor Required to Apply Slate Roofing

The labor operations required for slate roofing will include laying the felt, applying slates, and installing snow guards, if they are required. Production rates will vary considerably with the size and thickness of slates, slope of the roof, kind of pattern specified, and complexity of the areas.

Table 16-6 gives the labor-hours to apply a square of felt and slate under various conditions. Use the lower rates for plain roofs, such as gable roofs, and the higher rates for steep roofs with valleys, hips, and dormers.

BUILT-UP ROOFING

Building up roofing consists of applying alternate layers of roofing felt and hot pitch or asphalt over the area to be covered, with gravel, crushed stone, or slag applied uniformly over the top layer of pitch or asphalt. This type of roofing may be applied to wood sheathing, concrete, poured gypsum, precast concrete tiles, precast gypsum blocks, book tile, and approved insulation.

The quality of built-up roofing is designated by specifying the weight and number of plies of felt, the weight and number of applications of pitch or asphalt, and the weight of gravel or slag used. The unit of area is a square.

Felt

Pitch-impregnated felt should be used with pitch cement and asphalt-impregnated felt with asphalt. Felt weighing 15 or 30 lb per square may be used, with the 15-lb felt being more commonly specified. Felt is available in rolls 36 in wide with gross areas of 108, 216, and 432 ft^2.

Fiberglass felt is sometimes used in place of organic felt for built-up roofing. Fiberglass, type IV, is available in 540-lb rolls.

Pitch and Asphalt

Pitch should be applied at a temperature not exceeding 400°F and asphalt at a temperature not exceeding 450°F. Applications are made with a mop to the specified thickness or weight. Pitch and asphalt will weight about 10 lb/gal. Pitch is purchased in 200-lb cartons, and asphalt is purchased in 100-lb cartons.

Gravel and Slag

After the final layer of pitch or asphalt is applied, and while it is still hot, gravel or slag is spread uniformly over the area. The aggregate should be $\frac{1}{4}$ to $\frac{5}{8}$ in in size and thoroughly dry. Application rates are about 400 lb per square for gravel and 300 lb per square for slag.

Laying Built-up Roofing on Wood Decking

While specifications covering the roofing laid on wood decking will vary, the method described below is representative of common practice.

1. Over the entire surface lay two plies of 15-lb asphalt felt, lapping each sheet 19 in over the preceding one and turning these felts up not less than 4 in along all vertical surfaces. Nail as often as necessary to secure, until the remaining felt is laid.
2. Over the entire surface embed in asphalt two plies of asphalt felt, lapping each sheet 19 in over the preceding one, rolling each sheet immediately behind the mop to ensure a uniform coating of hot asphalt, so that in no place shall felt touch felt. Each sheet shall be nailed 6 in from the back edge at intervals of 24 in. These felts shall be cut off at the angles of the roof deck and all walls or vertical surfaces.
3. Over the entire surface spread a uniform coating of asphalt into which, while hot, embed not less than 400 lb of gravel or 300 lb of slag per 100 ft^2 of area. Gravel or slag must be approximately $\frac{1}{4}$ to $\frac{5}{8}$ in in size, dry and free from dirt. If the roofing is applied during cold weather or the gravel or slag is damp, it should be heated and dried immediately before application.

Not less than the following quantities of materials should be used for each 100 ft^2 of roof area:

Material	Weight, lb	
	Gravel	Slag
Four plies of 15-lb asphalt felt	60	60
Roofing asphalt	100	100
Gravel or slag	400	300
Total weight	560	460

The application previously specified is designated as four-ply roofing, with two plies dry and two plies mopped.

Sometimes specifications require that the decking first be covered with a single thickness of sheathing paper, weighing not less than 5 or 6 lb per square, with the edges of the sheets lapped at least 1 in.

Laying Built-up Roofing on Concrete

Before built-up roofing is applied on a concrete deck, the concrete should be cleaned and dried, after which a coat of concrete primer should be applied cold at a rate of about 1 gal per square. After the primer has dried, embed in hot asphalt two, three, four, or five plies of asphalt felt, lapping each sheet enough to give the required number of plies. Roll each sheet immediately behind the mop to ensure a uniform coating of hot asphalt, so that in no place shall felt touch felt. Over the entire area spread a uniform coating of hot asphalt into which, while hot, embed not less than 400 lb of dry gravel or 300 lb of dry slag per square.

For a three-ply roofing the quantities of materials given in the following table might be used per square.

Material	Weight, lb	
	Gravel	Slag
Concrete primer	10	10
Three plies of 15-lb asphalt felt	45	45
Roofing asphalt	125	125
Gravel or slag	400	300
Total weight	580	480

Cost of Materials

Table 16-7 gives representative costs of materials required for built-up roofing.

TABLE 16-7
**Representative costs of materials
required for built-up roofing**

Material	Unit	Cost per unit
Sheathing paper	Square	$2.15
Asphalt felt, 15 lb	Square	2.60
Asphalt felt, 30 lb	Square	5.10
Concrete primer	Square	3.05
Roofing asphalt	100-lb cartons	9.20

Labor Laying Built-up Roofing

The operations required to lay built-up roofing will vary with the type of roofing specified. If the time required to perform each operation is estimated, the sum of these times will give the total time for a roof or for completing a square. If a building is more than two or three stories high, additional time should be allowed for hoisting materials.

A typical crew for laying roofing on buildings up to three stories high would include one person each tending the kettle, handling hot asphalt or pitch, laying felt, rolling felt, and mopping, with a foreman supervising all operations. On some jobs one person may be able to tend the kettle and handle the asphalt or pitch.

Table 16-8 gives the labor-hours per square required to perform each operation and to complete all operations for various types of built-up roofing. Use

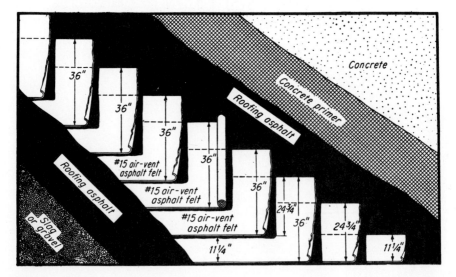

FIGURE 16-2
Method of laying built-up roofing on concrete.

TABLE 16-8
Labor-hours required to perform operations and lay a square of built-up roofing

Operation	Labor-hours†
Apply sheathing paper on wood deck	0.10–0.15
Apply primer on concrete	0.10–0.15
Lay 1 ply of roofing felt	0.10–0.15
Apply asphalt with mop	0.10–0.15
Apply asphalt and gravel	0.50–0.75
Apply 2-ply roofing on wood deck	1.3–2.0
Apply 3-ply roofing on wood deck	1.5–2.2
Apply 4-ply roofing on wood deck	1.7–2.5
Apply 5-ply roofing on wood deck	1.8–2.7
Apply 3-ply roofing on concrete	1.6–2.3
Apply 4-ply roofing on concrete	1.8–2.6
Apply 5-ply roofing on concrete	1.9–2.8

† The labor-hours for the last eight applications include three workers on the roof and two workers heating and supplying hot asphalt.

the lower values for large plain roofs and the higher values for roofs with irregular areas.

FLASHING

Flashing is installed to prevent water from passing into or through areas such as valleys and hips on roofs, where roofs meet walls, or where openings are cut through roofs. Materials used for flashing include sheets of copper, tin, galvanized steel, aluminum, lead, and sometimes mopped layers of roofing felt.

Flashing is usually measured by the linear foot for widths up to 12 in and by the square foot for widths greater than 12 in.

Metal Flashing

When metal flashing requires nails or other metal devices to hold it in place, it is essential that the fastener and the flashing be of the same metal; otherwise, galvanic action will soon destroy one of the metals.

Flashing Roofs at Walls

Where a roof and a parapet wall join, it is common practice to extend the layers of built-up roofing 4 to 8 in up the wall, with mopping applied to the wall and the felt, with no metal flashing underneath the felt. A metal counter flashing, whose upper edge is bent and inserted in a raggle or slot in the mortar joint between bricks about

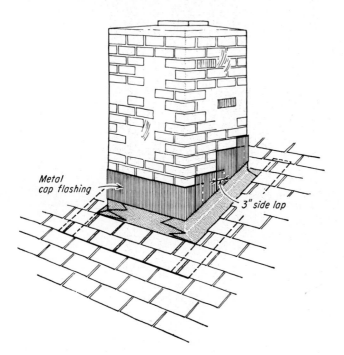

FIGURE 16-3
Metal cap flashing laid over base flashing.

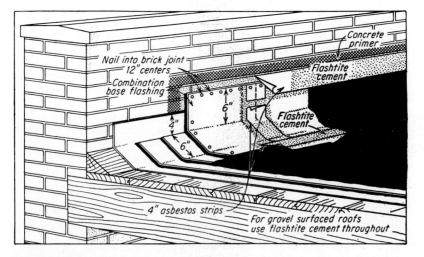

FIGURE 16-4
Method of flashing metal roof at parapet wall.

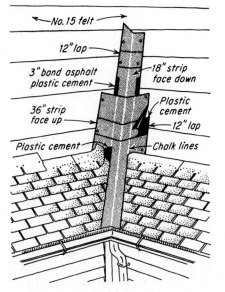

FIGURE 16-5
Use of roll roofing for valley flashing.

12 in above the roof, is installed to cover the portion of the felt flashing attached to the wall. If metal is installed under the felt and extended up the wall, it is called *base flashing*. Base and counter flashing should be soldered at all end joints.

Flashing Valleys and Hips

When roofs are covered with shingles, it is necessary to flash the valleys and hips, usually with metal flashing 10 to 30 in wide. The flashing may be soldered at end joints, or it may be lapped enough to eliminate danger of leakage. The quantity is measured by the linear foot.

Labor Required to Install Flashing

Flashing is installed by tinners. The labor required will vary with the material used, type of flashing required, and specifications covering the installation. Table 16-9 gives the labor-hours required to install flashing.

TABLE 16-9
Labor-hours required to install flashing

Class of work	Unit	Labor-hours
Metal base around parapet wall	100 lin ft	5.0– 6.0
Metal counter flashing around walls	100 lin ft	5.0– 6.0
Metal along roof and wood walls	100 lin ft	4.0– 5.0
Metal valleys and hips	100 lin ft	5.0– 6.0
Metal shingles along chimneys	100 each	8.0–10.0

TABLE 16-10
Costs of materials used for flashing per 100 lin ft

Galvanized steel:	
Plain ridge roll, 8 in wide	$ 25.92
Plain ridge roll, 10 in wide	31.68
Plain ridge roll, 12 in wide	36.64
Roll flashing, weight per 100 lin ft:	
8 in wide, 45 lb	21.60
10 in wide, 56 lb	26.40
12 in wide, 67 lb	30.54
14 in wide, 78 lb	34.82
16 in wide, 89 lb	40.04
24 in wide, 134 lb	58.96
Flashing shingles, 5 × 7 in, per 1,000	60.50
Aluminum flashing:	
4 in wide	13.56
5 in wide	15.44
6 in wide	17.64
20 in wide	56.28
Copper flashing, valley, etc., 16 oz	
12 in wide	180.00
16 in wide	225.00
20 in wide	315.00

Cost of Flashing

Table 16-10 lists prices for certain types of metal flashing materials.

PROBLEMS

For these problems use the national average wage rates and the costs of materials given in this book.

16-1. Estimate the total cost and the cost per square for furnishing and laying 15-lb asphalt felt and three-tab strip asphalt shingles, 12 by 36 in, on a wood roof decking whose area is 3,436 ft^2, for a building one-story high. The roof will have numerous valleys, hips, and dormers. Use carpenters to lay the roofing. The nails will cost $1.15 per pound.

16-2. Estimate the total cost and the cost per square for furnishing and laying 12- by 36-in fiberglass asphalt shingles weighing 340 lb per square. Lay one ply of 15-lb roofing felt over the wood decking prior to applying the shingles. Use carpenters to apply the felt and shingles.

The roof, whose area is 2,640 ft^2, is a simple gable type with a $\frac{1}{4}$ pitch. The nails will cost $1.15 per pound.

CHAPTER
17

PLUMBING

Although general contractors usually subcontract the furnishing of materials and the installation of plumbing in a building, they should have a reasonably good knowledge of the costs of plumbing. The fact that general contractors do not prepare detailed estimates for plumbing does not eliminate the need for estimating; it simply transfers the preparation to another party.

Plumbing involves the furnishing of materials, equipment, and labor to bring gas and water to a building; the furnishing and installation of fixtures; and the removal of the water and waste from the building. Water usually is obtained from a water main, and the waste is discharged into a sanitary sewer line. Materials include various types of pipe, fittings, valves, and fixtures, which will be more fully described later. Consumable supplies, such as lead, oakum, gasoline, etc., must be included in an estimate.

Plumbing Requirements

Some plans and specifications clearly define the types, grades, sizes, and quantities and furnish other information required for a complete plumbing installation, while other plans and specifications furnish limited information and place on the plumbing contractor the responsibility for determining what is needed to satisfy the owner and the local plumbing ordinance.

All cities have ordinances which require that plumbing installations conform with the plumbing code for the city in which the project will be constructed. A contractor must obtain a plumbing permit prior to starting an installation, and the work is checked by a plumbing inspector for compliance with the code. Although plumbing codes in different localities are similar, there will be variations which make it necessary for a plumbing contractor to be fully cognizant of the requirements of the code which will apply to a given project.

Plumbing Code

Some of but not all the requirements of the plumbing code for a major city are given here. The information is intended to serve as a guide in demonstrating the requirements of city codes for plumbing. When preparing an estimate of the cost of furnishing and installing items for plumbing services, the estimator should use the appropriate code for the area in which the project will be constructed.

Permit fee   A fee, which is generally paid by the plumbing contractors and included in their estimate, is charged by a city for issuing a plumbing permit. This fee may vary with the type and size of the project and with the location.

Plumbing-fixture facilities   Table 17-1 lists the minimum requirements for plumbing fixture facilities for a representative city. The requirements may differ in other cities.

Minimum-size trap and outlet   Table 17-2 gives the minimum-size trap and outlet permitted for the indicated fixture.

Fixture-unit values   Table 17-2 gives the value of each fixture unit for determining the relative load factors of different kinds of plumbing fixtures and estimating the total load carried by soil and waste pipe.

Sizes of Soil and Waste Pipe

Table 17-3 gives the maximum number of fixture units which may be connected to any given size of drain, soil, or waste pipe serving a building.

Steel Pipe

Black and galvanized steel pipes are available in standard, extra-strong, and double-extra-strong weights, with standard weights most commonly used for plumbing purposes. Table 17-4 gives the dimensions and weights of standard steel pipe.

Brass and Copper Pipe

Brass pipe and copper pipe are frequently used instead of steel pipe for water. They are available in standard and extra-strong grades. The external diameters and

TABLE 17-1
Minimum requirement for plumbing-fixture facilities (one fixture for each designated group)

Type of building	Water closet	Urinal	Lavatory	Drinking fountain	Shower	Bathtub	Kitchen sink
Dwellings and apartment houses	Each family	—	Each family	—	Choice of 1 per family		Each family
Places of employment, such as mercantile and office buildings, workshops, and factories where 5 or more persons work	25 males 20 females	25 males	15 persons	75 persons			
Foundries, mines, and places where exposed to dirty or skin-irritating materials where 5 or more persons work	25 males 20 females	25 males	5 persons	75 persons	15 males 15 females		
Schools	20 males 15 females	25 males	20 persons	75 persons			
Dormitories	10 males 8 females	25 males	6 persons	50 persons	8 males 10 females	40† males 35† females	

† Half may be additional showers.

TABLE 17-2

Fixture-unit values and minimum-size traps and outlets required

Fixture	Fixture-unit value as load factor	Minimum-size trap and outlet connection, in
Bathroom group consisting of water closet, lavatory and bathtub or shower stall and		
Tank with water closet	6	3
Flush-valve water closet	8	3
Bathtub with or without overhead shower	2	$1\frac{1}{2}$
Bidet	3	$1\frac{1}{2}$
Combination sink and tray	3	$1\frac{1}{2}$
Combination sink and tray with food disposal unit	4	$1\frac{1}{2}$
Dental unit or cuspidor	$\frac{1}{2}$	$1\frac{1}{4}$
Dental lavatory	1	$1\frac{1}{4}$
Drinking fountain	$\frac{1}{2}$	1
Dishwasher, domestic	2	$1\frac{1}{2}$
Floor drains	1	2
Kitchen sink, domestic	2	$1\frac{1}{2}$
Kitch sink, domestic, with food disposal unit	3	$1\frac{1}{2}$
Lavatory	2	$1\frac{1}{2}$
Lavatory, barber, beauty parlor	2	$1\frac{1}{2}$
Lavatory, surgeon's	2	$1\frac{1}{2}$
Laundry tray, 1- or 2- compartment	2	$1\frac{1}{2}$
Shower stall, domestic	2	2
Showers, group, per head	3	Varies
Sinks:		
Surgeon's	3	$1\frac{1}{2}$
Flushing rim, with valve	8	3
Service, standard trap	3	3
Service, with P trap	2	2
Pot, scullery, etc.	4	$1\frac{1}{2}$
Urinal, pedestal, syphon jet blowout	8	3
Urinal, wall lip	4	$1\frac{1}{2}$
Urinal, stall, washout	4	2
Urinal trough, each 2-ft section	2	$1\frac{1}{2}$
Wash sink, circular or multiple, each set of faucets	2	$1\frac{1}{2}$
Water closet:		
Tank-operated	4	3
Valve-operated	8	3

TABLE 17-3
Maximum number of fixture units permitted for soil and waste pipe

Pipe size, in	Fall or slope, in/ft			
	$\frac{1}{16}$	$\frac{1}{8}$	$\frac{1}{4}$	$\frac{1}{2}$
$2\frac{1}{2}$	—	2	24	31
3	—	20†	27‡	36‡
4	—	180	216	250
5	—	390	480	575
6	—	700	840	1,000
8	1,400	1,600	1,920	2,300
10	2,500	2,900	3,500	4,200
12	3,900	4,800	5,600	6,700

† Water closet not permitted.
‡ Not over two water closets permitted.

TABLE 17-4
Dimensions and weights of black and galvanized standard steel pipe

Size, in	Diameter, in		Thickness, in	Internal area, in^2	Weight, lb/lin ft	Threads per in
	External	Internal				
$\frac{1}{8}$	0.405	0.269	0.068	0.057	0.24	27
$\frac{1}{4}$	0.540	0.364	0.088	0.104	0.42	18
$\frac{3}{8}$	0.675	0.493	0.091	0.191	0.56	18
$\frac{1}{2}$	0.840	0.622	0.109	0.304	0.84	14
$\frac{3}{4}$	1.050	0.824	0.113	0.533	1.12	14
1	1.315	1.049	0.133	0.861	1.67	$11\frac{1}{2}$
$1\frac{1}{4}$	1.660	1.380	0.140	1.496	2.25	$11\frac{1}{2}$
$1\frac{1}{2}$	1.900	1.610	0.145	2.036	2.68	$11\frac{1}{2}$
2	2.375	2.067	0.154	3.356	3.61	$11\frac{1}{2}$
$2\frac{1}{2}$	2.875	2.467	0.203	4.780	5.74	8
3	3.500	3.066	0.217	7.383	7.54	8
$3\frac{1}{2}$	4.000	3.548	0.226	9.886	9.00	8
4	4.500	4.026	0.237	12.730	10.67	8
5	5.563	5.045	0.259	19.985	14.50	8
6	6.625	6.065	0.280	28.886	18.76	8
7	7.625	7.023	0.301	38.734	23.27	8
8	8.625	7.981	0.322	50.021	28.18	8
9	9.625	8.937	0.344	62.72	33.70	8
10	10.750	10.018	0.336	78.82	40.07	8
12	12.750	12.000	0.375	113.09	48.99	8

TABLE 17-5

Dimensions and weights of standard brass and copper pipe

Size, in	Diameter, in		Thickness, in	Weight, lb/lin ft	
	External	Internal		Brass	Copper
$\frac{1}{8}$	0.405	0.281	0.0620	0.246	0.259
$\frac{1}{4}$	0.540	0.375	0.0825	0.437	0.460
$\frac{3}{8}$	0.675	0.494	0.0905	0.612	0.644
$\frac{1}{2}$	0.840	0.625	0.1075	0.911	0.959
$\frac{3}{4}$	1.050	0.822	0.1140	1.235	1.299
1	1.315	1.062	0.1265	1.740	1.831
$1\frac{1}{4}$	1.660	1.368	0.1460	2.558	2.692
$1\frac{1}{2}$	1.900	1.600	0.1500	3.038	3.196
2	2.375	2.062	0.1565	4.018	4.228
$2\frac{1}{2}$	2.875	2.500	0.1875	5.832	6.136
3	3.500	3.062	0.2190	8.316	8.75
$3\frac{1}{2}$	4.000	3.500	0.2500	10.85	11.42
4	4.500	4.000	0.2500	12.30	12.94
$4\frac{1}{2}$	5.000	4.500	0.2500	13.74	14.46
5	5.563	5.062	0.2500	15.40	16.20
6	6.625	6.125	0.2500	18.45	19.41
7	7.625	7.062	0.2815	23.92	25.17
8	8.625	8.000	0.3125	30.06	31.63
9	9.625	8.937	0.3440	36.95	38.88
10	10.750	10.019	0.3655	43.93	46.22
12	12.750	12.000	0.3750	53.71	56.51

number of threads per inch are the same as for steel pipe. Table 17-5 gives the dimensions of brass and copper pipe.

Copper Tubing

Copper tubing is frequently used instead of steel pipes for water. It is furnished in both hard and soft tempers. Both tempers are available in straight 20-ft lengths, and the soft temper is furnished in 30-, 45-, and 60-ft coils in sizes to $1\frac{1}{4}$-in diameter. The tubes are joined by sweating the ends into special fittings, using a solder or brazing alloy, which is heated with a blowtorch. Because of the flexibility, the tubing can be bent easily to change directions, thus reducing the number of fittings and the amount of labor required. This tubing is frequently used instead of lead pipe for connections from water mains into buildings.

Table 17-6 gives the dimensions and weights of soft- and hard-temper copper tubing.

TABLE 17-6
Dimensions and weights of soft- and hard-temper copper tubing

Nominal size, in	Outside diameter, in	Type K soft		Type M hard	
		Wall thickness, in	Weight, lb/lin ft	Wall thickness, in	Weight, lb/lin ft
$\frac{1}{4}$	$\frac{3}{8}$	0.032	0.133	0.025	0.106
$\frac{3}{8}$	$\frac{1}{2}$	0.049	0.269	0.025	0.144
$\frac{1}{2}$	$\frac{5}{8}$	0.049	0.344	0.028	0.203
$\frac{5}{8}$	$\frac{3}{4}$	0.049	0.418		
$\frac{3}{4}$	$\frac{7}{8}$	0.065	0.641	0.032	0.328
1	$1\frac{1}{8}$	0.065	0.839	0.035	0.464
$1\frac{1}{4}$	$1\frac{3}{8}$	0.065	1.04	0.042	0.681
$1\frac{1}{2}$	$1\frac{5}{8}$	0.072	1.36	0.049	0.940
2	$2\frac{1}{8}$	0.083	2.06	0.058	1.46
$2\frac{1}{2}$	$2\frac{5}{8}$	0.095	2.92	0.065	2.03
3	$3\frac{1}{8}$	0.109	4.00	0.072	2.68
$3\frac{1}{2}$	$3\frac{5}{8}$	0.120	5.12	0.083	3.58
4	$4\frac{1}{8}$	0.134	6.51	0.095	4.66
5	$5\frac{1}{8}$	0.160	9.67	0.109	6.66
6	$6\frac{1}{8}$	0.192	13.87	0.122	8.91
8	$8\frac{1}{8}$	0.271	25.90	0.170	16.46
10	$10\frac{1}{8}$	0.388	40.26	0.212	25.57
12	$12\frac{1}{8}$	0.405	57.76	0.254	36.69

PVC Water-Supply Pipe

This pipe, made of polyvinyl chloride (PVC), is a rigid plastic pipe capable of withstanding internal water pressures up to 160 or 200 lb/in^2. It is lightweight and highly resistant to corrosion, has low resistance to the flow of water, and can be installed rapidly and economically.

Joints between the pipe and fittings are made by applying a coating of solvent cement around the end of the pipe and then inserting it into the fitting.

This pipe is not suitable for use with hot water.

Table 17-7 lists representative sizes of PVC pipe and costs. Table 17-8 lists representative PVC fittings for use with the pipe, with current costs.

Indoor CPVC Plastic Water Pipe

This pipe, made of chlorinated polyvinyl chloride (CPVC), may be used inside a building for hot or cold water, if local building codes permit. It will withstand internal water pressures up to 100 lb/in^2 or more, at temperatures up to 180°F.

TABLE 17-7
Representative costs of PVC plastic water pipe

Pipe size, in	Cost per lin ft
$\frac{1}{2}$	$0.11
$\frac{3}{4}$	0.13
1	0.18
$1\frac{1}{4}$	0.24
$1\frac{1}{2}$	0.29
2	0.37

Joints between pipes and fittings are made by applying coatings of solvent cements to the ends of the pipes and then inserting them into the fittings.

Labor Installing Plastic Water Pipe

Installing plastic water pipe involves cutting standard lengths of pipe to the desired lengths, if necessary using a hacksaw, applying a solvent cement to the ends, and then inserting the ends into plastic fittings. This forms a solid joint, which should be free of any leakage.

Connections between plastic pipes and standard steel pipes can be made by using plastic female or male adapters, which are threaded at one end to match the threads of the steel pipe.

Table 17-9 gives representative times required to perform various operations in installing plastic pipe.

TABLE 17-8
Representative costs of fittings for PVC plastic water pipe

Fitting	Price each by size, in				
	$\frac{1}{2}$	$\frac{3}{4}$	1	$1\frac{1}{4}$	$1\frac{1}{2}$
90° elbow	$0.18	$0.19	$0.24	$0.46	$0.65
45° elbow	0.23	0.37	0.44	0.63	0.88
Tee	0.18	0.24	0.42	0.58	0.74
Coupling	0.10	0.12	0.22	0.26	0.43
Male adapter	0.13	0.15	0.25	0.30	0.43
Female adapter	0.15	0.20	0.24	0.36	0.41

TABLE 17-9
Representative hours required to install plastic water pipe and fittings

	Size of pipe, in				
Operation	$\frac{1}{2}$	$\frac{3}{4}$	1	$1\frac{1}{4}$	$1\frac{1}{2}$
Cut pipe	0.08	0.10	0.11	0.12	0.13
Apply cement and join pipe and fittings†	0.05	0.05	0.06	0.07	0.08
Join plastic pipe to steel pipe with adapter	0.20	0.20	0.25	0.25	0.30

† This is the time required for making a joint. A coupling and an elbow require two joints, and a tee requires three joints.

Soil, Waste, and Vent Pipes

The pipes which convey the discharge liquids from plumbing fixtures to the house drain are called soil and waste pipes. Pipes which receive the discharge from water closets are called *soil pipes*, while the pipes that receive the discharge from other fixtures are called *waste pipes*. Pipes which provide ventilation for a house drainage system and prevent siphonage of water from traps are called *vent pipes*.

Cast-iron soil pipe is furnished in 5-ft lengths, with single or double hubs, in standard and extra-heavy weights. The weights are given in Table 17-10.

The joints for cast-iron soil pipe are usually made with lead and oakum. The oakum, which is wrapped around the spigot end of the pipe, should be well caulked into the hub of the joint to center the spigot and to prevent molten lead from flowing into the pipe. An asbestos runner is wrapped around the pipe adjacent to the hub, with an opening at the top, into which molten lead is poured to fill the joint in one operation. After the lead solidifies, the runner is removed, and the lead is heavily caulked.

TABLE 17-10
**Approximate weights of soil pipe,
lb/lin ft**

Size, in	Standard, single hub	Extra heavy	
		Single hub	Double hub
2	3.6	5.0	5.2
3	5.2	9.0	9.4
4	7.0	12.0	12.6
5	9.0	15.0	15.6
6	11.0	19.0	20.0
8	17.0	30.0	31.4
10	23.0	43.0	45.0
12	33.0	54.0	57.0

TABLE 17-11
Approximate quantities of lead and oakum required for cast-iron soil pipe

Size pipe, in	Weight per joint, lb	
	Lead	Oakum
2	1.5	0.13
3	2.5	0.16
4	3.5	0.19
5	4.3	0.22
6	5.0	0.25
8	7.0	0.38
10	9.0	0.50
12	11.0	0.70

Table 17-11 gives the approximate quantity of lead and oakum required for a joint in cast-iron soil pipe.

House Drain Pipe

The cast-iron soil pipe must extend a specified distance outside a building, depending on the plumbing code. The balance of the drain pipe extending to the sanitary sewer main may be vitrified clay, concrete sewer pipe, or sometimes a composition pipe of fiber and bitumen. Joints for clay and concrete pipes may be made with cement mortar or an asphaltic jointing compound.

Fittings

Fittings for black and galvanized-steel pipe should be malleable iron. Fittings for copper and brass pipe and tubing should be copper and brass, respectively, threaded or sweat type. Fittings for plastic pipe should be of the same material as the pipe. If different types of plastic are used, the solvent cement will not be effective on both types of materials. Fittings for cast-iron pipe should be cast iron.

Adapters are available which may be used in joining one type of pipe or fitting to another type.

Valves

Valves are installed to shut off the flow of water. Several types are used, including globe, gate, check, sill cocks, drain, etc. They are usually made from bronze and brass.

Traps

Traps are installed below the outlets from fixtures to retain water as a seal to prevent sewer gases from entering a building. Vent pipes installed into the drain pipes below the traps prevent the water in the traps from being siphoned out.

Roughing in Plumbing

Roughing in includes the installation of all water pipes from the meter into and through a building, soil pipe, waste pipe, drains, vents, traps, plugs, cleanouts, etc., but does not include the installation of plumbing fixtures, which is called *finish plumbing*.

Estimating the Cost of Roughing in Plumbing

In preparing a detailed estimate covering the cost of furnishing materials, equipment, and labor for roughing in the plumbing, a comprehensive list of items required should be used as a check and for establishing all costs. Each item should be listed separately by description, grade, size, quantity, and cost, both unit and total. If there are other costs such as tapping fees, permits, or removing and replacing pavement, then they must be included in an estimate. Table 17-12 may be used as a guide in preparing a list for estimating purposes.

Cost of Materials for Rough Plumbing

While the costs of materials for rough plumbing will vary with grades, quantities, locations, and time, the costs given in the tables in this book may be used as a guide in preparing an approximate estimate. The actual costs that will apply should be determined and used in preparing an estimate for bid purposes.

Cost of Lead, Oakum, and Solder

Ingot or pig lead used for joints with cast-iron pipe will cost about $1.05 per pound. Oakum will cost about $1.65 per pound. Wiring solder used to sweat joints for copper pipe and tubing will cost about $4.20 per pound.

Plastic Drainage Pipe and Fittings

Many building codes permit the use of plastic drainage pipe and fittings, such as ABS-DWV ASTM-2661-68, or later, for use in draining liquids from buildings. These pipes are generally available in sizes $1\frac{1}{4}$ in in diameter and larger and in lengths of 10 or 20 ft or more. Joints are made by applying a solvent cement to the spigot end of a pipe before inserting it into a fitting. Adapters are available to permit this pipe to be connected into cast-iron pipe.

TABLE 17-12
Checklist for estimating the cost of roughing in plumbing

Description	Quantity	Unit cost	Total cost
Permits			
Tapping fees			
Removing and replacing pavement			
Water meter and box, if required			
Steel pipe			
Copper pipe and tubing			
Pipe fittings			
Valves			
Soil pipe			
Water pipe			
Vent pipe and stacks			
Traps, cleanouts			
Cast-iron fittings			
Floor drains			
Roof flashings			
Vitrified-clay sewer pipe and fittings			
Concrete sewer pipe and fittings			
Catch basins and covers			
Manholes and covers			
Lead, solder, oakum, gasoline, etc.			
Cement and sand			
Brackets, hangers, supports, etc.			
Other items			

Figure 17-1 illustrates an assembly of pipes and fittings representing a typical drainage system for a home or other building. Table 17-13 lists the items illustrated in Fig. 17-1, together with representative costs of the items.

Labor Required to Rough in Plumbing

Plumbers frequently work in teams of two, a plumber and a helper. However, since this arrangement is not always followed, the labor time given in the tables is expressed in labor-hours, which includes the combined time for a plumber and a helper.

The operations required to install water pipe will include cutting and threading the pipe and screwing it together with the appropriate fittings. Cutting and threading may be done with hand or power tools. The time required will vary with the size of the pipe, tools used, and working conditions at the job. If power tools can be moved along with the work, a minimum amount of time will be required; but if the pipe must be carried some distance to the tools, a great deal of

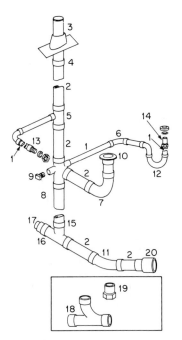

FIGURE 17-1
Representative piping and fittings for a drain system.

time will be consumed in walking. If the pipe is suspended from hangers, additional time will be required to install the hangers. Some specifications require that all pipe cut on the job be reamed to remove burrs. This will consume additional time.

Copper tubing is furnished in coils up to 60 ft long in all sizes to $1\frac{1}{4}$ in. Since the tubing is longer than steel pipe and changes in direction are obtained by bending the tubing, fewer fittings are required and the labor time will be less than for steel pipe. Joints usually are made by sweating the tubing into fittings, using solder and a blowtorch. Some cutting of the tubing with a hacksaw will be necessary.

Cast-iron soil, waste, and vent pipes are joined with molten lead and oakum joints. Many of these joints can be made with the pipe or fittings standing in an upright position, which will reduce the time required. Line joints are made by pouring the lead with the pipe in place, using an asbestos runner to hold the lead in the joint. After the lead has solidified, it must be caulked. Time should be estimated by the number of joints.

Vitrified-clay sewer pipe is manufactured in lengths varying from $2\frac{1}{2}$ to 4 ft. Joints are made with oakum and cement mortar or a heated asphaltic joint compound.

Many plumbing contractors add a percentage of the cost of materials to cover the cost of labor required to rough in plumbing. The rates used vary from 40 to 80 percent of the cost of materials. Although this is a simple operation, it does not necessarily produce an accurate estimate.

TABLE 17-13
Representative costs of PVC plastic drainage and waste pipe and fittings

Number	Item	Size, in	Unit	Unit cost
1	PVC pipe	$1\frac{1}{2}$	Lin ft	$0.29
	PVC pipe	2	Lin ft	0.37
2	PVC pipe	3	Lin ft	0.89
3	Roof flashing, neoprene	$1\frac{1}{2}$	Each	3.75
		2	Each	3.90
		3	Each	3.90
4	Coupling	$1\frac{1}{2}$	Each	0.17
		2	Each	0.21
		3	Each	0.55
5	Sanitary tee	$1\frac{1}{2}$	Each	0.42
		2	Each	0.65
		3	Each	1.54
6	90° elbow	$1\frac{1}{2}$	Each	0.27
		2	Each	0.37
7	90° elbow	3	Each	1.16
8	Sanitary tee with two $1\frac{1}{2}$-in side outlets	3	Each	1.98
9	Slip plug	$1\frac{1}{2}$	Each	0.35
10	Reducing closet flange	4×3	Each	3.00
11	45° elbow	$1\frac{1}{2}$	Each	0.24
		2	Each	0.34
		3	Each	0.98
12	P trap with union	$1\frac{1}{2}$	Each	0.90
		2	Each	1.75
13	Trap adapter, $1\frac{1}{2}$-in pipe to $1\frac{1}{4}$-in waste	$1\frac{1}{4}$	Each	0.48
14	Trap adapter, $1\frac{1}{2}$-in pipe to $1\frac{1}{2}$-in waste	$1\frac{1}{2}$	Each	0.45
15	Wye branch	$1\frac{1}{2}$	Each	0.72
		2	Each	0.92
		3	Each	1.65
16	Cleanout adapter	$1\frac{1}{2}$	Each	0.35
		2	Each	0.59
		3	Each	1.10
17	Threaded cleanout plug	$1\frac{1}{2}$	Each	0.25
		2	Each	0.27
		3	Each	0.47
18	Combination Y and $\frac{1}{8}$ bend	3	Each	2.82
19	Male iron pipe adapter	$1\frac{1}{2}$	Each	0.26
		2	Each	0.45
20	Adapter, plastic to iron hub	$1\frac{1}{2} \times 2$	Each	0.90
		2×2	Each	0.90
	Adapter, plastic to plastic hub	3×4	Each	1.85

TABLE 17-14
Representative cost of standard-weight steel pipe

Pipe size, in	Cost per linear foot	
	Black	Galvanized
$\frac{1}{2}$	$0.32	$0.36
$\frac{3}{4}$	0.40	0.46
1	0.60	0.71
$1\frac{1}{4}$	0.79	0.96
$1\frac{1}{2}$	0.95	1.15
2	1.25	1.50
$2\frac{1}{2}$	1.99	2.39
3	2.61	3.13
4	3.79	4.54

Table 17-23 gives representative time in labor-hours required to rough in plumbing. The values are based on an analysis of the rates reported by a substantial number of plumbing contractors whose individual rates varied considerably. The rates given in the table include the combined time for plumbers and helpers. Thus, if a plumber and a helper work together, one-half of the labor-hours should be assigned to each person.

TABLE 17-15
Representative cost of malleable fittings for steel pipe per fitting

Fitting	Size, in					
	$\frac{1}{2}$	$\frac{3}{4}$	1	$1\frac{1}{4}$	$1\frac{1}{2}$	2
Black						
Couplings	$0.52	$0.57	$0.88	$1.09	$1.37	$1.97
Elbows	0.32	0.39	0.74	1.22	1.60	2.35
Reducers	0.54	0.57	0.88	1.26	1.46	1.97
Tees	0.43	0.63	1.16	1.89	2.33	5.40
Unions	1.43	1.47	2.16	3.00	3.63	4.62
Galvanized						
Couplings	0.52	0.60	0.95	1.21	1.47	2.11
Elbows	0.35	0.29	0.83	1.28	1.69	2.45
Reducers	0.51	0.60	0.95	1.24	1.47	2.11
Tees	0.43	0.74	1.24	1.96	2.45	3.51
Unions	1.52	1.56	2.30	3.13	3.82	4.75

TABLE 17-16

Cost of copper pipe and tubing per linear foot

Pipe size, in	Type L, hard or soft	Type K, hard or soft
$\frac{3}{8}$	$0.92	$1.40
$\frac{1}{2}$	1.36	1.74
$\frac{3}{4}$	2.08	2.56
1	2.84	3.14
$1\frac{1}{4}$	3.14	4.24
$1\frac{1}{2}$	4.24	4.98
2	7.12	8.58

TABLE 17-17

Cost of copper solder fittings, per fitting

Fittings	Size, in					
	$\frac{1}{2}$	$\frac{3}{4}$	1	$1\frac{1}{4}$	$1\frac{1}{2}$	2
Couplings	$0.11	$0.18	$ 0.43	$ 0.70	$ 1.06	$ 1.51
Elbows	0.11	0.27	0.60	1.18	1.48	2.79
Tees	0.20	0.43	1.64	2.40	3.44	5.39
Unions	1.07	1.62	2.59	3.50	4.91	8.36
Gate valves	6.96	8.70	11.09	16.17	19.25	25.30

TABLE 17-18

Cost of 125-lb brass screwed valves, each

Valves	Size, in					
	$\frac{1}{2}$	$\frac{3}{4}$	1	$1\frac{1}{4}$	$1\frac{1}{2}$	2
Globe and angle	$6.96	$8.70	$11.08	$16.17	$19.25	$25.30
Gate and check	6.96	9.06	12.00	18.90	22.00	29.80
Hose bib	1.98	2.18				

TABLE 17-19

Cost of cast-iron soil pipe per joint

Pipe size, in	Single hub, 10 lin ft		Double hub, 5 lin ft	
	Standard	Extra heavy	Standard	Extra heavy
2	$15.29	$ 21.65	$13.16	$17.87
3	21.23	29.72	14.86	23.35
4	27.60	39.06	19.46	29.25
5	38.64	51.80	30.57	37.36
6	46.28	58.59	36.52	45.01
8	77.28	104.88		

TABLE 17-20
Cost of single-hub fittings for cast-iron soil pipe, per fitting

Fitting	Size, in				
	2	3	4	6	8
Quarter-bend, standard	$2.51	$ 4.63	$ 6.88	$11.89	$ 36.52
Extra heavy	2.80	5.86	8.75	14.63	36.13
Eight-bend, standard	1.95	4.63	6.88	9.89	26.33
Extra heavy	2.46	4.76	6.88	11.17	28.79
T and Y branch, standard	6.11	9.13	12.65	29.25	71.37
Extra heavy	7.64	12.74	16.60	32.10	72.18
Cross, standard	5.90	10.45	13.93	30.23	64.54
Extra heavy	9.64	14.18	19.96	32.69	77.28
Running trap, with hub vent		19.91	27.69	38.94	116.77
Extra heavy					
Plain trap, S and P, standard	8.24	14.36	26.50	63.69	

TABLE 17-21
Cost of vitrified-clay sewer pipe, per linear foot

Pipe size, in	ASTM C-700-71	ASTM C-425-71, rubber seal
4	$ 1.56	$ 1.56
6	2.84	2.32
8	3.75	3.45
10	5.10	5.00
12	6.65	6.56
15	12.82	11.64
18	16.75	16.44

TABLE 17-22
Cost of vitrified-clay sewer pipe fittings with rubber seals, per fitting

Pipe size, in	Wyes and tees	$\frac{1}{8}, \frac{1}{4}, \frac{1}{2}$ bends	Cleanout wyes	Stoppers
4	$ 5.64	$ 5.36		$1.48
6	12.86	9.24	$14.82	1.96
8	16.58	14.75	21.96	2.84
10	24.52	36.86		3.74
12	39.58	36.74		5.24

TABLE 17-23
Labor-hours required for roughing in plumbing

Class of work	Labor-hours
Hand cut, thread, and install steel pipe, per joint:	
$\frac{1}{2}$- and $\frac{3}{4}$-in pipe	0.6–0.7
1- and $1\frac{1}{4}$-in pipe	0.8–0.9
$1\frac{1}{2}$- and 2-in pipe	1.3–1.5
$2\frac{1}{2}$- and 3-in pipe	1.8–2.4
4-in pipe	2.2–2.5
Machine cut, thread, and install steel pipe, per joint:	
$\frac{1}{2}$- and $\frac{3}{4}$-in pipe	0.4–0.7
1- and $1\frac{1}{4}$-in pipe	0.5-0.6
$1\frac{1}{2}$- and 2-in pipe	0.7–0.8
$2\frac{1}{2}$- and 3-in pipe	0.9–1.0
4-in pipe	1.2–1.4
Install copper tubing, per joint:	
$\frac{1}{2}$- and $\frac{3}{4}$-in tubing	0.3–0.4
1- and $1\frac{1}{4}$-in tubing	0.4–0.5
$1\frac{1}{2}$- and 2-in tubing	0.5–0.6
$2\frac{1}{2}$- and 3-in tubing	0.6–0.7
4-in tubing	0.7–0.8
Install cast-iron soil pipe and fittings, per joint:	
2-in diameter	0.3–0.4
3-in diameter	0.4–0.5
4-in diameter	0.5–0.6
6-in diameter	0.7–0.9
8-in diameter	1.0–1.3
Install vitrified-clay or concrete pipe, per joint:	
4-in diameter	0.2–0.3
6-in diameter	0.2–0.3
8-in diameter	0.3–0.4
10-in diameter	0.5–0.6
12-in diameter	0.7–0.9
Rough in for fixtures:	
Bathtub	10–18
Bathtub with shower	16–24
Floor drain	4–6
Grease trap	5–10
Kitchen sink, single	8–14
Kitchen sink, double	10–16
Laundry tub, 2-compartment	8–12
Lavatory	10–15
Shower with stall	12–18
Slop sink	8–12
Urinal, pedestal type	8–12
Urinal with stall	10–14
Urinal, wall type	7–10
Water closet	9–18
Water heater, 30 to 50-gal automatic	10–12
Water heater, 50- 100-gal automatic	12–15

FINISH PLUMBING

Finish plumbing includes fixtures such as lavatories, bathtubs, shower stalls, water closets, sinks, urinals, water heaters, etc., and the accessories which usually are supplied with the fixtures. The prices of finish plumbing vary a great deal with the type, size, and quality specified. No book of this kind can give a complete listing of fixtures and prices. An estimator should list each item required for a building and then refer to a current catalog for prices.

Table 17-24 gives representative prices for plumbing fixtures in the medium-grade range; these should be used as a guide only.

Labor Required to Install Fixtures

The labor required to install plumbing fixtures will vary with the kind and quality of fixture, the kind of accessories used, the type of building, and to some extent the building code. The simplest fixtures will require the least time, and the more elaborate fixtures will generally require the most time.

A lavatory to be attached to the wall will be delivered to a job packed in a crate, complete with wall brackets, faucets, pop-up stopper, and other accessories, unassembled. It is necessary to remove it from the crate, attach the brackets, then attach and level the lavatory, connect the hot- and cold-water lines, and connect the trap and drain into the waste pipe. Additional time will be required to install legs and towel bars.

More time is required to install a combination bathtub with a shower than to install a bathtub only. Also considerably more time is required to install a built-in bathtub, especially in a bathroom with tile floors and walls, than to install a tub

TABLE 17-24

Representative costs of plumbing fixtures and accessories

Fixture	Cost per unit
Bathtub, porcelain on cast iron, flat bottom	
5 ft 0 in	$164.00
5 ft 6 in	390.00
Kitchen sink, enamel on cast iron	
Single, $24 \times 21 \times 8$ in	69.00
Double, $32 \times 21 \times 8$ in	89.00
Kitchen sink, stainless steel	
Single, $24 \times 21 \times 8$ in	38.00
Double, $33 \times 22 \times 8$ in	36.00
Lavatory, porcelain on cast iron, 18×24 in	54.00
Urinal stall, vitreous china	50.00
Water closet, vitreous china, close-coupled	189.00
Garbage disposal unit, with $\frac{1}{2}$-hp motor	134.00

TABLE 17-25
Labor-hours required to install plumbing
fixtures and accessories

Fixtures	Labor-hours
Bathtub, leg type	10–16
Bathtub, flat bottom, no shower	14–20
Bathtub, flat bottom, with shower	18–26
Kitchen sink, enamel, single	6–10
Kitchen sink, enamel, double	8–12
Laundry tub, double	10–14
Lavatory, wall type	5–8
Lavatory, pedestal type	6–9
Urinal, wall type	8–12
Urinal, pedestal type	9–13
Urinal with stall	16–20
Water closet	6–9
Garbage disposal unit	6–8
Drinking fountain	5–8

supported on legs. All factors that affect the rates of installation must be considered in preparing an estimate.

Table 17-25 gives representative labor-hours required to install plumbing fixtures, based on reports furnished by a substantial number of plumbing contractors. Use the lower rates for simple fixtures installed under favorable conditions and the higher rates for more complicated fixtures installed under more difficult conditions.

PROBLEMS

For these problems use the national average wage rates and the costs of materials given in this book. Use a plumber and a helper as a team to perform the work.

17-1. Estimate the direct cost of furnishing and installing 286 lin ft of 1-in galvanized steel pipe, whose average length per joint prior to cutting will be 20 ft 0 in. It will be necessary to cut and thread the pipe for installing eight galvanized tees, six galvanized elbows, four galvanized unions, and three brass gate valves. The pipe will be cut and threaded by hand.

17-2. Estimate the direct cost of furnishing and installing the pipe, fittings, and valves of Prob. 17-1 when the pipe is cut and threaded by machines.

17-3. Estimate the cost of furnishing and installing 468 lin ft of $\frac{3}{4}$-in type-K soft-copper tube, including 14 couplings, 26 tees, 16 elbows, and 4 unions, using sweated joints.

17-4. Estimate the cost of furnishing and installing 426 lin ft of $1\frac{1}{2}$-in type-L copper pipe, including 16 couplings, 12 tees, and 6 ells, using sweated joints.

17-5. Estimate the cost of furnishing and installing 256 lin ft of 4-in single-hub standard cast-iron pipe, including six wyes, eight standard quarter bends, and three P traps, all standard fittings, under average conditions. All joints will be made with oakum and lead.

17-6. Estimate the cost of furnishing and installing 280 lin ft of $1\frac{1}{2}$-in-diameter ABS drain pipe supplied in 20-ft lengths. The following fittings will be required, with each requiring that the pipe be cut: six sanitary tees and five 90° elbows. The cost of a 1-pt can of solvent cement will be $4.20. This is enough cement to make 50 joints with $1\frac{1}{2}$-in pipe.

CHAPTER
18

ELECTRIC
WIRING

Electric wiring is generally installed by a subcontractor who specializes in this type of work, with all materials and labor furnished under the contract. This procedure does not eliminate the need for estimating a job; it simply transfers the operation from the general to the electrical contractor.

Approximate estimates, and sometimes estimates for bid purposes, are prepared by counting the number of outlets required for a job and then applying a unit price to each outlet to arrive at the total cost. While this method will permit estimates to be prepared quickly, it will not always give an accurate cost for a job, because conditions may vary a great deal between two jobs. When this method is used, each switch, plug, and fixture served will count as an outlet.

Factors which Affect the Cost of Wiring

The cost of wiring a building will be affected by the type, size, and arrangement of the building and the materials used in constructing the building. Wiring for a frame building may be attached to the framing members or run through holes drilled in the members. Wiring for a building constructed of masonry may be placed in conduit attached to or concealed within the walls. Wiring for concrete structures is placed in conduit installed in the concrete, which will require the installation of the conduit prior to placing the concrete.

The cost per outlet will be affected by the lengths of runs between outlets, kind of installation, and sizes of wires required. Job specifications may require no conduit, rigid or flexible conduit, armored cable, or nonmetallic cable. The costs of these materials and the labor required to install them may vary considerably. Unless an estimator is able to apply a dependable cost factor, based on job conditions, to a particular job, the outlet method may result in a substantial variation from the true cost.

The best method of preparing a dependable estimate is to prepare an accurate material takeoff which lists the items separately by type, size, quantity, quality, and cost. When you are working from plans, it is good practice to indicate on the plans, using a colored pencil, each item as it is transferred to the material list. This procedure will reduce the danger of omitting items or counting them more than once. The application of appropriate costs for materials and labor to these items will give a dependable estimate.

Items Included in the Cost of Wiring

When you are preparing a detailed estimate covering the cost of wiring a project, it is good practice to start with the first item required to bring electric service to the project. This may be the service wires from the transmission lines to the building, unless they are furnished by the utility. The listing should proceed from the first item to and including the last ones, which may be the fixtures. The following list may be used as a guide:

1. Service wires, if required
2. Meter box
3. Entrance switch box and fuses or circuit breakers
4. Wire circuits
   a. Conduit
   b. Wires
5. Junction boxes
6. Outlets for
   a. Switches, single-pole, double-pole, three-way
   b. Plugs, wall and floor
   c. Fixtures
7. Bell systems, with transformers
8. Sundry supplies, such as solder, tape, gasoline, etc.

Types of Wiring

The types of wiring used in buildings may be goverened by the National Electrical Code (NEC), published by the National Board of Fire Underwriters, or in some locations by state or municipal codes, which specify minimum requirements for materials and workmanship. Codes require that electrical wiring be protected against physical damage and possible short circuits by one or more methods such as installing them in conduit or by the use of wires that are furnished with factory-wrapper coverings.

Rigid Conduit

Rigid conduit is made of light galvanized pipe. Heavy conduit, which may be used for outside or inside installations, is threaded at each end. Since the walls of light conduit are too thin for threads, it is necessary to use special clamping fittings for making joints or entering boxes. The conduit extends into outlet or other boxes, where it is securely fastened with locknuts and bushings or with clamps. The conduit must be adequately supported and fastened in position along its run. The electric wires are pulled through the conduit at any desirable time after the conduit is installed.

Table 18-1 gives the sizes of conduit required for the indicated sizes and numbers of wires, as specified by the NEC.

Flexible Metal Conduit

This conduit consists of an interlocking spiral steel armor, which is constructed in such a manner that it permits considerable flexibility. It may be used where

TABLE 18-1
Size conduit required for rubber-covered electric wires

Size of wire, B & S gage	Maximum no. of wires in one conduit								
	1	2	3	4	5	6	7	8	9
	Size conduit required, in								
14	$\frac{1}{2}$	$\frac{1}{2}$	$\frac{1}{2}$	$\frac{1}{2}$	$\frac{3}{4}$	$\frac{3}{4}$	$\frac{3}{4}$	1	1
12	$\frac{1}{2}$	$\frac{1}{2}$	$\frac{1}{2}$	$\frac{3}{4}$	$\frac{3}{4}$	1	1	1	$1\frac{1}{4}$
10	$\frac{1}{2}$	$\frac{1}{2}$	$\frac{3}{4}$	$\frac{3}{4}$	1	1	1	$1\frac{1}{4}$	$1\frac{1}{4}$
8	$\frac{1}{2}$	$\frac{3}{4}$	1	1	$1\frac{1}{4}$	$1\frac{1}{4}$	$1\frac{1}{4}$	$1\frac{1}{4}$	$1\frac{1}{2}$
6	$\frac{1}{2}$	1	$1\frac{1}{4}$	$1\frac{1}{4}$	$1\frac{1}{2}$	$1\frac{1}{2}$	2	2	2
5	$\frac{1}{2}$	$1\frac{1}{4}$	$1\frac{1}{4}$	$1\frac{1}{4}$	$1\frac{1}{2}$	2	2	2	2
4	$\frac{1}{2}$	$1\frac{1}{4}$	$1\frac{1}{4}$	$1\frac{1}{2}$	2	2	2	2	$2\frac{1}{2}$
3	$\frac{3}{4}$	$1\frac{1}{4}$	$1\frac{1}{4}$	$1\frac{1}{2}$	2	2	2	$2\frac{1}{2}$	$2\frac{1}{2}$
2	$\frac{3}{4}$	$1\frac{1}{4}$	$1\frac{1}{4}$	$1\frac{1}{2}$	2	2	$2\frac{1}{2}$	$2\frac{1}{2}$	$2\frac{1}{2}$
1	$\frac{3}{4}$	$1\frac{1}{2}$	$1\frac{1}{2}$	2	2	$2\frac{1}{2}$	$2\frac{1}{2}$	3	3
0	1	$1\frac{1}{2}$	2	2	$2\frac{1}{2}$	$2\frac{1}{2}$	3	3	3
00	1	2	2	$2\frac{1}{2}$	$2\frac{1}{2}$	3	3	3	$3\frac{1}{2}$
000	1	2	2	$2\frac{1}{2}$	3	3	3	$3\frac{1}{2}$	$3\frac{1}{2}$
0000	$1\frac{1}{4}$	2	2	$2\frac{1}{2}$	3	3	$3\frac{1}{2}$	$3\frac{1}{2}$	4

frequent changes in direction are necessary. End connections must be made with special fittings.

Armored Cable

Armored cables or conductors are made with or without a lead sheath under the armor. Special fittings are required to connect this cable into boxes, outlets, etc.

Nonmetallic Cable

The wires of this cable are covered with a flexible insulated braiding impregnated with a moisture-resisting compound. It is furnished in wire sizes 14 to 4 gage. It is very popular for use in residences.

Electric Wire

Electric wires are usually made of copper, either solid or standard. The wires may be rubber-covered for use indoors or weatherproof for use outdoors. The size of a wire is designated by specifying the Brown and Sharpe (B & S) gage number, or for sizes larger than B & S gage no. 0000, by the number of circular mils.

Table 18-2 gives the resistance in ohms per 1,000 feet and the electric current capacity in amperes, as specified by the NEC, for copper wire.

TABLE 18-2
Resistance and current capacity of copper wire

| B & S gage | Resistance, ohms/1,000 ft | Capacity, A | |
		Rubber-covered	Weather-proof
18	6.374	3	5
16	4.009	6	10
14	2.527	15	20
12	1.586	20	25
10	0.997	25	30
8	0.627	35	50
6	0.394	50	70
5	0.313	55	80
4	0.248	70	90
3	0.197	80	100
2	0.156	90	125
1	0.124	100	150
0	0.098	125	200
00	0.078	150	225
000	0.062	175	275
0000	0.049	225	325

Accessories

In addition to the conduit and wire, numerous accessories will be needed to install wiring. Among the items needed will be the following:

1. Meter box
2. Entrance cap and conduit
3. Entrance box with switch and fuses or circuit breaker
4. Couplings, elbows, locknuts, bushings, and straps for conduit
5. Special fittings and clamps for flexible conduit and cable
6. Outlet boxes
7. Junction boxes
8. Light sockets
9. Rosettes
10. Solder, tape, straps, etc.

Cost of Materials

The cost of materials for electric wiring will vary with the location, quality, quantities purchased, and time. Representative costs in effect in 1988 are given in Tables 18-3 to 18-8. The costs given in the tables should be used as a guide only. When estimating the cost of a project for contract purposes, the estimator should obtain current prices applicable to the particular project.

Labor Required to Install Electric Wiring

The labor required to install electric wiring is estimated by at least three methods: by assuming a certain cost per outlet, by assuming that labor will cost a certain

TABLE 18-3
Representative costs of rigid conduit

Size, in	Cost per 100 lin ft	
	Heavy	Light
$\frac{1}{2}$	$ 47.45	$ 13.89
$\frac{3}{4}$	57.28	19.90
1	77.10	30.02
$1\frac{1}{4}$	101.37	42.06
$1\frac{1}{2}$	119.06	48.10
2	162.07	62.10
$2\frac{1}{2}$	256.24	140.15
3	342.37	176.39

TABLE 18-4
Representative costs per fittings for rigid conduit

Fitting	Size, in					
	$\frac{1}{2}$	$\frac{3}{4}$	1	$1\frac{1}{4}$	$1\frac{1}{2}$	2
For heavy conduit						
Couplings	$0.62	$0.78	$1.10	$ 1.38	$ 1.74	$ 2.70
Elbows	1.83	2.22	3.28	4.74	5.90	8.69
Locknuts	0.09	0.15	0.26	0.32	0.49	0.71
Bushings	0.10	0.15	0.26	0.35	0.47	0.86
Straps, 2 holes	0.08	0.09	0.13	0.22	0.25	0.35
Condulets, LB etc.	3.14	3.77	5.66	9.80	12.79	21.06
Condulets, T	3.93	4.72	7.07	10.38	13.83	21.83
Condulets, X	5.19	6.13	8.49	11.95	15.41	29.34
For thin-wall conduit						
Elbow			1.76	2.41	3.05	4.88
Setscrew connectors	0.19	0.31	0.50	0.91	1.37	1.84
Setscrew couplings	0.21	0.34	0.54	1.03	1.49	1.98

percentage of the cost of materials, and by assuming the time required to perform each operation of the work.

The first two methods may produce a sufficiently accurate estimate if dependable cost data are available from similar work previously done. If an estimate is prepared for a project that is unlike work previously done, it may be difficult to assume costs accurately.

The third method requires the preparation of a list of all materials needed, by quantity and quality. If the time required to install each item is known, the total time for the job can be estimated. Electric wiring is usually installed by two workers, an electrician and a helper, who work together as a team. The time per operation may be based on labor-hours or team-hours.

TABLE 18-5
Representative costs of flexible metal conduit

Size, in	Cost per 100 lin ft
$\frac{1}{2}$	$ 19.80
$\frac{3}{4}$	27.50
1	60.50
$1\frac{1}{4}$	79.20
$1\frac{1}{2}$	91.30
2	136.40

TABLE 18-6
Representative costs of nonmetallic cable

B & S gauge	No. wires	Cost per 1,000 lin ft	
		Without ground wires	With ground wires
14	2	$ 60.90	$ 63.65
12	2	72.57	78.00
10	2	140.25	169.17
8	2	246.30	366.81
14	3	113.82	122.10
12	3	157.14	167.07
10	3	243.27	259.32
8	3	468.36	545.64
6	3	695.69	756.26

If the estimate is started with the service wires that enter a building, the estimate should include any work required to bring the wires to the building. Frequently the utility company will bring the service wires to a building, and the contractor will install an entrance cap, a conduit, and wires to the meter box, then install conduit and wires from the meter to a service fuse panel, with the main switch located within the building. From this panel, as many separate circuits as are necessary are installed throughout the building. Frequently a single conduit may be used for several circuits for at least a portion of the building, with individual circuits coming out of junction boxes installed in the main conduit line.

Wires must be installed from a circuit to each outlet, switch, and plug.

TABLE 18-7
Representative costs of rubber-covered copper wire

B & S gauge	Cost per 1,000 lin ft
14 THHN solid	$ 20.95
12 THHN solid	28.60
10 THHN solid	48.13
8 THHN stranded	96.03
6 THW stranded	134.30
4 THW stranded	209.50
2 THW stranded	310.60
1 THW stranded	408.77
0 THW stranded	481.50

TABLE 18-8
Representative costs of electrical accessories

Item	Cost
Service fuse panel with switch:	
4 branches, ea.	$11.50
6 branches, ea.	13.00
8 branches, ea.	22.84
Outlet boxes, 4 in, ea.	0.38
Switch boxes, ea.	0.28
Wall plates, ea.	0.37
Switches:	
Toggle, single-pole, ea.	0.70
Toggle, three-way, ea.	1.30
Mercury, single-pole, ea.	2.60
Receptacles	0.56

Heavy rigid conduit is installed with threaded couplings, elbows, and locknuts and bushings where it enters boxes. It should be supported with pipe straps or in some other approved manner, depending on the type of construction used for the building. This conduit is furnished in 10-ft lengths.

Lightweight rigid conduit is installed with special couplings and connectors.

Flexible conduit, armored, and nonmetallic cable may be bent easily as it is installed. Usually only end connectors are required where it enters boxes. It can be installed more rapidly than rigid conduit.

Where conduit, boxes, and accessories are installed in a building under construction, especially a concrete building, it may be necessary to keep one or more electricians on the job most of the time, even though they are unable to work continuously. This possibility should be considered by an estimator.

When 100 ft of straight rigid conduit is installed, the operations will include connecting 10 joints into couplings, plus the installation of any outlet boxes required. The installation of an outlet box requires the setting of a locknut and a bushing for each conduit connected to the box. More time will be required for large than for small conduit. Table 18-9 gives the labor-hours required for installing conduit.

The installation of a switch, fixture outlet, or plug requires a two-wire circuit from the main circuit to the outlet. The time required will vary with the length of the run and the type of building. For a frame building it may be necessary to drill one or more holes through lumber, whereas for a masonry building it may be necessary to install the conduit through cells in concrete blocks, then chip through the wall of a block to install the outlet box.

The wires usually are pulled through the conduit just before the building is completed.

TABLE 18-9
Labor-hours required to install electric wiring per 100 lin ft

Class of work	Electrician	Helper
Install service entrance cap and conduit†	0.5–1.0	0.5–1.0
Install conduit and fuse panel†	0.5–1.0	0.5–1.0
Install heavy rigid conduit with outlet boxes:		
$\frac{1}{2}$ and $\frac{3}{4}$ in	5.0–10.0	5.0–10.0
1 and $1\frac{1}{4}$ in	7.0–11.0	7.0–11.0
$1\frac{1}{2}$ in	9.0–13.0	9.0–13.0
2 in	12.0–17.0	12.0–17.0
$2\frac{1}{2}$ in	15.0–21.0	15.0–21.0
3 in	20.0–28.0	20.0–28.0
4 in	25.0–35.0	25.0–35.0
Install thin-wall conduit with outlet boxes:		
$\frac{1}{2}$ and $\frac{3}{4}$ in	4.0–6.0	4.0–6.0
1 in	4.3–7.0	4.3–7.0
$1\frac{1}{4}$ in	4.5–7.5	4.5–7.5
$1\frac{1}{2}$ in	5.5–9.0	5.5–9.0
Install flexible conduit with outlet boxes:		
$\frac{1}{2}$ and $\frac{3}{4}$ in	3.0–5.0	3.0–5.0
1 and $1\frac{1}{4}$ in	4.0–6.0	4.0–6.0
Install nonmetallic cable with outlet boxes:		
14/2, 12/2, 10/2, and 8/2‡	3.0–6.0	3.0–6.0
14/3, 12/3, 10/3, and 8/3‡	3.5–6.5	3.5–6.5
6/3 and 4/3	4.0–7.0	4.0–7.0
Pull wire through conduit and make end connections, per circuit:		
14, 12, and 10 gauge	0.5–1.0	0.5–1.0
8 and 6 gauge	1.0–1.5	1.0–1.5
4 gauge	1.5–2.0	1.5–2.0
2 gauge	2.2–3.2	2.2–3.2
1 gauge	2.7–3.7	2.7–3.7

† The unit in this case is "each."
‡ The numbers 8/2 designate two wires, each no. 8 gauge in a cable.

FINISH ELECTRICAL WORK

Under roughing in electrical work, the conduit, wires, and outlet boxes are installed. After the building is completed and the surfaces are painted, the fixtures are installed. The costs of fixtures vary so much that a list of prices in this book would be of little value to an estimator. The costs should be obtained from a current catalog or jobber.

Labor Required to Install Electric Fixtures

Installing a fixture such as a switch or a plug involves connecting it to the wires which are already in the outlet box and attaching a cover plate. Installing a ceiling

TABLE 18-10
Labor-hours per fixture required to install electric fixtures

Class of work	Electrician	Helper
Install ceiling fixture	0.2–0.4	0.2–0.4
Install wall light	0.2–0.4	0.2–0.4
Install base or floor plug	0.1–0.2	0.1–0.2
Install wall switch	0.1–0.2	0.1–0.2
Install three-way switch	0.15–0.25	0.15–0.25

or wall fixture such as a light involves securing the fixture to a bracket and connecting it to the wires. Table 18-10 gives representative labor-hours required to install electrical fixtures.

PROBLEMS

For these problems use the representative wage rates and the costs of materials listed in this book. Use an electrician and a helper to do the work.

18-1. Estimate the cost of furnishing and installing 420 lin ft of $1\frac{1}{2}$-in heavy-duty conduit, including 36 couplings, 12 elbows, and 18 T-type Condulets under average conditions.

18-2. Estimate the cost of furnishing and installing 680 lin ft of 1-in heavy rigid conduit, including 38 couplings, 20 elbows, and 14 T-type Condulets, when the installation is made in a concrete building under construction. The conditions are such that the electrician and the helper can work at only about 75 percent of normal efficiency.

18-3. Estimate the cost of furnishing and installing the following items in a frame residence:

1 service fuse panel with switch for six branches
180 lin ft of $\frac{3}{4}$-in light conduit
130 lin ft of $\frac{1}{2}$-in light conduit
23 junction boxes
28 switch boxes
12 single-pole toggle switches
16 flush receptacles
8 $\frac{3}{4}$-in elbows
12 $\frac{3}{4}$-in couplings
24 $\frac{3}{4}$-in connectors
18 $\frac{1}{2}$-in couplings
12 $\frac{1}{2}$-in elbows
18 $\frac{1}{2}$-in couplings
48 $\frac{1}{2}$-in connectors
28 wall plates
320 lin ft of no. 10 gauge rubber-covered copper wire, two wires per circuit
280 lin ft of no. 12 rubber-covered copper wire, two wires per circuit

18-4. Estimate the cost of furnishing and installing the wiring for a residence, including the following items:

        1 $1\frac{1}{4}$-in cap, cost $5.85

        20 lin ft of $1\frac{1}{4}$-in heavy rigid conduit

        4 $1\frac{1}{4}$-in elbows

        2 $1\frac{1}{4}$-in couplings

        4 $1\frac{1}{4}$-in locknuts and bushings

        1 service panel with switch for six branches

        980 lin ft of no. 12 gauge two-wire nonmetallic cable with ground wire

        320 lin ft of no. 14 gauge two-wire nonmetallic cable with ground wire

        26 junction boxes, 4 in

        36 switch boxes

        36 wall plates

        20 flush receptacles, wall plugs

        16 single-pole mercury switches

CHAPTER
19

STEEL
STRUCTURES

Types of Steel Structure

Steel is used to erect such structures as multistory buildings, auditoriums, gymnasiums, theaters, churches, mill buildings, roof trusses, stadiums, bridges, wharfs, towers, etc. In addition to steel structures, steel members are frequently used for columns, beams, and lintels and for other purposes.

Materials Used in Steel Structures

Insofar as it is possible, steel structures should be constructed with members fabricated from standard shapes, such as H columns, I beams, WF beams, channels, angles, and plates. Members made from standard rolled shapes are usually more economical than fabricated members. However, if standard shapes are not available in sufficient sizes to supply the required strength, it is necessary to fabricate the members from several parts, such as standard shapes and plates or lattices.

Connections for Structural Steel

In fabricating standard shapes to form the required members or in connecting the members into the structure, two types of connections are used: bolts and welds.

Each type of connection has its place in the field of structural-steel construction and will be discussed in greater detail later in this chapter.

Estimating the Cost of Steel Structures

In estimating the cost of structural steel for a job, a contractor will submit a set of plans and specifications for the structure to a commercial steel fabricator for quotations. The fabricator will make a quantity takeoff, including main members, details, and miscellaneous items, to which he or she will apply shop costs for fabricating, welding, painting, overhead, and profit as a basis for submitting a quotation to the general contractor. The cost of transporting the steel to the project must be added to the cost of the finished products at the shop. This procedure will establish the cost of the fabricated steel delivered to the job.

Most general contractors who erect buildings and similar structures subcontract the erection of the steel to contractors who specialize in this work. This practice is justified because the erection of steel is a highly specialized operation which should be performed by a contractor with suitable equipment and a well-trained erection crew. Because of these conditions, the general contractor can usually have the erection done more economically by a subcontractor than with her or his own equipment and employees. The charge for the erection is generally based on an agreed price per ton of steel in place, including bolting or welding the connections.

When estimating the cost of structural steel in place, a building contractor will include in the estimate the cost of the steel delivered to the project, the cost of erection, and the cost of field painting as required. To these costs the contractor will add his or her cost for job overhead, general overhead, and profit.

Items of Cost in a Structural-Steel Estimate

The items of cost which should be considered in preparing a comprehensive detailed estimate for a steel structure include the following:

1. Cost of structural-steel shapes at the fabricating shop
2. Cost of preparing drawings for use by the shop in fabricating the steel
3. Cost of handling and fabricating the steel shapes into finished members
4. Cost of shop painting, if required
5. Cost of shop overhead, sales, and profit
6. Cost of transporting the steel to the job
7. Cost of erecting the steel, including equipment, labor, bolts, or welding
8. Cost of field painting the steel structure
9. Cost of job overhead, general overhead, insurance, taxes, and profit

The cost of any one or all of these items may vary considerably between two projects, and consequently the cost of each item must be estimated for a particular project.

Cost of Structural-Steel Shapes at the Fabricating Shop

Structural-steel shapes are made in many sizes by rolling mills. These shapes are purchased by shops which specialize in fabricating steel members. The base price of the shapes at a fabricating shop varies with the price charged by the mills and the cost of transporting the steel to the shop.

If the steel members are furnished by a fabricating shop, the job estimator does not need to consider the mill price and the cost of transportation from the mill to the shop separately. Only the base price at the fabricating shop will be of concern.

The actual price per pound of structural-steel shapes varies at a given mill or shop with the size and weight of the shape and the quantity of steel required.

Extra Charges for Size and Section

As indicated in Table 19-1, the rolling mills which supply steel shapes to fabricating shops assess charges that are added to the base price of steel. Table 19-1 illustrates representative extra charges for only selected sizes and sections. A complete list of the extra charges may be obtained from mills which produce structural-steel shapes.

Extra Charges for Quantity

Quantity extra charges are determined by the total theoretical weight of the individual size, weight, gage, or thickness of a structural section ordered of one grade or analysis on the same order, released and accepted for one mode of shipment to one destination at one time. Table 19-2 gives the extra charges for quantity effective as of January 1, 1988.

Example 19-1. Determine the cost per cwt for 3,160 lb of 6- by 6- by $\frac{1}{2}$-in structural-steel angles, including the base price, size extra, and quantity extra charges only.

    Base price at shop   = $23.00 per cwt
    Size extra charge   =   3.10 per cwt
    Quantity extra charge   =   0.75 per cwt
    Total cost   = $26.85 per cwt

TABLE 19-1
Representative size extra charges over base prices for structural-steel shapes

Size, in	Weight, lb/ft	Size extras per cwt
Wide-flange shapes		
W36	135–300	$3.60
W30	99–211	3.60
W24	55–162	3.60
W21	44–147	3.60
W18	35–119	2.10
W16	57–100	3.60
W16	36–50	2.10
W14	61–426	3.60
W12	53–190	3.60
W12	27–50	2.10
W10	49–112	3.60
W10	15–30	2.10
W8	18–67	2.10
W6	12–16	2.10
Standard beams		
S24	80–121	$4.10
S20	66–96	4.10
S18	54.7–70	4.10
S15	42.9–50	4.10
S12	31.8–50	4.10
S10	25.4–35	2.10
S8	18.4–23	2.10
S6	12.5–17.25	2.10
S4	7.7–9.5	2.10
Standard channels		
C15	All	3.10
C10, C12	All	2.10
C8, C9	All	2.10
C4, C6	All	2.10
Angle sections		
All	All	3.10

TABLE 19-2
Quantity extra charges for structural steel

Quantity, lb	Extra charge per cwt
10,000 and over	None
6,000–10,000	$0.25
4,000–6,000	0.50
2,000–4,000	0.75
1,000–2,000	2.50
Under 1,000	5.00

Estimating the Weight of Structural Steel

In estimating the probable weight of structural steel for a job, the estimator should determine from the plans the total number of linear feet for each shape by size or weight. Structural-steel handbooks give the nominal weights of all sections. However, variations in weights amounting to $2\frac{1}{2}$ percent above or below the nominal weights are permissible and may occur. The purchaser is charged for the actual weight furnished, provided that the weight does not fall outside the permissible variation.

The weight of the details for connections should be estimated and priced separately if a detailed estimate is desirable. In estimating the weight of a steel plate of irregular shape, the weight of the rectangular plate from which the shape is cut should be used. Steel weighs 490 lb/ft³.

Cost of Preparing Shop Drawings

In preparing plans for a steel structure, the engineer or architect does not furnish drawings in sufficient detail to permit the shop to fabricate the members without additional information. Fabricating shops maintain drafting departments, which prepare shop drawings in sufficient detail to permit the shops to fabricate the members. One job may require as few as one or two sheets, while another job may require more than 100 sheets. The cost of preparing the drawings is based on the complexity of the detailing and the number of sheets required. Since the total cost of the drawings is charged to the steel supplied for a job, the cost per unit weight of steel will vary with the total cost of the drawings and the quantity of steel supplied. The cost per ton for lightweight roof trusses will be high, while the cost per ton for large beams fabricated from rolled sections will be considerably lower.

Shop drawings are usually prepared on 24- by 36-in sheets. The cost of preparing a sheet may vary from $100.00 to $200.00, depending on the amount of detailing required. The cost of preparing a sheet must be charged to the finished steel members that are fabricated from the information given on the sheet. If only

TABLE 19-3
Representative cost of shop drawings for structural steel

	Bolted connections		Welded connections	
	Per ton	Per cwt	Per ton	Per cwt
Office buildings	$50.00–$90.00	$2.50–$4.50	$50.00–$90.00	$2.50–$4.50
Industrial buildings	80.00–150.00	4.00–7.50	80.00–170.00	4.00–8.50
Roof trusses	50.00–350.00	12.50–17.50	250.00–350.00	12.50–17.50
Plate girders	36.00–70.00	1.80–3.50	40.00–90.00	2.00–4.50
Highway bridges	40.00–50.00	2.00–2.50	40.00–50.00	2.00–2.50

one member is fabricated, the cost per member is high; if a great many members are fabricated, the cost per member will be much lower.

Table 19-3 gives the approximate cost of shop drawings for various types of steel structures.

Cost of Shop Handling and Fabricating Structural Steel

The cost of shop handling and fabricating structural steel will vary considerably with the operations performed, sizes of the members, and extent to which the operations are duplicated on similar members.

For bolted connections the fabricating operations include cutting, punching, milling, planing, and marking each member. For welded connections the fabricating operations include cutting, some punching for temporary bolt connections, milling, beveling, planing, and shop welding.

Tables 19-4 and 19-5 give ranges in the costs of various fabricating operations for structural steel. The cost per ton or per hundredweight is based on the weight of the finished members, including details, cutting, punching, welding, etc. The lower costs should be used when identical operations are performed on several members, while the higher costs should be used when the operations are performed on only a few members. This is necessary to provide for the time required to set up the fabricating equipment, which is fairly constant regardless of the number of operations performed.

Cost of Applying a Coat of Shop Paint to Structural Steel

Specifications for structural steel frequently require the fabricator to apply a coat of paint after the fabricating is completed. A gallon of paint should cover about 400 ft^2 of surface. Spray guns are generally used to apply the paint.

TABLE 19-4
Representative costs of fabricating structural steel, using bolted connections

Operation	Cost per ton	Cost per cwt
Punching beams and channels:		
6 to 10 in	$56.00–$80.00	$2.80–$4.00
18 in and larger	42.00–60.00	2.10–3.00
Punching and framing beams and channels with end connections:		
6 to 16 in	104.00–144.00	5.20–7.20
18 in and larger	70.00–104.00	3.50–5.20
Framing beams with plates and channels;		
6 to 16 in	140.00–210.00	7.00–10.50
18 in and larger	88.00–122.00	4.40–6.10
Fabricating built-up girders:		
Weight up to 200 lb/lin ft	230.00–260.00	11.50–13.00
Weight 200 to 300 lb/lin ft	190.00–240.00	9.50–12.00
Weight 300 to 400 lb/lin ft	174.00–230.00	8.70–11.50
Weight 400 to 600 lb/lin ft	158.00–192.00	7.90–9.60
Fabricating H columns, including bases, splices, and end milling:		
Up to 8 in	160.00–230.00	8.00–11.50
Larger than 8 in	122.00–174.00	6.10–8.70
Fabricating roof trusses:		
Weight up to 1,200 lb	300.00–370.00	15.00–18.50
Weight 1,200 to 2,400 lb	280.00–336.00	14.00–16.80
Weight 2,400 to 3,600 lb	230.00–280.00	11.50–14.00
Weight over 3,600 lb	174.00–230.00	8.70–11.50

Table 19-6 gives the approximate cost for applying a coat of paint to structural steel for various types of members and structures. A painter, using a spray gun, should paint 1 to 2 tons/h, depending on the size of the sections.

Cost of Shop Overhead and Profit

If the prices given previously are for the actual costs of steel, shop drawings, and fabricating and painting steel, it will be necessary to add to these costs the cost of overhead and profit in order to determine the probable cost of the fabricated steel at the shop. The combined cost of overhead and profit will vary from 15 to 25 percent, depending on the size of the order.

Cost of Transporting Structural Steel

The cost of transporting structural steel from the fabricating shop to the job will vary with the quantity of steel, method of transporting it, and distance from the

TABLE 19-5

Representative costs for fabricating structural steel, using welded connections

Operation	Cost per ton	Cost per cwt
Punching and framing beams and channels with end connections:		
6 to 18 in	$260.00–$400.00	$13.00–$20.00
18 in and larger	175.00–280.00	8.75–14.00
Framing beams with angles:		
6 to 18 in	260.00–400.00	13.00–20.00
18 in and larger	175.00–280.00	8.75–14.00
Fabricating plate and angle columns:		
Weight up to 40 lb/lin ft	600.00–775.00	30.00–38.75
Weight 40 to 80 lb/lin ft	390.00–575.00	19.50–28.75
Weight over 80 lb/lin ft	305.00–435.00	15.25–21.75
Fabricating built-up plate girders:		
Weight up to 200 lb/lin ft	480.00–600.00	24.00–30.00
Weight 200 to 300 lb/lin ft	420.00–520.00	21.00–26.00
Weight 300 to 400 lb/lin ft	380.00–480.00	19.00–24.00
Weight 400 to 600 lb/lin ft	353.00–420.00	17.65–21.00
Fabricaing H columns, including bases, splices, beam connections, and end milling:		
Up to 8 in	435.00–525.00	21.75–26.25
Larger than 8 in	350.00–480.00	17.50–24.00
Fabricating roof trusses:		
Weight up to 1,200 lb	800.00–1,000.00	40.00–50.00
Weight 1,200 to 2,400 lb	725.00–900.00	36.25–45.00
Weight 2,400 to 3,600 lb	600.00–740.00	30.00–37.00

TABLE 19-6

Cost of applying a shop coat of paint to structural steel

Member or structure	ft²/ton	Cost per ton			Cost per cwt
		Paint	Labor	Total	
Beams	200–250	$49.00	$ 8.33	$57.33	$2.87
Girders	125–200	27.30	7.20	34.50	1.73
Columns	200–250	49.00	8.33	57.33	2.87
Roof trusses	275–300	54.60	10.80	65,40	3.27
Bridge structures	200–250	49.00	8.33	57.33	2.87

shop to the job. If it is possible to haul the steel the entire distance by trucks instead of by a combination of railroad and trucks, then the cost of an intermediate handling will be eliminated. A truck should haul approximately 20 tons per load.

The estimator should determine the freight or truck cost per ton or per hundredweight for the particular project in order to include the correct amount in the estimate.

Cost of Fabricated Structural Steel Delivered to a Project

The following example illustrates a method of estimating the cost of structural steel delivered to a project.

Example 19-2. Estimate the total cost and the cost per cwt of structural steel for the columns, beams, and details for a framed steel building. All members are to be fabricated at a shop for bolted connections with high-tensile-strength bolts for connections.

The list of members and details is given in the accompanying table:

No. of pieces	Description	Length each	Weight, lb/lin ft	Total weight, lb
18	Columns, W10 × 89	24 ft 9 in	89	39,649
12	Columns, W10 × 112	24 ft 9 in	112	33,264
18	Columns, W10 × 54	21 ft 6 in	54	20,898
12	Columns, W10 × 72	21 ft 6 in	72	18,576
Total weight of columns				112,387
30	16- × 16- × $1\frac{1}{2}$-in base plates			3,267
120	6- × 18- × $\frac{3}{4}$-in splice plates			2,763
120	∟ 3 × 3 × $\frac{3}{8}$	0 ft 8 in	7.2	575
420	∟ $2\frac{1}{2}$ × 2 × $\frac{3}{16}$	0 ft 8 in	2.75	577
Total weight of column details				7,182
68	Beams, W14 × 48	17 ft 6 in	48	57,120
54	Beams, W14 × 34	19 ft 9 in	34	36,261
24	Beams, W12 × 32	16 ft 6 in	32	12,672
18	Beams, W12 × 28	22 ft 6 in	28	11,340
Total weight of beams				117,393
492	∟ 3 × 3 × $\frac{3}{8}$	1 ft 0 in	7.2	3,542
164	∟ 3 × 3 × $\frac{1}{4}$	0 ft 10 in	4.9	670
Total weight of angles				4,212
Total weight of fabricated steel, lb				241.174
Total weight, tons				120.59

The total cost for the job will be:

Base cost for the fabricating shop:
Columns, 1,123.87 cwt @ $23.00 = $25,849.01
Beams, 1,173.93 cwt @ $23.00   = 27,000.39
Plates, 60.30 cwt @ $22.00   =   1,326.60
L3 × 3, 47.87 cwt @ $24.00   =   1,148.88
L2½ × 2, 5.77 cwt @ $24.00   =   138.48
Size extra charges:
Columns, 1,123.87 cwt @ $3.60 =   4,045.93
Beams, 571.20 cwt @ $3.60   =   2,056.32
Beams, 362.61 cwt @ $2.10   =   761.48
Beams, 240.12 cwt @ $2.10   =   504.25
L3 × 3, 47.87 cwt @ $3.10   =   148.40
L2½ × 2, 5.77 cwt @ $3.10   =   17.89
Quantity extra charges:
L3 × 3, 6.7 cwt @ $5.00   =   33.50
L2½ × 2, 5.77 cwt @ $5.00   =   28.85

Working drawings:
Total weight, 120.59 tons @ $70.00   = $ 8,441.30
Fabricating cost:
Columns, 56.20 tons @ $148.00   =   8,317.60
Beams, 58.70 tons @ $105.00   =   6,163.50
Shop painting:
Total weight, 120.59 tons @ $57.33   =   6,913.42
Total cost of materials   = $ 92,895.80
Shop overhead, 10% of $92,895.80   =   9,289.58
Profit, 10% of $92,895.80   =   9,289.58
Total cost f.o.b. shop   = $111,474.96
Trucking cost to job, 134 mi @ $1.75 =   234.50
Total cost delivered to job   = $111,709.46
Cost per ton, $111,709.46 ÷ 120.59 = 926.36
Cost per cwt, $926.36 ÷ 20   =   46.32

Erecting Structural Steel

When a structural-steel building is erected, the columns are erected first usually on previously prepared concrete foundations, with the anchor bolts already in place. The bases for the columns are supported above the concrete foundations by suitable means, such as with steel shims or steel wedges, to permit the bases to be set at the required elevations and grouted to these elevations. After the columns are erected, the beams are installed for the first tier of floors, usually two floors. The connections between the columns and the beams are temporarily bolted through holes, by using two or more bolts per connection. After the tier of columns and beams is in place, it is necessary to plumb the structure before installing the permanent bolts. This operation is repeated for subsequent tiers until the erection of the structure is completed.

Equipment for Erecting Structural Steel

The equipment used for erecting steel structures depends on the type of structure, the size of the structure and its component parts, and the location.

Roof trusses are usually delivered to the job partly or completely assembled and hoisted directly from the delivery trucks into place by power cranes. The mobility of a crane makes it very useful for such operations.

Multistoried framed steel buildings may be erected with cranes if the height is not excessive, usually about four stories. If a building is so tall that a crane can not be used, the steel members may be placed with one or more guy derricks, or a tower crane may be used.

Labor Erecting Structural Steel

The cost of labor erecting structural steel will vary with the type of structure, the kind of equipment used, sizes of members, kind of connections, climatic conditions, and prevailing wage rates.

A crew of steel erectors, classified as ironworkers, may vary from five to eight or more persons, excluding the individuals who bolt or weld the connections.

If a power crane is used to erect a steel structure, with relatively lightweight members, the crew might include the indicated persons, at the specified wage rates.

Classification	Hourly wage rate	Total wages
1 foreman	$18.30	$ 18.30
1 crane operator	17.25	17.25
1 crane oiler	11.25	11.25
4 ironworkers	15.80	63.20
	Total hourly cost = $110.00	

One of the ironworkers will hook the lifting line to the members to be lifted, while the other two will make the temporary or permanent bolt connections.

If a tower crane is used to handle the steel members, one or two additional crew members may be required. This crew will set the members in place, temporarily bolt the connections, and plumb the structure for the final bolting or welding of the connections.

Table 19-7 gives the approximate rates of erecting steel for various types of structures.

Labor Bolting Structural Steel

When roof trusses are erected, one worker will be required at each end of the truss to install at least enough bolts to secure the truss to the supporting structure. One or two other workers may be required on the floor of the building to assist in

TABLE 19-7
Approximate rates of erecting structural steel (does not include final bolting or welding)

Type of structure	Type of equipment	Size of crew	Crew-hour per ton
Roof trusses:			
Up to 1,200 lb	Crane	5	1.6
1,200–2,400 lb	Crane	5	1.3
2,400–3,600 lb	Crane	5	1.0
3,600–4,800 lb	Crane	6	0.8
4,800–6,000 lb	Crane	6	0.7
Purlins and braces	Crane	5	1.2
Framed steel structures:			
Up to 4 stories high	Crane	7	0.5
Up to 8 stories high	Tower crane	7	0.5
8 to 18 stories high	Tower crane	8	0.5
Mill buildings, factories, etc.,			
columns and beams	Crane	6	0.8
Churches, theaters, etc.	Crane	5	1.2
Plate girders:			
5–10 tons	Crane	6	0.4
10–20 tons	2 cranes	8	0.3

FIGURE 19-1
Two power cranes erecting steel roof trusses.

moving the ends of the trusses into position for bolting. As the trusses are erected, some of or all the purlins should be attached to the trusses to ensure adequate rigidity for the trusses.

After the trusses are erected and plumbed, two ironworkers should be able to complete the permanent bolting. One helper might be required on the floor.

When a tower crane or a guy derrick is used to hoist the members and place them in position, it is common practice to install enough bolts in the connections to permit the structure to be plumbed and braced as it is erected. If this procedure is followed, the rest of the permanent bolts will be installed later. Calibrated torque wrenches, selected to produce the desired tension in the bolts, may be used in tightening the nuts on the bolts. The size crew required for installing the bolts will vary with the number of bolts needed and the ease or difficulty in getting at the bolts.

FIGURE 19-2
Tower crane erecting structural steel.

WELDED STRUCTURES

With the improvements in the techniques of welding which have been developed in recent years, it is becoming increasingly common practice to use arc welding in fabricating and erecting steel structures. Arc welding has passed beyond the experimental stage and is now accepted as a safe and satisfactory method of connecting steel members, both in the shop and in the field. Laboratory tests may be applied to determine whether or not welders are qualified and welded connections are satisfactorily made.

Advantages of Welded Connections

The use of arc-welded connections is in many instances more desirable than that of bolted connections. The advantages begin with the design of the structure and continue through the erection.

If a structure is designed for welded connections, it is possible to reduce the total weight of steel by as much as 15 to 25 percent. It is not necessary to punch bolt holes in members at points of critical stress for connection purposes. Where a few holes are required for temporarily bolting members in place, the holes may be punched at noncritical points, thus permitting the use of the full strength of a member at joints. Since the welded joints can be made as strong as the full sections of the members, it is possible to design beams as continuous members over several supports, thus reducing the critical bending-moment stresses. As a result of these conditions, lighter beams may be used in a structure, which will reduce the total weight of steel required. Smaller and less expensive foundation costs should result from reduction in deadweight of the structure.

The use of welded connections will simplify the preparation of plans for a steel structure, because less work is required in preparing the details for joint connections. It is not necessary to determine the number and spacing of bolts or to design gusset plates, for they are not used.

It is frequently less expensive to fabricate steel members for welded connections. The accurate location of bolt holes for shop and field assembling is eliminated. A somewhat greater freedom in dimensioning members is permissible since bolt holes for connecting members, which are punched separately, do not have to be matched exactly in assembling members. Members which are to be field-welded require fewer shop operations, thus reducing the costs and the fabricating time.

An experienced crew should be able to erect a welded-steel structure at no greater cost than for a bolted structure. This is especially true if the members are designed and fabricated in a manner which will facilitate welding, such as providing for down-hand welding of all joints where possible.

Erection Equipment Required for Welded-Steel Structures

In erecting a welded-steel structure, one or more electric generators and welding equipment are substituted for the air compressor and bolting equipment. Otherwise, the erection equipment is essentially the same as for a bolted structure.

Arc welding may be satisfactorily accomplished with either ac or dc electricity. To obtain good welds, it is necessary to provide electric energy with the desirable characteristics, amperes, and volts for the particular job. The generator selected should be capable of meeting the demands that will be made on it. If this precaution is not observed, it will be difficult to obtain welds of the desired quality.

If alternating current is used for welding, the current may be obtained from commercial sources with a transformer. If direct current is used, the current may be provided with a generator. An electric-motor-driven generator is usually more satisfactory than a gasoline- or diesel-engine-driven generator.

Erecting Steel Structures with Welded Connections

In erecting a steel structure with welded connections, the same general procedure is followed as for a structure with bolted connections. A few holes are punched in the members, at noncritical points, so that bolts may be installed for temporarily connecting the members. After a unit of the structure is plumbed or brought to the correct position, welding of the connections is started. To eliminate undesirable distortion of a structure, resulting from unequal heating at the connections, a definite pattern for welding should be established and rigidly followed. If beams are to be welded to opposite flanges of a column, the two welds should be performed simultaneously to eliminate unequal expansion of the two sides of the column, which would result from welding the beams separately.

In welding girders to columns or beams to columns and girders, the top flanges should be welded first and allowed to cool to the atmospheric temperature prior to welding the bottom flanges. As the welds for the bottom flanges cool, the bottom flanges at the ends of the beams will be subjected to tension. Since this stress is opposite the stress to which the beams will be subjected under loaded conditions, the effect of welding is to reduce the ultimate bending stresses in the beams.

The correct procedure in welding beams to columns is illustrated in Fig. 19-3.

Types of Welds

Names identifying the various types of welds have been standardized and should be used in specifying or referring to welds. Figure 19-4 illustrates various types of structural welds.

Arc-welding Terminology

Below are given definitions of terms associated with arc welding.

Kilowatts   Amperes × volts ÷ 1,000

Kilowatthours   Kilowatts × time, in hours.

Efficiency of a transformer   The ratio of the output divided by the input for a transformer, expressed as a percentage.

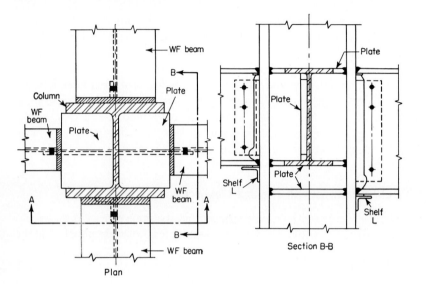

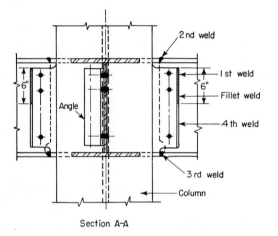

FIGURE 19-3
Details of beam-to-column welds.

Electrodes   Electrodes are steel rods of various sizes and types, which are melted by the electric current to form the welding metal.

Melt-off rate of electrode   The rate at which the electric current melts an electrode, expressed in inches per minute.

Arc speed   The speed at which an electrode is moved along a weld, without interrruptions, usually expressed in feet per hour.

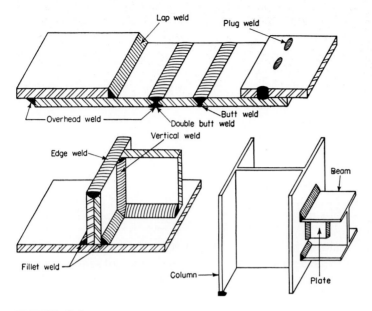

FIGURE 19-4
Typical welds.

Arc time   The time required to produce a given unit of welding, such as 100 ft, usually expressed in hours.

Interruption   A temporary stoppage in welding, as in changing electrodes, in moving from one weld to another, etc.

Operating factor   The percentage of time actually devoted to welding. For example, if one-half the total time is devoted to actual welding while the other half is consumed by interruptions, the operating factor is 50 percent.

Floor-to-floor speed   The net length of weld produced by a welder, in feet per hour, considering all interruptions.

Fit-up   The manner in which the component parts of a joint are brought together prior to welding, such as the width of the gap at a joint.

Passes   The number of times an electrode must be moved over a joint to complete a weld.

Deposition efficiency   The ratio of the amount of metal deposited in the weld divided by the amount of electrode consumed, expressed as a percentage.

Methods of Producing the Most Economical Welds

To obtain the most economical welds, the procedures given below should be followed where possible:

1. Travel as fast as possible within the limits of good weld appearance by always keeping the electrode just ahead of the molten pool.
2. Use the largest electrode practical.
3. Use the highest current practical.
4. Use as short an arc as possible, dragging the electrode coating where practical.
5. Use a proper plate preparation and fit-up.
6. Keep the buildup to a minimum.

Electrodes

Metallic electrodes are manufactured in sizes varying from $\frac{1}{16}$ to $\frac{3}{8}$ in in diameter or larger and in lengths of 14 and 18 in. For ordinary welding, a 2-in stub end will be wasted for each electrode.

Electrodes are available for use with either direct or alternating current, and with different coatings to satisfy the demands for the particular jobs for which they will be used. The proper electrode should be used to obtain the best results.

Cost of Welding

The cost of welding steel structures includes the cost of electrodes, electricity, labor, and job overhead. The effect on the cost of welding resulting from changes in the sizes of electrodes is illustrated in Table 19-8. The quantities and costs given are based on the following:

> The deposition efficiency is $66\frac{2}{3}\%$.
> The operating factor is 50%.
> The cost of electric energy is $0.06 per kWh.
> A welder is paid $14.66 per h.
> The cost of job overhead is $12.60 per h.

Tables 19-9 and 19-10 give information on the cost of labor, electricity, and electrodes for fillet and lap welds. The costs are based on the following:

> The cost of a welder is $14.66 per h.
> The cost of electricity is $0.06 per kWh.
> The cost of electrodes is $0.55 per lb.
> Horizontal and flat welds are downhand.

Table 19-11 gives information on the cost of labor, electricity, and electrodes for butt welds. The joints will be made with a 60° groove, a $\frac{1}{8}$-in shoulder, and a $\frac{1}{32}$-in gap. The cost of a welder, electricity, and electrodes will be the same as for Table 19-8.

TABLE 19-8
Effect of sizes of electrodes on cost of welding

	Size of electrode, in					
	$\frac{1}{8}$	$\frac{5}{32}$	$\frac{3}{16}$	$\frac{1}{4}$	$\frac{5}{16}$	$\frac{3}{8}$
Cost of electrode per lb	$0.55	$0.56	$0.55	$0.57	$0.56	$0.56
Amperage required, A	110	130	150	250	320	425
Arc volts required, V	24	25	26	30	34	38
Kilowatts at arc, kW	2.6	3.2	3.9	7.5	11.1	16.1
Efficiency of set, %	47	50	51	55	59	59
Kilowatt input, kW	5.6	6.5	7.6	13.6	18.8	27.3
Electrode consumed, lb/h	1.3	1.7	2.0	3.8	5.4	8.1
Metal deposited, lb/h	0.87	1.13	1.33	2.53	3.60	5.40
Kilowatthour/lb of metal deposited	3.20	2.90	2.85	2.70	2.60	2.55
Interruptions per lb of metal deposited	18	12	8	5	3	2
Cost per lb of metal deposited						
Labor	$16.961	$13.011	$10.995	$ 5.803	$4.072	$2.688
Overhead	14.574	11.172	9.450	4.767	3.507	2.331
Electricity	0.192	0.174	0.171	0.162	0.156	0.153
Electrode	0.825	0.840	0.807	0.836	0.822	0.822
Interruptions	0.526	0.346	0.230	0.146	0.084	0.052
Total cost per lb	$33.078	$25.543	$21.653	$11.714	$8.641	$6.046

TABLE 19-9
Cost of fillet and lap welds for horizontal and flat positions

Gage size of fillet, in	Electrode size, in	Current, A	Passes per weld	Electrodes, lb/ft of weld	Cost, ¢/ft of joint		
					Electrode	Electricity	Total
$\frac{5}{32}$	$\frac{3}{16}$	225	1	0.105	5.62	1.80	7.42
$\frac{3}{16}$	$\frac{1}{4}$	275	1	0.160	8.56	2.58	11.14
$\frac{1}{4}$	$\frac{1}{4}$	325	1	0.195	10.43	3.15	13.58
$\frac{5}{16}$	$\frac{1}{4}$	350	1	0.210	11.22	3.39	14.61
$\frac{3}{8}$	$\frac{5}{16}$	400	1	0.290	15.22	4.53	19.75
$\frac{7}{16}$	$\frac{5}{16}$	425	2	0.540	28.93	8.43	37.36
$\frac{1}{2}$	$\frac{5}{16}$	425	2	0.700	36.48	10.92	47.40
$\frac{5}{8}$	$\frac{5}{16}$	425	4	1.100	58.67	17.16	75.83
$\frac{7}{8}$	$\frac{5}{16}$	425	7	2.200	117.74	34.35	152.09
Labor cost							
Welding speed, ft/h	40	30	20	15	10	8	6	4
Labor cost, ¢/ft of joint	37	49	53	98	147	183	244	356

TABLE 19-10
Cost of fillet and lap welds, vertical position

Gage size of fillet, in	Electrode size, in	Current, A	Passes per weld	Electrodes, lb/ft of weld	Cost, ¢/ft of joint		
					Electrode	Electricity	Total
$\frac{5}{32}$	$\frac{5}{32}$	140	1	0.08	4.28	1.38	5.66
$\frac{3}{16}$	$\frac{3}{16}$	150	1	0.14	7.50	2.40	9.90
$\frac{1}{4}$	$\frac{3}{16}$	170	1	0.17	9.06	2.91	11.97
$\frac{5}{16}$	$\frac{3}{16}$	170	1	0.26	13.87	4.44	18.31
$\frac{3}{8}$	$\frac{3}{16}$	170	1	0.38	20.37	6.48	26.85
$\frac{7}{16}$	$\frac{3}{16}$	170	2	0.52	27.70	8.88	36.58
$\frac{1}{2}$	$\frac{3}{16}$	170	3	0.67	35.81	11.46	47.27
$\frac{5}{8}$	$\frac{3}{16}$	170	Varies	1.10	58.67	24.45	83.12
$\frac{7}{8}$	$\frac{3}{16}$	170	Varies	2.10	112.01	46.65	158.65

Labor cost

Welding speed, ft/h	20	15	12	10	8	6	4	2
Labor cost, ¢/ft of joint	73	98	122	147	183	244	367	733

TABLE 19-11
Cost of butt welds

Plate thickness, in	Electrode size, in	Current, A	Passes per weld	Electrodes, lb/ft of weld	Cost, ¢/ft of joint		
					Electrode	Electricity	Total
$\frac{1}{4}$	$\frac{3}{16}$	220	2	0.25	12.22	4.29	16.51
$\frac{5}{16}$	$\frac{3}{16}$	220	2	0.27	13.14	4.62	17.76
$\frac{3}{8}$	$\frac{1}{4}$	350	2	0.48	23.32	7.77	31.09
$\frac{7}{16}$	$\frac{1}{4}$	350	3	0.62	30.15	10.02	40.17
$\frac{1}{2}$	$\frac{1}{4}$	350	3	0.82	39.72	13.26	52.98
$\frac{5}{8}$	$\frac{1}{4}$	350	4	1.25	60.70	20.22	80.92
$\frac{3}{4}$	$\frac{1}{4}$	350	5	1.83	88.81	29.22	118.03
1	$\frac{1}{4}$	350	7	3.13	151.76	50.64	202.40

Labor cost

Welding speed, ft/h	20	15	10	8	6	4	2
Labor cost, ¢/ft of joint	73	98	147	183	244	367	611

Quantity of Details and Welds for a Structural-Steel Frame Building

The relation between the quantities of structural-steel members, details, and welds is illustrated by the information given below. This structure was arc-welded in accordance with the method shown in Fig. 19-3.

Height of structure	16 stories
Average story height	12 ft $5\frac{1}{2}$ in
Floor area	253,000 ft^2
Volume of building	3,153,000 ft^3
Total weight of steel	2,274 tons
Weight of steel per ft^2	17.98 lb
Weight of steel per ft^3	1.44 lb
Weight of columns	745 tons
Weight of column details	101.5 tons
Weight of column base plates	47.9 tons
Weight of beams and girders	1,379.6 tons
Total weight of shop electrodes	14,844 lb
Total weight of field electrodes	9,210 lb
Ratios of weights of parts to total weight:	
Columns	32.76%
Column base plates	2.11%
Column details	4.46%
Columns, base plates, and details	39.33%
Beams and girders	60.67%
Job details	6.48%
Electrodes, shop	6.52 lb/ton
Electrodes, field	4.05 lb/ton

Analysis of the Cost of Welded Connections Compared with Bolted Connections

To compare the cost of welded joints with bolted joints for a representative framed steel building, a bay with one column, girders, and beams was analyzed. The height of one story was 12 ft.

The total lengths of main members were as follows:

Member	Size	Length	Weight, lb
Column	W14 × 211	12 ft 0 in	2,532
Girder	W24 × 131	16 ft 4 in	2,140
Girder	W24 × 117	13 ft 6 in	1,580
Beam	W18 × 71	24 ft 6 in	1,740
Beam	W18 × 55	39 ft 0 in	2,145
Beam	W18 × 50	39 ft 0 in	1,950
	Total weight of main members =		12,087

List of details for bolted joints:

4 beams W18 × 119 × 1 ft $2\frac{1}{2}$ in	575
4 beams W18 × 106 × 1 ft 0 in	424
4 ∟s 8 × 8 × $\frac{3}{4}$ × 0 ft $8\frac{1}{2}$ in	110
12 ∟s 4 × $3\frac{1}{2}$ × $\frac{3}{8}$ × 0 ft $11\frac{1}{2}$ in	105
Bolts	288
Total weight of details	1,502

List of details for welded joints:

12 ∟s 4 × $3\frac{1}{2}$ × $\frac{3}{8}$ × 0 ft $11\frac{1}{2}$ in	105
4 plates 8 × 1 × 1 ft $0\frac{1}{2}$ in	113
2 plates 8 × $\frac{7}{8}$ × 1 ft $0\frac{1}{2}$ in	50
2 plates 12 × $\frac{1}{2}$ × 1 ft 5 in	58
1 plate 4 × $\frac{5}{8}$ × 1 ft 7 in	14
1 plate 4 × $\frac{1}{2}$ × 1 ft 7 in	12
Electrodes	34
Total weight of details	386

For bolted joints:

Weight of main members	=	12,087
Weight of details	=	1,502
Total weight	=	13,589

Ratio weight of details to total weight, 1,502 ÷ 13,589 = 11.05%

For welded joints:

Weight of main members	=	12,087
Weight of details	=	386
Total weight	=	12,473

Ratio weight of details to total weight, 386 ÷ 12,473 = 3.10%

Reduction in total weight:

Bolted joints	=	13,589
Welded joints	=	12,473
Reduction in weight	=	1,116

Ratio of reduction, 1,116 ÷ 13,589 = 8.21%

If a structure is designed for welded joints, it is frequently possible to reduce the sizes of main members. Thus the total reduction in the weight of steel will exceed the reduction due to joints only.

Painting Structural Steel

The cost of applying coats of paint in the field to structural steel will vary with the type of structure, sizes of members to be painted, and ease or difficulty of gaining access to steel members. The cost of painting a ton of steel for a roof truss will be considerably higher than for a frame steel building because of the greater area of steel per ton for the truss and also because of the difficulty of moving along the truss.

Paint is usually applied with a spray gun unless local restrictions ban the use of guns. Two field coats are usually applied.

TABLE 19-12
Cost of applying a field coat of paint to steel structures, using a spray gun

Member or structure	ft²/ton	Cost, $/ton		
		Paint	Labor	Total
Beams	200–250	53.10	14.50	67.60
Girders	125–200	31.95	12.56	44.51
Columns	200–250	53.10	14.50	67.60
Roof trusses	275–350	63.90	18.10	82.00
Bridge structures	200–250	53.10	15.70	68.80

Table 19-12 gives the approximate cost of applying a field coat of paint, by using a spray gun, to structural steel for various types of members and structures. The costs are based on paying $13.50 per gallon for paint and $16.10 per hour to the painter. A painter should paint $\frac{3}{4}$ to $1\frac{1}{2}$ tons/h, depending on the structure.

CHAPTER
20

WATER DISTRIBUTION SYSTEMS

Cost of Water Distribution Systems

The cost of a water distribution system will include the materials, equipment, labor, and supervision to accomplish some of or all the following:

1. Clear the right-of-way for the trench.
2. Remove and replace pavement.
3. Excavate and backfill trenches.
4. Relocate utility lines.
5. Install pipe.
6. Install fittings.
7. Install valves and boxes.
8. Install fire hydrants.
9. Install service connections, meter, and meter boxes.
10. Drill holes under roads and pavements, and install casings for pipeline.
11. Test and disinfect water pipe.

Pipelines

The following types of pipes are used for water systems:

1. Cast iron
   a. Bell and spigot
   b. Mechanical joint
   c. Push-on joint, or gasket-seal joint
   d. Threaded
   e. Cement-lined
2. Steel
   a. Threaded, black or galvanized
   b. Welded, plain or cement-lined
3. Reinforced concrete
   a. Prestressed
   b. Nonprestressed
4. Brass and copper
5. Lead
6. Plastic

Bell-and-Spigot Cast-Iron Pipe

Bell-and-spigot pipe is cast with a bell on one end and a spigot on the other. Joints are made by inserting the spigot end into the bell of an adjacent pipe, caulking one or two encircling strands of yarning material into the bell, then filling the balance of the bell with a specified jointing material, such as lead, cement, or an approved substitute. Fittings and valves are installed in a pipeline in the same manner. A joint of pipe may be cut if necessary to permit a fitting or valve to be installed at a designated location.

The inside surface of cast-iron pipe may be lined with portland cement to reduce tuberculation.

Several types of cast-iron pipe are available. Table 20-1 gives the weights of bell-and-spigot and push-on joint pipe cast in metal molds, for water and other liquids.

Push-on Joint Cast-Iron Pipe

This pipe, which may be identified also as *gasket-type joint*, or *gasket-seal joint*, has a built-in rubber-type gasket in the bell or hub which produces a watertight joint when the spigot end of the joining pipe is forced into the bell. This is now the most widely used type of cast-iron pipe in the water service.

The American Water Works Association (AWWA) and the American National Standards Institute (ANSI) specify the same laying lengths, properties, and weights for this pipe as for bell-and-spigot cast-iron pipe for service under the

TABLE 20-1
Weight of bell-and-spigot and push-on joint cast-iron centrifugally cast in metal molds, AWWA C106-70 or ANSI A21.6-1970, lb/lin ft

Pipe size, in	Internal water pressure 100 lb/in²		Internal water pressure 150 lb/in²		Internal water pressure 200 lb/in²		Internal water pressure 250 lb/in²	
	18-ft pipe	20-ft pipe	18-ft pipe	20-ft pipe	18-ft pipe	20-ft pipe	18-ft pipe	20-ft pipe
3	12.0	11.9	12.0	11.9	12.0	11.9	12.0	11.9
4	16.1	16.0	16.1	16.0	16.1	16.0	16.1	16.0
6	25.6	25.6	25.6	25.6	25.6	25.6	25.6	25.6
8	36.9	36.8	36.9	36.8	36.9	36.8	36.9	36.8
10	49.0	48.7	49.0	48.7	49.0	48.7	49.0	48.7
12	63.4	63.1	63.4	63.1	63.4	63.1	68.3	67.9
14	78.2	77.8	78.2	77.8	83.8	83.4	89.5	89.0
16	94.5	94.0	94.5	94.0	100.9	100.4	109.0	108.4
18	113.9	113.3	113.9	113.3	122.8	122.2	131.8	131.1
20	134.9	134.2	134.9	134.2	144.9	144.2	154.8	154.0
24	177.3	176.4	189.4	188.4	203.6	202.6	203.6	202.6

same water pressure. Thus the information appearing in Table 20-1 applies to this pipe.

Fittings for Bell-and-Spigot Cast-Iron Pipe

Fittings for bell-and spigot cast-iron pipe include tees, crosses, bends, reducers, etc. The fittings may be purchased with any desired combination of bells and spigots. The joints are made with the same materials which are used in joining the pipelines.

Table 20-2 gives the approximate weight of lead and jute required per joint for cast-iron pipe and fittings.

Mechanical-Joint Cast-Iron Pipe

A joint for this pipe is made by inserting the plain end of one pipe into the socket of an adjoining pipe, then forcing a gasket ring into the socket by means of a cast-iron gland which is drawn to the socket by tightening bolts through the gland and socket. Fittings are installed in a similar manner. Joints may be made quickly with an unskilled crew, an ordinary ratchet wrench being the only tool required.

Table 20-3 gives the weights of mechanical-joint cast-iron pipe for working pressures of 100, 150, 200, and 250 lb/in^2 for 18- and 20-ft lengths. In some locations 16-ft lengths are available.

TABLE 20-2
Approximate weights of lead and jute required per joint for cast-iron pipe and fittings

Size pipe, in	Weight of lead, lb				Weight of jute, lb		
	Per joint	Per lin ft		Per joint	Per lin ft		
		12-ft pipe	18-ft pipe			12-ft pipe	18-ft pipe
3	6.50	0.54	0.36	0.18	0.015	0.010
4	8.00	0.67	0.45	0.21	0.018	0.012
6	11.25	0.94	0.63	0.31	0.026	0.017
8	14.50	1.21	0.81	0.44	0.037	0.025
10	17.50	1.46	0.97	0.53	0.044	0.030
12	20.50	1.71	1.14	0.61	0.051	0.034
14	24.00	2.00	1.33	0.81	0.068	0.045
16	33.00	2.75	1.84	0.94	0.078	0.052
18	36.90	3.07	2.05	1.00	0.083	0.056
20	40.50	3.37	2.25	1.25	0.104	0.070
24	52.50	4.38	2.92	1.50	0.125	0.084
30	64.75	5.40	3.60	2.06	0.172	0.115
36	77.25	6.45	4.30	3.00	0.250	0.167
42	104.25	8.70	5.80	3.50	0.292	0.195
48	119.00	9.93	6.62	4.00	0.334	0.222

TABLE 20-3
Weights of mechanical-joint cast-iron pipe centrifugally cast in metal molds, AWWA C106-70 or ABSI A21.6-1970, lb/lin ft

Pipe size, in	Internal water pressure 100 lb/in²		Internal water pressure 150 lb/in²		Internal water pressure 200 lb/in²		Internal water pressure 250 lb/in²	
	18-ft pipe	20-ft pipe	18-ft pipe	20-ft pipe	18-ft pipe	20-ft pipe	18-ft pipe	20-ft pipe
3	12.0	11.9	12.0	11.9	12.0	11.9	12.0	11.9
4	16.2	16.1	16.2	16.1	16.2	16.1	16.2	16.1
6	25.5	25.4	25.5	25.4	25.5	25.4	25.5	25.4
8	36.4	36.2	36.4	36.2	36.4	36.2	36.4	36.2
10	48.2	48.0	48.2	48.0	48.2	48.0	48.2	48.0
12	62.6	62.3	62.6	62.3	62.6	62.3	67.4	67.1
14	78.2	77.8	78.2	77.8	83.8	83.4	89.4	89.0
16	94.5	94.0	94.5	94.0	100.9	100.3	108.9	108.3
18	113.9	113.2	113.9	113.2	122.8	122.2	131.7	131.0
20	134.9	134.2	134.9	134.2	144.9	144.2	154.8	154.1
24	177.2	176.2	189.2	188.2	203.5	202.6	203.5	202.6

Valves

Valves for cast-iron water pipes are usually cast-iron body, bronze-mounted, bell or hub type. Gate valves should be used. A cast-iron adjustable-length valve box should be installed over each wrench-operated valve to permit easy access when it is necessary to operate the valve.

Service Lines

Service lines are installed from the water pipes to furnish water to the customers. These lines usually include a bronze corporation cock, which is tapped into the water pipe, copper pipe extending to the property line or meter, a bronze curb cock, and a meter set in a box. Customers install the service pipe from the meter to their house or business.

Fire Hydrants

Fire hydrants are specified by the type of construction, size of valve, sizes and number of hose connections, size of hub for connection to the water pipe, and depth of bury. It is good practice to install a gate valve between each hydrant and the main water pipe, so that the water may be shut off in the event repairs to the hydrant are necessary.

Tests of Water Pipes

Specifications usually require the contractor to subject the water pipe to a hydrostatic test after it has been laid, prior to backfilling the trenches. If any joints show excessive leakage, they must be recaulked. It is common practice to lay several blocks of pipe, install a valve temporarily, and subject the section to a test. If a test satisfies the specifications, the valve is removed and the trench is immediately backfilled. Additional lengths are laid and tested. This procedure is repeated until the system is completed.

Sterilization of Water Pipes

Prior to placing a water distribution system in service, it should be thoroughly sterilized. Chlorine is most frequently used to sterilize water pipes. It should be fed continuously into the water which is used to flush the pipe lines. After the pipes are filled with chlorinated water, the water is permitted to remain in the pipes for the specified time; then it is drained, and the pipes are flushed and placed in service.

Cost of Cutting Cast-Iron Pipe

Cast-iron pipe may be cut with chisels or with chain cutters. Chisel cutting is done by two or more laborers using a steel chisel with a wood handle and a 6- to 8-lb

FIGURE 20-1
Power-driven pipe cutter.

hammer. Chain cutters may be operated by hand for pipes up to 12 in diameter, but for larger pipes a power-driven cutter should be used.

Figure 20-1 illustrates a power-driven cutter, which is operated by compressed air. This machine can cut 10- to 60-in diameter pipe. It requires 60 to 70 ft^3/min of air at a pressure of 85 lb/in^2. The saw is a portable milling machine on wheels, which travels around the pipe under two silent-type chains, which hold the machine to the pipe and act as a flexible ring gear for the feed sprockets. The machine moves, while the chains remain stationary. A complete cut is made in 1 revolution (r) around the pipe. It will require 1 min of cutting time for each inch of pipe diameter. Thus a 24-in pipe is cut in 24 min of cutting time. An experienced crew of two individuals can install the machine on a pipe in approximately 15 min.

The total time allowed for cutting cast-iron pipe should include measuring, supporting on skids, if necessary, and cutting. In cutting pipe larger than 24 to

TABLE 20-4
Approximate time required to cut cast-iron pipe

| Size pipe, in | Hand cutting | | | | Power cutting, h | |
| | With chisel | | With chain cutter | | | |
	Skilled	Laborer	Skilled	Laborer	Skilled	Laborer
4	0.30	0.30	0.25	0.25		
6	0.45	0.45	0.40	0.40		
8	0.60	0.60	0.55	0.55		
10	0.90	0.90	0.80	0.80	0.60	0.60
12	0.95	1.90	0.85	1.70	0.63	1.26
14	1.10	2.20	—	—	0.66	1.32
16	1.25	2.50	—	—	0.70	1.40
18	1.40	2.80	—	—	0.75	1.50
20	1.50	4.50	—	—	0.80	2.40
24	1.65	6.60	—	—	0.85	3.40
30†	1.25	3.75	—	—	0.95	2.85
36†	1.50	4.50	—	—	1.00	3.00
42†	2.00	6.00	—	—	1.10	3.30
48†	2.40	7.20	—	—	1.20	3.60
54†	2.80	8.40	—	—	1.30	3.90
60†	3.20	9.60	—	—	1.50	4.50

† For pipes larger than 24 in diameter, include the cost of a crane and an operator to handle the pipe.

30 in, it may be necessary to use a crane to handle the pipe. Table 20-4 gives the approximate time in hours required to cut various sizes of cast-iron pipe. The time given includes measuring, setting up, and cutting for average conditions.

Labor Required to Lay Cast-Iron Pipe

The installation of bell-and-spigot cast-iron pipe will include some of or all the following operations:

1. Cutting the pipe, if necessary
2. Lowering the pipe into the trench
3. Inserting the spigot into the bell
4. Yarning the bell
5. Attaching a runner and pouring the lead
6. Removing the runner and caulking the lead

Each joint of pipe is lowered into the trench by hand or with a crane, tractor-mounted side boom, or other suitable equipment. After the spigot end is forced into the bell to full depth, two or more strands of yarning, which completely encircle the pipe, are caulked into the bell to center the pipe and to prevent molten lead from flowing into the pipe. An asbestos runner is placed around the pipe against the bell, with an opening near the top to permit molten lead to be poured into the joint. The lead for a joint should be poured in one continuous operation, without interruption. After the lead cools to the temperature of the pipe, the runner is removed and the lead is caulked by hand or with a pneumatic caulking hammer.

The size of crew required to lay the pipe and the rate of laying will vary considerably with the following factors:

1. Class of soil
2. Extent of groundwater present
3. Depth of trench
4. Extent of shoring required
5. Extent of obstruction such as utilities, sidewalks, pavement
6. Size of pipe
7. Method of lowering pipe into trench
8. Extent of cutting required for fittings and valves

The crew required to dig the trenches, lay the pipe, and backfill the trenches for 12-in pipe furnished in 18-ft lengths, using a tractor-mounted side boom to lower the pipe into the trench, in trenches 3 to 6 ft deep in firm earth, with no groundwater and no shoring needed, might include the following:

    1 trenching machine operator
    2 laborers on bell holes
    1 tractor operator
    2 laborers on pipe
    2 workers centering pipe and installing yarning
    1 person melting and supplying lead
    1 person installing runners and pouring lead
    2 workers caulking lead joints by hand†
    1 driver for utility truck
    1 bulldozer operator backfilling trench
    1 foreman

A crew should install four to six joints per hour, either pipe or fittings. The length of pipe laid will vary from 72 to 108 ft/h, with 90 ft/hr a fair average for the conditions specified.

Table 20-5 gives representative labor-hours required to lay cast-iron pipe in trenches 3 to 6 ft deep in firm soil with little or no shoring required, and no groundwater, using caulked lead joints. For pipe laid under other conditions, the

---

† If the joints are caulked with a pneumatic hammer, use only one person. It will be necessary to include the cost of an air compressor, hose, and hammer.

TABLE 20-5
Labor-hours required to lay 100 lin ft of cast-iron pipe with lead joints

Size pipe, in	Pipe laid, ft/h	Trenching machine operator	Tractor operators	Truck driver	Kettle operator	Pipe layers†	Laborers‡	Foreman
					Labor-hours			
4	120	0.83	0.83	0.83	0.83	1.66	4.15	0.83
6	110	0.91	1.82	0.91	0.91	2.73	4.55	0.91
8	100	1.00	2.00	1.00	1.00	3.00	6.00	1.00
10	95	1.05	2.10	1.05	1.05	3.15	6.30	1.05
12	90	1.10	2.20	1.10	1.10	4.40	6.60	1.10
14	85	1.17	2.34	1.17	1.17	4.68	7.02	1.17
16	80	1.25	2.50	1.25	1.25	6.25	7.50	1.25
18	70	1.43	2.86	1.43	1.43	7.15	8.58	1.43
20	60	1.67	3.34	1.67	1.67	8.35	10.02	1.67
24	50	2.00	4.00	2.00	2.00	10.00	12.00	2.00

† Includes workers pouring lead and caulking joints by hand. If the joints are caulked with a pneumatic hammer, reduce the time by one-third.

‡ Includes the workers digging bell holes and handling and centering the pipe. Five workers are used for the 4- and 6-in pipe and six for the other sizes.

labor-hours should be altered to fit the conditions for the particular project. If a crane is used to lower the pipe, replace the tractor operator with a crane operator.

Labor Required to Lay Cast-Iron Pipe with Mechanical Joints

The operations required to lay cast-iron pipe with mechanical joints include lowering the pipe into the trench, installing the gasket and gland on the spigot, centering the pipe in the bell, pushing the gasket into the bell, and setting the gland by tightening the bolts. The nuts may be tightened with a hand or pneumatic wrench.

The crew required to dig the trench, lay 12-in pipe furnished in 18-ft lengths, and backfill the trenches for trenches 3 to 6 ft deep in firm earth with no shoring needed might include the following:

1 trenching machine operator
2 laborers on bell holes
1 tractor operator
2 laborers on pipe
2 workers installing gland and gasket and centering pipe
1 person tightening nuts
1 driver for utility truck
1 bulldozer operator backfilling trench
1 foreman

TABLE 20-6
Labor-hours required to lay 100 lin ft of cast-iron pipe with mechanical joints

Size pipe, in	Pipe laid, ft/h	Trenching machine operator	Tractor operators	Truck driver	Pipe layers	Laborers	Foreman
				Labor-hours			
4	140	0.72	1.44	0.72	1.44	3.60	0.72
6	130	0.77	1.54	0.77	1.54	3.85	0.77
8	120	0.83	1.66	0.83	1.66	4.15	0.83
10	110	0.91	1.82	0.91	1.82	4.55	0.91
12	100	1.00	2.00	1.00	2.00	5.00	1.00
14	90	1.11	2.22	1.11	2.22	5.55	1.11
16	85	1.18	2.36	1.18	2.36	5.90	1.18
18	80	1.25	2.50	1.25	2.50	6.25	1.25
20	75	1.33	2.66	1.33	2.66	6.65	1.33
24	65	1.54	3.08	1.54	3.08	7.70	1.54
30	55	1.82	3.64	1.82	3.64	9.10	1.82

A crew should lay five to seven joints per hour, either pipe or fittings. The length of pipe laid will vary from 90 to 136 ft/h, with 100 ft/h a fair average.

Table 20-6 gives representative labor-hours required to lay cast-iron pipe with mechanical joints in trenches 3 to 6 ft deep in firm soil with little or no shoring required and no groundwater. For pipe laid under other conditions, the labor-hours should be altered to fit the conditions of the particular job. If a crane is used to lower the pipe, replace the tractor operator with a crane operator.

Labor Required to Lay Push-on Joint Cast-Iron Pipe

The operations required to lay cast-iron pipe and fittings with push-on joints include cutting the pipe to length, if necessary, lowering the pipe and fittings into the trench, using a tractor with a side boom or a power crane, then forcing the spigot end of the pipe being laid into the bell end of the pipe previously laid. Bell holes should be dug in the trench for proper bedding of the pipe and joints.

The crew required to dig the trench, lay the pipe, and backfill the trench will vary with the requirements of the job, including the method of backfilling the earth around and over the top of the pipe.

Consider a project which requires the laying of 12-in push-on joint cast-iron water pipe for a pressure of 150 lb/in^2 in 18-ft lengths, in a trench whose depth will vary from 4 to 6 ft in a soil that will not require shores or side support for the trenches. The pipe will be laid in a new subdivision of a city, under average conditions.

The trench, which will be 30 in wide, will be dug with a gasoline-engine-operated wheel type of trenching machine. The backfill must be of select earth, free of rocks, hand-placed, and compacted to a depth of 6 in above the top of the pipe. The rest of the backfill may be placed with a bulldozer and compacted by running the bulldozer along the trench over the backfill, requiring an estimated three or four passes of the bulldozer.

The compaction around and 6 in over the top of the pipe will be obtained from using a hand-operated gasoline-engine type of self-contained tamper.

The pipe will be lowered into the trench with a tractor-mounted side boom. The pipe will be laid along one side of the trench prior to digging the trench.

The crew should lay four to seven joints per hour, for either pipe or fittings. For pipes furnished in 18-ft lengths, the total length laid should vary from 72 to 136 ft/h with 100 ft/h representing a reasonable average rate.

The crew might include the following:

1 trenching machine operator
2 laborers cleaning trench and digging bell holes
1 tractor operator lowering pipe into trench
1 laborer assisting tractor operator
2 laborers in trench handling pipe
1 pipe layer with pipe and fittings
2 laborers backfilling trench
1 laborer operating tamper
1 bulldozer operator backfilling trench
1 foreman

Table 20-7 gives representative labor-hours required to lay push-on joint cast-iron pipe in trenches whose depths average around 6 ft, in firm soil with little or no shoring required, with no groundwater, and no major obstacles to delay the progress.

In some locations it may be necessary to reclassify some of the individuals of the crew into semiskilled or skilled ratings.

Cost of a Cast-Iron Pipe Water Distribution System

In estimating the cost of installing a water distribution system, the estimator must consider the many variables which will influence the cost of the project. No two projects are alike. For one project there may be very favorable conditions, such as a relatively level terrain, free of trees and vegetation, out in the open with no obstructions, no rocks, no groundwater, no utility pipes, and no pavement, and little rain to delay the project. The equipment may be in good physical condition. The construction gang may be well organized and experienced. The specifications may not require rigid exactness in construction methods, tests, and cleanup. For another project the conditions may be entirely different, with rough terrain, restricted working room (as in alleys), considerable rock, pavement, and unmarked

TABLE 20-7
Representative labor-hours required to lay 100 lin ft of push-on joint cast-iron pipe†

Size pipe, in	Pipe laid, ft/h	Labor-hours					
		Trenching machine operator	Tractor operators	Truck driver	Pipe layers	Laborers	Foreman
4	140	0.72	1.44	0.72	0.72	5.70	0.72
6	130	0.77	1.54	0.77	0.77	6.15	0.77
8	120	0.83	1.66	0.83	0.83	6.65	0.83
10	110	0.91	1.82	0.91	0.91	7.25	0.91
12	100	1.00	2.00	1.00	1.00	8.00	1.00
14	90	1.11	2.22	1.11	1.11	8.88	1.11
16	85	1.18	2.36	1.18	1.18	9.44	1.18
18	80	1.25	2.50	1.25	1.25	10.00	1.25
20	75	1.33	2.67	1.33	1.33	10.64	1.33
24	65	1.54	3.08	1.54	1.54	12.32	1.54

† If for a given job the estimated number of feet of pipe laid per hour differs from the values given in this table, the information in the table can be used. If the estimated rate of laying 8-in pipe is 100 ft/h, the labor-hours for a rate of 100 ft/h should be used.

utility pipes to contend with, as well as groundwater and rain. The equipment may be in poor physical condition, and the construction gang may be poorly organized, with inexperienced workers. The specifications may be very rigid regarding construction methods, tests, and cleanup. As a result of the effect of these variable factors, an estimator should be very careful about using cost data from one project as the basis of estimating the probable cost of another project, especially if the conditions are appreciably different. The following example is intended to illustrate a method of estimating the cost of a water distribution system, but it should not be used as the basis of preparing an estimate without appropriate modifications to fit the particular project.

> **Example 20-1.** Estimate the cost per unit in place for a cast-iron pipe water distribution system consisting of the indicated quantities for each item.
> The cast-iron pipe will be AWWA push-on joint for 150-lb/in² water pressure, in 18-ft lengths. The fittings will be push-on joint AWWA class D.
> The quantities will be:
>
> 3,964 lin ft of 12-in pipe
> 8 each 12 × 12 × 8 × 8 all bell crosses
> 6 each 12 × 12 × 6 × 6 all bell crosses
> 9 each 12 × 12 × 6 all bell tees
> 2 each cast-iron body bronze-mounted 12-in gate valves
> 2 each cast-iron valve boxes for 3-ft 0-in depth

84 each domestic water services including for each service:
1 corporation cock, $\frac{3}{4}$ in
30-lin-ft average length of $\frac{3}{4}$-in type-K copper pipe
2 copper-to-steel pipe couplings, $\frac{3}{4}$ in
1 brass curb cock, $\frac{3}{4}$ in
1 water meter, $\frac{5}{8}$-in disk type
1 cast-iron meter box with cover

   The trenches for the pipe will be dug to an average depth of 3 ft 6 in and 30 in wide, by using a wheel-type trenching machine. The soil is firm, sandy clay which will not require shores or trench supports. In backfilling the trench around and over the pipe, the earth will be placed by hand and compacted to a depth of 6 in above the top of the pipe, by using a self-contained gasoline-engine-operated tamper.
   The pipe and fittings will be distributed along the trenches prior to digging the trenches. The cost of the pipe and fittings includes furnishing and delivering the materials along the trenches. Assume that the job conditions will permit the pipe to be laid at a rate of 80 lin ft/h.
   Use Table 20-7 to estimate the cost of labor.
   The cost per unit will be:

12-in cast-iron pipe:
Pipe delivered to job, per lin ft	= $	10.250
Trenching machine, $18.20 per h ÷ 80 ft/h	=	0.227
Tractor lowering pipe, $16.90 per h ÷ 80 ft/h	=	0.211
Bulldozer backfilling trench, $17.86 per h ÷ 80 ft/h	=	0.223
Pipe cutter, 36 cuts × 0.25 h per cut = (9 h × $6.80)/3,964 =		0.015
Utility truck, $11.62 ÷ 80 ft/h	=	0.145
Earth tamper, $3.22 per h ÷ 80 ft/h	=	0.040
Trenching machine operator, $16.65 ÷ 80 ft/h	=	0.208
Tractor operator, $15.80 per h ÷ 80 ft/h	=	0.198
Bulldozer operator, $15.80 per h ÷ 80 ft/h	=	0.198
Truck driver, $8.10 per h ÷ 80 ft/h	=	0.101
Pipe layer, $15.56 per h ÷ 80 ft/h	=	0.195
Laborers, 8 × $8.10 per h ÷ 80 ft/h	=	0.810
Foreman, $19.65 per h ÷ 80 ft/h	=	0.246
Testing and sterilizing pipe	=	0.150
Total direct cost per lin ft	= $	13.217
Total cost for pipe, 3,964 lin ft × $13.217 per lin ft	=	52,392.19

Cast-iron fittings, per unit, delivered to the job:
12- × 12-in fitting crosses or tees	= $	296.00
Trenching machine, 4 lin ft × $0.227 per lin ft	=	0.91
Trenching machine operator, 4 lin ft × $0.208 per lin ft	=	0.83
Assume that 0.5 h is required to place fitting in trench		
Tractor lowering fitting in trench, 0.5 h × $16.90 per h	=	8.45
Tractor operator 0.5 h × $15.80 per h	=	7.90
Labor helping place fitting, 0.5 h × $8.10 per h	=	4.05
Pipe layer installing fitting, 0.5 h × $15.56 per h	=	7.78
Helper for pipe layer, 0.5 h × $8.10 per h	=	4.05

Foreman, 0.5 h × $19.65 per h	=	9.83
Testing and sterilizing fittings	=	0.90
Utility truck and tools, 0.5 h × $11.62 per h	=	5.81
Total direct cost per fitting	=	$346.51

12-in gate valves, delivered to the job, per valve	= $	592.00
Labor digging 2 bell holes, 0.25 h × $8.10 per h	=	2.02
Pipe layer installing valve, 0.5 h × $15.56 per h	=	7.78
Laborer helping pipe layer, 0.5 h × $8.10 per h	=	4.05
Labor backfilling around box, 1 h × $8.10 per h	=	8.10
Foreman, 0.25 h × $19.65 per h	=	4.91
Utility truck and tools, 0.5 h × $11.62 per h	=	5.81
Total direct cost per valve	=	624.67

Valve boxes, delivered to the job, per valve:

Valve box with lid	= $	43.00
Labor installing box, 0.5 h × $8.10 per h	=	4.05
Labor backfilling around box, 0.5 h × $8.10 per h	=	4.05
Foreman, 0.25 h × $19.65 per h	=	4.91
Utility truck and tools, 0.25 h × $11.62 per h	=	2.90
Total direct cost per box	=	$58.91

Domestic water services, per service:

Corporation cock, $\frac{3}{4}$ in	= $	14.80
Copper pipe, $\frac{3}{4}$ in, 30 lin ft × $7.00 per lin ft	=	210.00
Copper-to-steel pipe couplings, 2 × $3.10	=	6.20
Curb cock, $\frac{3}{4}$ in	=	10.10
Water meter, $\frac{5}{8}$ in	=	25.00
Meter box with cover	=	14.60
Hauling material to job	=	4.80
Power-operated trenching machine, 0.5 h × $5.20 per h	=	2.60
Trenching machine operator, 0.5 h × $12.60 per h	=	6.30
Labor tapping cast-iron pipe and installing corporation cock, 1 h × $8.10 per h	=	8.10
Pipe layer installing copper pipe, meter, and meter box, 1.5 h × $15.56 per h	=	23.34
Labor backfilling trench, 1 h × $8.10 per h	=	8.10
Helper for pipe layer, 1.5 h × $8.10 per h	=	12.15
Foreman, 1 h × $16.95 per h	=	16.95
Utility truck and tools, 0.5 h × $11.62 per h	=	5.81
Total direct cost per service	= $	368.85

Compilation of costs:

Pipe, 3,964 lin ft × $13.217 per lin ft	= $	52,392.19
Cast-iron fittings, 23 × $346.51 each	=	7,969.73
Gate valves, 2 × $624.67 each	=	1,249.34
Valve boxes, 2 × $58.91 each	=	117.82
Domestic water services, 84 × $368.85 each	=	30,983.40
Total direct cost of project	=	$92,712.48

CHAPTER

21

SEWERAGE SYSTEMS

Items Included in Sewerage Systems

Sewerage systems include one or more types of pipes and other appurtenances necessary to permit the system to collect waste water and sewage and transmit them to a suitable location where they may be disposed of in a satisfactory manner.

A sewerage system may be classified as one of two or more types depending on its primary function:

1. Sanitary sewer
2. Storm sewer
3. Other

This chapter is devoted to the construction of sanitary sewerage systems. Among the items included in this system are the following:

1. Sewer pipe and fittings
2. Manholes
3. Service connections
4. Cleanout services

Sewer Pipes

Sewer pipes may be made of one of the following materials:

1. Vitrified clay
2. Ductile iron
3. PVC (polyvinyl chloride)

The properties and other qualities of these pipes may be designated by such agencies as the American Society for Testing Materials (ASTM), the American Water Works Association (AWWA), or another appropriate agency.

Vitrified-clay pipe. This pipe is available in variable diameters, lengths, and wall thicknesses to permit it to be used under different load conditions. The pipe has a rubberlike gasket installed in the bell end which engages the spigot of the inserted pipe to produce a flexible watertight joint.

Ductile iron pipe. This pipe is available in various diameters, lengths, and wall thicknesses. Joints are made by inserting the spigot end of one pipe into the bell end of another pipe. An O-ring or a rubberlike gasket is used to produce a watertight flexible joint.

PVC pipe. This pipe is available in various sizes, lengths, and wall thicknesses. Joints are made by inserting the spigot end of one pipe into the bell end of the connecting pipe, to produce a flexible watertight joint.

Fittings for Sewer Pipe

Fittings for sewer pipe include single and double wyes, tees, curves, increasers, decreasers, and stoppers or plugs.

Construction Operations

When a sewerage system is constructed, the operations will include, but may not be limited to the following:

1. Clearing the right-of-way, if necessary
2. Excavating the trench for the pipe
3. Laying the pipe
4. Excavating the hole for the manhole
5. Placing the manhole in the hole
6. Installing service lines
7. Installing cleanout boots, if required
8. Backfilling with earth over the pipe and manhole
9. Removing surplus earth, if necessary

Equipment Required

The equipment required to install a sewerage system will include, but is not limited to, the following items:

1. Trenching machines
   a. Wheel-type trencher
   b. Ladder-type trencher
   c. Backhoe
   d. Dragline
2. Combination front-end loader, dozer, pipe handler
3. Truck to dispose of surplus earth
4. Power-type earth compactor
5. Trench braces or shoring, if necessary
6. Dewatering pumps, if necessary
7. A power-operated pipe saw
8. A general utility truck
9. A laser beam generator

Constructing a Sewerage System

In constructing a sewerage system it is common practice to begin at the outlet end of the pipeline and proceed to the other end. As the excavation progresses, shores or bracings are installed, if necessary, to protect against earth cave-ins.

The depth of the trench as excavation progresses is determined by using a laser beam, which produces a light spot shining on a target mounted on a depth pole set in the trench from time to time, as required (Fig. 21-1). This beam is set at a slope parallel to the slope of the bottom of the trench, and it can be used from one set up for several hundred feet of trench.

Depending on their weights, the pipe and fittings may be lowered into the trench manually or by a machine. The spigot end of a pipe should be lubricated with a specified compound, to permit easy insertion into the bell of the connecting pipe and to provide a watertight joint.

Manholes

Manholes are installed along sewer lines, usually 300 to 400 ft apart, to permit access to the lines for inspection and cleaning. Also, manholes should be installed at intersections with laterals, at changes in the grade or directions, and for changes in the size of the pipe.

At one time manholes were constructed with bricks laid in place. But brick manholes have largely been replaced with precast (Fig. 21-2) or cast-in-place concrete manholes. When precast manholes are used, the base may be cast at the bottom of the hole first; then the remainder of the precast manhole will be set on

FIGURE 21-1
Generator for a laser beam used to establish the depth and slope of the bottom of a pipeline trench.

the base to complete the installation. Another method of constructing a manhole is to cast the concrete base monolithically with a lower section of the manhole. This section of the manhole is set in place; then one or more sections are added to produce the desired depth. At the time the manholes are cast, holes are cast through the walls to permit pipes of the sewer line to enter or leave the manhole. Cast-in-place rubberized gaskets are installed to produce watertight seals for the pipes. The manhole may be cast with an invert at the bottom to permit the sewage to flow more effectively.

Cleanout Boots

Cleanout boots are frequently installed at the upper end of sewer lines to permit easy access to the lines for cleaning and flushing purposes. The boots are made of cast iron, with a base, throat, and cover, usually set in concrete for stability. To provide this installation, the last joint of the sewer line ends with a wye, pointing upward. Sewer pipe of the desired size is extended upward to the boot. Figure 21-4 illustrates a cleanout boot installation.

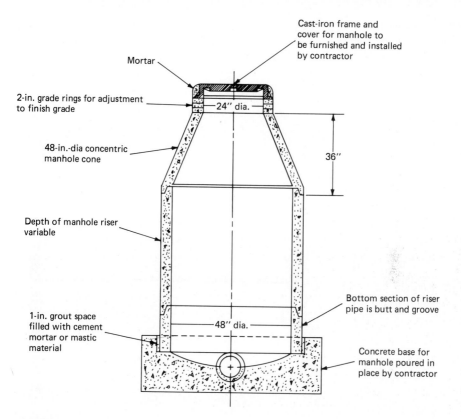

Cast-iron frame and
cover for manhole to
be furnished and installed
by contractor

Mortar

2-in. grade rings for adjustment
to finish grade

24″ dia.

48-in.-dia concentric
manhole cone

36″

Depth of manhole riser
variable

Bottom section of riser
pipe is butt and groove

1-in. grout space
filled with cement
mortar or mastic
material

48″ dia.

Concrete base for
manhole poured in
place by contractor

FIGURE 21-2
A representative precast manhole.

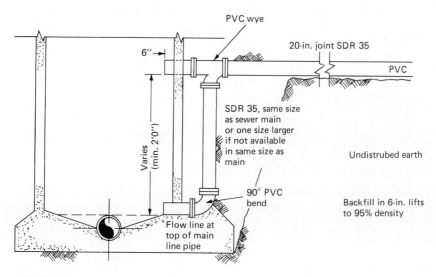

PVC wye

6″

20-in. joint SDR 35

PVC

Varies
(min. 2′0″)

SDR 35, same size
as sewer main
or one size larger
if not available
in same size as
main

Undistrubed earth

90° PVC
bend

Backfill in 6-in. lifts
to 95% density

Flow line at
top of main
line pipe

FIGURE 21-3
Drop connection detail for a manhole.

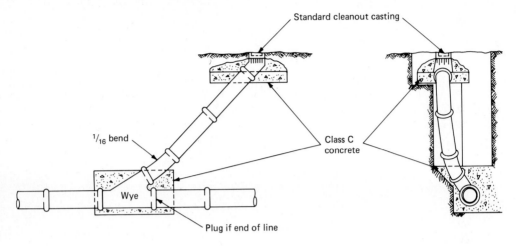

FIGURE 21-4
Cleanout detail for a sewer line.

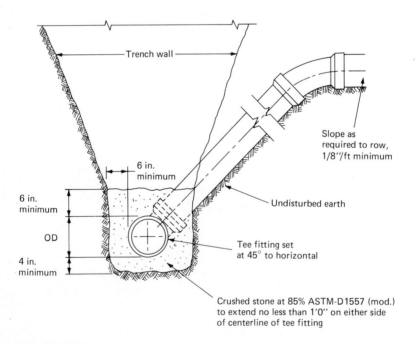

FIGURE 21-5
Service connection detail for a sewer line.

Service Branches

When the main pipeline is installed, it may be desirable to provide access to the pipe for future connections from users of the facility. This provision may be made by installing wye branches in the main line at the time it is installed. Pipes of desired sizes may be extended from the wye branch to locations near the surface of the ground. The ends of the pipe are closed with stoppers, which permits easy access to the services. Figure 21-5 illustrates representative service line installation.

TABLE 21-1

Labor-hours required to lay PVC sewer pipe, using joints 20 ft long

Size pipe, in	Pipe laid, lin ft/h	Labor-hours per 100 lin ft	
		Pipe layers	Laborers
4	70	1.5	1.5
6	60	1.7	1.7
8	60	1.7	1.7
10	40	2.5	2.5
12	40	2.5	5.0
15	40	2.5	7.5

TABLE 21-2

Labor-hours required to backfill and tamp earth by hand to 12 in above the top of the sewer pipe

Size of pipe, in	Labor-hours per 100 lin ft†
4	7
6	9
8	10
10	11
12	12
15	14

† If the backfill is placed by machine but tamped by hand, the number of labor-hours may be reduced by 50 percent.

TABLE 21-3
Machine-hours and labor-hours required to excavate the earth for manholes, using a backhoe

Depth of manhole, ft	Hours per manhole		
	Machine	Operator	Labor
Up to 6	0.75	0.75	1.5
6–9	1.00	1.00	2.0
9–12	1.25	1.25	3.0
12–15	1.50	1.50	4.0

TABLE 21-4
Labor-hours required to install service inlets to sewer pipe

Size of inlet, in	Length of inlet, ft	Labor-hours		
		Pipe layers	Helper	Backhoe operator
4	10–15	1.00	1.00	0.50
4	15–20	1.50	1.50	0.75
4	20–30	2.00	2.00	1.00
6	10–15	2.25	2.00	0.50
6	15–20	2.75	2.75	0.75
6	20–30	3.25	3.25	1.00

TABLE 21-5
Labor-hours required to install a cleanout boot

Depth of trench, ft	Labor-hours			
	Pipe layers	Helpers	Backhoe	
			Operator	Helper
4–6	1.00	1.00	0.50	0.50
6–8	1.50	1.50	0.75	0.75
8–12	2.00	2.00	1.00	1.00
12–15	2.75	2.75	1.25	1.25

Example 21-1. Estimate the direct cost of installing a sewerage system requiring the laying of PVC pipe, fittings, manholes, service connections, and cleanout boots for the given conditions.

> Length of project, 4,290 lin ft
> Size of pipe, 8 in
> Maximum depth of trench, 12.4 ft
> Minimum depth of trench, 6.5 ft
> Average depth of trench, 8.0 ft
> Number of manholes, 14
> Average depth of manholes, 8.0 ft

The project will be constructed in an open and clear area with no obstructions, land clearing, utility lines, or other items to delay progress. Surplus earth will be hauled to a disposal site approximately 2 mi distant.

The soil will be dense clay, with no groundwater. The trenches will be excavated with a $\frac{3}{4}$-yd^3 backhoe, which will permit the trench walls to assume normal slopes, with no shoring required. After the pipe is laid, the backfill will be select earth from along the trench. The backfill around the pipe will be hand-tamped to a depth of 12 in above the top of the pipe. The balance of the backfill will be placed by a bulldozer-loader in 12-in layers and compacted with a power-operated tamper.

The manholes will be precast concrete of a type illustrated in Fig. 21-2.

Service inlets will be spaced at 40-ft intervals along the pipeline. Each inlet will require a wye fitting.

Four cleanout boots will be required.

The quantities of pipe, fittings, manholes, service connections, and cleanout boots will be as follows:

Total length of pipeline	= 4,290 lin ft
Deduct total diameters of manholes, requiring no pipe, 14 × 4 lin ft	= −56 lin ft
Subtotal length of pipe	= 4,234 lin ft
Number of wyes required, 4,290 ÷ 40 = 107	
Deduct length of wyes, 107 × 2 lin ft	= −214 lin ft
Subtotal length of pipe	= 4,020 lin ft
Add for breakage and waste, 2%	= +80 lin ft
Total length of pipe required	= 4,100 lin ft

The direct costs will be:

Sewer pipe, 4,100 lin ft @ $2.05	=	$ 8,405.00
Wyes, 4,290 lin ft ÷ 40 lin ft = 107 @ $17.50 =		1,872.50
Pipe stoppers, 107 @ $3.60	=	385.20
Joint lubricant, 2 gal @ $9.60 per gal	=	19.20
Backhoe, 4,290 lin ft ÷ 60 ft/h = 71 h @ $21.65 per h	=	1,537.15
Backhoe operator, 71 h @ $17.25 per h	=	1,224.75
Backhoe oiler, 71 h @ $11.25 per h	=	798.75

Pipe layer, 71 h @ $15.56 per h	=	1,104.76
Laborers, 2 × 71 h = 142 h @ $8.10 per h	=	1,150.20
Laborers tamping and backfilling trench, 4,234 lin ft × 5 h/100 lin ft = 211 h @ $8.10 per h	=	1,709.10
Bulldozer backfilling trench, 71 h @ $12.56	=	891.76
Bulldozer operator, 71 h @ $15.80 per h	=	1,121.80
Laser beam generator, 71 h @ $0.40 per h	=	28.40
Foreman, 71 h @ $19.65 per h	=	1,395.15
Utility truck, 71 h @ $4.26 per h	=	302.46
Total cost of pipeline	=	$21,946.18

Manholes, total no., 14:

Concrete base, 6-ft diameter, 9 in thick	= $	50.00
Manhole barrel, 8 ft deep @ $45.00 per ft	=	360.00
Cast-iron ring and cover	=	110.00
Cost per manhole	= $	520.00

Cost of manholes, 14 × $520.00	= $	7,280.00
Concrete for setting rings, 14 × 4 ft³/ring = 56 ft³ @ $1.60 per ft³	=	89.60
Backhoe excavating hole @ 1 h/hole = 14 h @ $21.65	=	303.10
Backhoe operator, 14 h @ $17.25	=	241.50
Backhoe setting manholes, 14 @ 1 h each = 14 h @ $21.65	=	303.10
Backhoe operator, 14 h @ $17.25 per h	=	241.50
Laborers, 2 × 14 h = 28 h @ $8.10 per h	=	226.80
Backhoe and loader backfilling earth around manhole, 14 @ 1 h each = 14 h @ $12.56	=	175.84
Laborers tamping earth around manholes, 14 @ 3 h each = 42 h @ $8.10	=	340.20
Power-operated earth tamper, 14 @ 1 h each = 14 h @ $1.85	=	25.90
Dump truck to haul surplus earth to disposal site, 14 trips @ 1 h/trip = 14 h @ $12.68	=	177.52
Total cost of manholes	= $	9,405.06

Service lines:

Total no. of service lines, 4,290 lin ft ÷ 40 lin ft per line = 107
Average cost per service:

20 ft of 4-in pipe @ $0.64 per lin ft	= $	12.80
4 in × ⅛ bend @ $3.07	=	3.07
4-in stopper	=	3.60
Backhoe excavating trench, ½ h @ $12.56	=	6.28
Backhoe operator, ½ h @ $17.25	=	8.68
Pipe layer, ½ h @ $15.56	=	7.78
Laborer, ½ h @ $8.10	=	4.05
Total cost per service line	=	$46.26

Total cost for 107 service lines, 107 × $46.26	= $	4,949.82

Cleanout boots:
  Total number, 14
  Average cost per boot:

8 lin ft of 6-in pipe @ $1.39 per lin ft	= $	11.12
6 in $\times \frac{1}{16}$ bend @ $5.91	=	5.91
Cleanout casting	=	6.96
6-in plug	=	2.12
$\frac{1}{4}$ yd^3 concrete @ $45.00 yd^3	=	11.25
Pipe layer, 1 h @ $15.56	=	15.56
Labor placing pipe and casting, 2 h @ $8.10	=	16.20
Trenching machine excavating trench, $\frac{1}{2}$ h @ $12.56	=	6.28
Trenching machine operator, $\frac{1}{2}$ h @ $17.25	=	8.63
Total cost per boot	= $	84.03
Total cost for 4 boots, 4 × $84.03	=	336.12

Total cost for the project:

Pipeline	=	$21,946.18
Manholes	=	9,405.06
Service lines	=	4,949.82
Cleanout boots	=	336.12
	=	$36,637.18

CHAPTER
22

TOTAL COST OF ENGINEERING PROJECTS

The previous chapters of this book have discussed methods of estimating the costs of constructing engineering projects. However, the cost of construction is not the only cost which the owner of the project must pay. The total cost to the owner may include, but is not necessarily limited to, the following items.

1. Land, right-of-way, easements
2. Legal expense
3. Bond expense, or cost of obtaining money to finance the project
4. Cost of construction
5. Engineering and/or architect's expense
6. Interest during construction
7. Contingencies

Cost of Land, Right-of-Way, and Easements

If it is necessary to purchase land or obtain rights to use land in constructing an engineering project, the owner of the project must provide the money to finance these acquisitions.

The land on which a project is to be constructed may be purchased by the owner. In general, the cost of acquiring the land should be included in the total cost of the project.

If the project includes the construction of pipelines, power lines, telephone lines, or other extended items, the owner of the project may prefer to obtain a continuing right to construct and maintain a facility through the property without actually purchasing it. This right may be defined as an *easement*, for which the owner of the project may pay the owner of the land.

Legal Expenses

The construction of a project frequently involves actions and services which require the employment or use of an attorney. The actions requiring an attorney may be the acquisition of land and easements, holding a bond election for a government agency, printing and obtaining approval of bonds to be sold for providing the money to finance the project, or assistance to a private corporation. The legal fees paid for these services should be included in the cost of the project.

Bond Expense

Before a project can be constructed by a government agency, it is usually necessary to hold a bond election for the purpose of permitting qualified voters to approve or reject the project. In the event voters approve the project, it is necessary to print, register, and sell the bonds, usually through a qualified underwriting broker, who charges a fee for these services. Private corporations frequently sell bonds to finance new construction. In any event, it is correct to charge to the project the costs of these services.

Cost of Construction

The cost of constructing a project is usually an estimate only, made in advance of receiving bids from contractors, prepared by an engineer or an architect. It is the amount that she or he believes the owner will have to pay for the construction of the project.

The estimate may be a lump-sum cost, such as for a building, or it may be a unit-price estimate, for a project that requires the construction of various items whose exact quantities are not exactly known in advance of construction. For example, the contract for constructing a highway may specify a given payment to the contractor for each ton of asphaltic material placed in the pavement.

For a project that involves the construction of units of work, the bid form will provide for the bidders to state the amounts which they will charge for constructing each specified unit.

Engineering Expense

The owner who desires to have a project constructed engages an engineer or an architect to make the necessary surveys and studies, to prepare the plans and specifications, and to assist in securing bids for the construction and supervising the construction of the project. The cost for this service is usually based on an agreed percentage of the cost of construction. The fee usually varies from 5 to 10 percent of the cost, depending on several factors.

Interest during Construction

Most construction contracts provide that the owner pay to the contractor at the end of each month during the period of construction a specified percentage of the value of the work completed during the month, frequently about 90 percent. In addition to the amounts that the owner must pay to the contractor, he or she likely will have to pay other costs prior to completion of the project. Thus the owner will have considerable funds invested in the project while it is under construction. Since the owner must pay interest on the money required to finance these costs, there will be an interest cost during construction, which should be included in the total cost of the project.

The amount of interest chargeable to the project during construction is usually estimated when the total cost of the project is estimated. A method sometimes used is to assume that one-half the cost of the project will require the payment of interest during the full period of construction. However, if the full cost of the project is secured in advance of the beginning of construction, and if interest is paid on the full amount during this period, then the total cost of interest should be included in the cost of the project.

Contingencies

If the exact cost of the project is not known in advance of raising funds to finance it, as is frequently the case, it is good practice to provide funds to cover any additional costs that may occur during construction. An estimator should rely on personal judgment to determine the amount of contingencies desirable.

Example Illustrating an Estimate for Total Cost of an Engineering Project

An example illustrating an estimate for the total cost of a project involves drilling three water wells and furnishing all materials, labor, and equipment required to provide additional sources of water for a city. The project will require acquiring land and easements, drilling wells, and installing pumps and cast-iron water pipes, fittings, and valves to bring the water to the city.

A bond election will be held to provide the money to finance the total cost of the project. The estimated total cost is determined as shown in the following table.

This is the minimum amount that should be included in a bond election for the project.

Item	Estimated cost
1. Land and easements	= $ 86,000
2. Legal expense	= 12,000
3. Bond expense	= 8,100
4. Cost of construction	
a. Water wells with pumps, 3 @ $106,625	= $319,875
b. Pump houses, 3 @ $4,500	= 13,500
c. Extend electric power lines to well,	
9,420 lin ft @ $11.36	= 107,011
d. 12-in class 250 ductile iron pipe,	
8,780 lin ft @ $15.63	= 137,231
e. Cast-iron fittings, 8.64 tons @ $1,003.11	= 8,667
f. 12-in gate valves, 4 @ $920.00	= 3,680
Total cost of construction	$589,964
5. Engineering expense, 7% of $589,964	= 41,298
6. Interest during construction, estimated to be	
10 mo @ 8%/yr, $\dfrac{\$589,964}{2} \times 0.08 \times \dfrac{10}{12}$	= 19,665
7. Contingencies	= 37,851
8. Estimated total cost	= $794,878

CHAPTER
23

CONCEPTUAL COST ESTIMATING

At the inception of a project by the owner, prior to any design, only limited information is known about a project. However, the owner must know the approximate cost in order to evaluate the economic feasibility of proceeding with the project. Thus, there is a need to determine the approximate cost of a project during its conceptual phase.

As discussed in Chap. 1, cost estimates may be divided into at least two different types, depending on the purposes for which they are prepared and the amount of information known when the estimate is prepared: approximate estimates (sometimes called preliminary, conceptual, or budget estimates) and detailed estimates (sometimes called final or definitive estimates). Each of these estimates may be subdivided.

The previous chapters of this book have discussed methods and procedures of estimating the construction costs of engineering projects for which detailed information is known. For example, the cost of excavation was determined for a known quantity and type of soil, size of excavator, and job conditions. The cost of concrete forms was determined based on specific sizes, shapes, and numbers of concrete columns, beams, walls, etc.

This chapter presents methods and procedures for estimating project costs during the preliminary or conceptual phase. Because there is little definition of a project at this stage, the accuracy of the estimate will be less than that of a detailed estimate. However, the conceptual cost estimate is important because the owner will examine this estimate before continuing with development of the project.

Accuracy of Conceptual Estimates

The accuracy of any estimate will depend on the amount of information known about a project. A conceptual cost estimate should be identified by the information from which the estimate was compiled. For example, a conceptual estimate which is prepared from the project scope (sometimes called the *project charter* or *project mission*), when there is little or no design, may be identified as *level I*, in terms of accuracy. *Level I accuracy* may be defined as accurate within $+40$ and -10 percent. A *level II* conceptual cost estimate may be defined as an estimate that is prepared upon completion of the preliminary design. Such estimates might be accurate within $+25$ and -5 percent. *Level III* conceptual estimates might be identified as estimates that are compiled upon completion of the final design. A level III conceptual cost estimate might be accurate within $+10$ and -3 percent.

It is important to classify the conceptual estimate by predefined levels of accuracy similar to those described above. Such classifications provide a measure of reliability to those who must use the information.

Liability of Conceptual Cost Estimates

Generally the designers are not obligated under standard-form contracts to guarantee the construction cost of a project. However, as a part of their design responsibility, designers prepare an estimate of the probable construction cost for the project for which they have prepared the design.

Major decisions are often made by the owner from information contained in the conceptual cost estimate. This places a responsibility and liability on the estimator. Although the initial estimate may be prepared from little-known information, it is the duty of the estimator to reestimate the project as additional information about the project becomes available. This is particularly important for a cost-plus construction contract where the designer has the responsibility to monitor costs during construction.

Preparation of Conceptual Estimates

There are many variations in the contract arrangements among the three principle parties in a project (owner, designer, and contractor). The conceptual estimate is generally prepared by the owner during the owner's feasibility study or by the designer during the design phase. It may be prepared by the contractor for negotiated work between an owner and a contractor.

Multibuilder owners are involved with the construction of projects on a continual basis. Examples are electric utilities, oil and gas firms, retail stores with nationwide locations, etc. To initiate a project, these firms generally conduct an owner's study. The owner's study consists of a technical feasibility study and an economic feasibility study of the proposed project. A conceptual cost estimate is prepared by the owner as part of the economic feasibility analysis.

Designers prepare conceptual cost estimates throughout the design process. The estimate is used in the selection of design alternatives and to keep the owner informed of forecast costs. It is the responsibility of the designer to develop a design that will produce a completed project within the amount of money authorized in the owner's budget.

Some projects are of an emergency nature and must be completed in the least time possible. For such projects the owner will negotiate a contract with a contractor. The contractor may be asked to prepare a conceptual cost estimate for the project for which there are no, or limited, plans and specifications.

The preparation of conceptual cost estimates requires knowledge and experience with the work required to complete the project. Cost information from previous projects of similar type and size is essential. The estimator must combine all known information with her or his personal experience and use considerable judgment to prepare a reliable conceptual estimate.

Broad-scope Conceptual Estimates

A broad-scope conceptual cost estimate of a proposed project is prepared prior to the design of the project. It is prepared from cost information on previously completed projects similar to the proposed project. The number of units, or size, of the project is the only known information, such as the number of square feet of building area, the number of cars in a parking garage, the number of miles of 345-kV transmission line, the number of barrels of crude oil processed per day, etc.

The best source of information for preparation of conceptual cost estimates is the cost records from previous projects. Although the range of costs will vary among projects, the estimator can develop unit costs to forecast the cost of future projects.

The unit cost should be developed from a weighting of the data that emphasizes the average value yet it should account for the extreme maximum and minimum values. Equation (23-1) can be used for weighting cost data from previous projects:

$$UC = \frac{A + 4B + C}{6} \qquad (23\text{-}1)$$

where UC = forecast unit cost
A = minimum unit cost of previous projects
B = average unit cost of previous projects
C = maximum unit cost of previous projects

The following example illustrates the weighting of the cost data from a previous project to determine the forecast unit cost of a proposed project.

Example 23-1. Cost information from eight previously constructed parking garage projects is shown below.

Project	Cost	No. cars
1	$466,560	150
2	290,304	80
3	525,096	120
4	349,920	90
5	259,290	60
6	657,206	220
7	291,718	70
8	711,414	180

Use the weighted unit cost to determine the conceptual cost estimate for a proposed parking garage that is to contain 135 parked cars.

The unit cost per car is calculated as shown below:

Project	Unit cost per car
1	$ 3,110.40
2	3,628.80
3	4,375.80
4	3,888.00
5	4,321.50
6	2,987.30
7	4,167.40
8	3,952.30
Total =	$30,431.50

The average cost per car is $30,431.50 ÷ 8 = $3,803.94. From Eq. (23-1) the forecast unit cost will be

$$UC = \frac{2,987.30 + 4(3,803.94) + 4,375.80}{6} = \$3,763.14$$

The conceptual cost estimate for 135 cars will be

$$135 \text{ cars @ } \$3,763.14 = \$508,023.00$$

The technique of conceptual cost estimating illustrated in Example 23-1 can be applied to other types of projects, for example, for apartment units, motel rooms, miles of electric transmission line, barrels of crude oil processed per day, square yards of pavement, etc.

It is necessary for the estimator to adjust the cost information from previously completed projects for use in the preparation of a conceptual cost estimate for a proposed project. There should be an adjustment for time, location, and size.

Time Adjustments for Conceptual Estimates

The use of cost information from a previous project to forecast the cost of a proposed project will not be reliable unless an adjustment is made proportional to the difference in time between the two projects. The adjustment should represent the relative inflation or deflation of costs with respect to time due to factors such as labor rates, material costs, interest rates, etc.

Various organizations publish indices that show the economic trends of the construction industry with respect to time. The Engineering News Record (ENR) annually publishes indices of construction costs. An index can be used to adjust previous cost information for use in the preparation of a conceptual cost estimate.

The estimator can use the change in value of an index between any two years to calculate an equivalent compound interest rate. This equivalent interest rate can be used to adjust past cost records, to forecast future project costs. The following example will illustrate the use of indices for time adjustments.

Example 23-2. Suppose the indices for building construction projects show the following economic trends:

Year	Index
3 yr ago	358
2 yr ago	359
1 yr ago	367
Current year	378

An equivalent compound interest can be calculated based on the change in the cost index during the 3-yr period:

$$\frac{378}{358} = (1 + i)^3$$

$$i = 1.83\%$$

Suppose that cost information from a $843,500 project completed last year is to be used to prepare a conceptual cost estimate for a project proposed for construction 3 yr from now. The cost of the proposed project should be adjusted for time as follows:

$$\text{Cost} = \$843,500 \times (1 + 0.0183)^4$$
$$= \$906,960$$

Adjustments for Location

The use of cost information from a previous project to forecast the cost of a proposed project will not be reliable unless an adjustment is made that represents the difference in cost between the locations of the two projects. The adjustment should represent the relative difference in costs of materials, equipment, and labor with respect to the two locations.

Various organizations publish indices that show the relative differences in construction costs with respect to geographic location. The ENR is an example of a publisher of location cost indices. The following example illustrates the use of indices for location adjustments.

Example 23-3. The location indices for construction costs show the following:

Location	Index
City A	1,025
City B	1,170
City C	1,260
City D	1,105
City E	1,240

Suppose that cost information from a $387,200 project completed in city A is to be used to prepare a conceptual cost estimate for construction of a proposed project in city D. The cost of the proposed project should be adjusted for location as follows:

$$\text{Cost} = \frac{1,105}{1,025} \times \$387,200 = \$417,420$$

Adjustment for Size

The use of cost information from a previous project to forecast the cost of a future project will not be reliable unless an adjustment is made that represents the difference in size of the two projects. In general, the cost of a project is directly proportional to its size. The adjustment is generally a simple ratio of the size of the proposed project to the size of the previous project from which the cost data are obtained.

Combined Adjustments

The conceptual cost estimate for a proposed project is prepared from cost records of a project completed at a different time and at a location with a different size. The estimator must adjust the previous cost information for the combination of time, location, and size.

The following example illustrates the use of combined adjustments for preparation of a conceptual cost estimate.

Example 23-4. Use the time and location indices presented in previous examples to prepare the conceptual cost estimate for a building with 62,700 ft² of floor area. The building is to be constructed 3 yr from now in city B. A similar type of building that cost \$2,197,540 and contained 38,500 ft^2 was completed 2 yr ago in city E. Estimate the probable cost of the proposed building.

Proposed cost

$$= \text{Previous cost} \times \text{Time adjustment} \times \text{Location adjustment} \times \text{Size adjustment}$$

$$= \$2,197,540 \times (1 + 0.0183)^5 \times \frac{1,170}{1,240} \times \frac{62,700}{38,500}$$

$$= \$2,197,540 \times 1.095 \times 0.944 \times 1.629$$

$$= \$3,700,360$$

Unit-cost Adjustments

Although the total cost of a project will increase with size, the cost per unit may decrease. For example, the cost of a 1,800-ft^2 house may be \$53.50 per square foot whereas the cost of a 2,200-ft^2 house of comparable construction may be only \$48.75 per square foot. Certain items, such as the kitchen appliances, garage, etc., are independent of the size of the project.

Size adjustments for a project are unique to the type of project. The estimator must obtain cost records from previous projects and develop appropriate adjustments for his or her particular project. The following example illustrates a method of size adjustments for preparation of a conceptual cost estimate.

Example 23-5. Cost records from previous projects show the following information:

Project	Total cost	Size, no. of units
1	\$2,250	100
2	1,485	60
3	2,467	120
4	2,730	150
5	3,401	190

The cost per unit can be calculated as shown:

Project	Cost per unit
1	\$22.50
2	24.75
3	20.56
4	18.20
5	17.90

A plot of the cost records can be prepared (see Fig. 23-1).

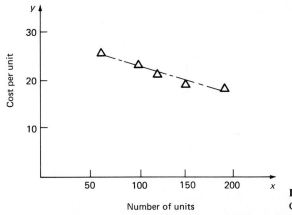

FIGURE 23-1
Comparison of size and cost per unit.

For a first-order relationship, the general equation for a straight line is

$$Y = b + mX$$

where $b =$ intercept of the line and $m =$ slope of the line. Substituting in values for b and m, we get

$$Y = 24.75 + \left(\frac{17.90 - 24.75}{190 - 60}\right)X$$

$$= 24.75 - 0.05269X \qquad \text{where } 60 < X < 190$$

The following equation can be written for the unit cost with respect to the number of units:

$$\text{Forecast unit cost} = 24.75 - 0.05269(S - 60)$$

where $S =$ the number of units in the proposed project.

The above equation represents the relationship between the unit cost and size for the five previously completed projects. This equation can be used to calculate the cost per unit for future projects whose sizes may range from 60 to 190 units. For example, the unit cost for a 170-unit project would be

$$\text{Unit cost} = 24.75 - 0.05269(S - 60)$$
$$= 24.75 - 0.05269(170 - 60)$$
$$= \$18.95$$

As illustrated in Example 23-5, the adjustment of unit costs based on the size of a project is unique and can only be obtained from previous cost records.

The cost data for some types of projects could be nonlinear, rather than linear as previously illustrated. For example, a second-order equation may better fit the data for some types of projects. The technique presented in Example 23-5 can also be applied for nonlinear data.

The estimator must evaluate his or her own particular cost records and develop a unit cost–size relationship. There are numerous methods of curve fitting, such as linear regression or least squares. Reference 1 provides graphical and statistical methods to develop equations for forecasting and establishing confidence levels.

Narrow-Scope Conceptual Cost Estimates

As the design of a project progresses and more information becomes known about the various components, a narrow-scope conceptual cost estimate can be prepared. For example, upon completion of the foundation design, the number of cubic yards of concrete will be known. When the structural-steel design is complete, the number of tons of structural steel will be known.

A narrow-scope conceptual estimate is prepared in a manner similar to that for a broad-scope estimate, except the project is subdivided into parameters. For a building construction project the parameters might be square yards of asphalt parking, cubic yards of concrete foundations, tons of structural steel, square feet of finished floor, number of doors, etc. For a steel-pole electric transmission line, the parameters might be acres of clearing land, cubic yards of concrete foundations, tons of steel pole, linear feet of conductor wire, number of insulator strings, etc.

The cost of a proposed project is prepared from historical cost records of previous projects, with an appropriate adjustment for time, size, and location, as previously discussed.

Factors Affecting Cost Records

The estimator must be cautious when using historical cost records from completed projects. A proposed project may have features significantly different from those of the completed project from which the cost records are obtained. For example, the cost per square foot for a building with a high ratio of perimeter to floor area will be significantly higher than for a building with a low ratio of perimeter to floor area. Other factors that could affect costs are span lengths, height between floors, quality of furnishings, quality of work, etc.

The estimator must compare the features of the proposed project with those of previous projects and make appropriate adjustments.

REFERENCES

1. Paul J. Ossenbruggen, *Systems Analysis for Civil Engineers: Technological and Economic Factors in Design*, Wiley, New York, 1984.
2. Jack R. Benjamin and C. Allin Cornell, *Probability, Statistics, and Decision for Civil Engineers*, McGraw-Hill, New York, 1970.
3. William Mendanhall and James E. Reinmuth, *Statistics for Management and Economics*, 2d ed., Duxbury Press, North Scituate, Mass., 1974.

PROBLEMS

23-1. Use the time and location indices presented in this chapter to estimate the cost of a building that contains 32,500 ft^2 of floor area. The building is to be constructed 2 yr from now in city A. The cost of a similar type of building that contained 48,300 ft^2 was completed last year in city C for a cost of $3,308,500.

23-2. Calculate the weighted unit cost per square foot for the project data shown below, and determine the cost of a 2,700-ft^2 project.

Project	Total cost	Size, ft^2
1	$147,300	2,580
2	153,700	2,900
3	128,100	2,100
4	118,400	1,850
5	135,700	2,300

23-3. Determine the relationship between unit cost and size for the project data shown in Prob. 23-2, to estimate the cost of a 2,200-ft^2 project.

APPENDIX

MASTERFORMAT—
Master List of Section Titles and Numbers

BIDDING REQUIREMENTS, CONTRACT FORMS*, AND CONDITIONS OF THE CONTRACT*

Document Number	Title		
		00200	**INFORMATION AVAILABLE TO BIDDERS**
00010	**PRE-BID INFORMATION**		
-020	Invitation to Bid	-210	Preliminary Schedules
-030	Advertisement for Bids	-220	Geotechnical Data
-040	Prequalification Forms		*Geotechnical Report*
			Soil Boring Data
00100	**INSTRUCTIONS TO BIDDERS**	-230	Existing Conditions
			Description of Existing Site
-120	Supplementary Instructions to Bidders		*Description of Existing Buildings*
			Property Survey
-130	Pre-Bid Conferences	-240	Project Financial Information

* Documents marked with an asterisk (*) have important legal consequences. Initiation or modifications without explicit approval and guidance from the owner or the owner's legal counsel is not recommended.

Reprinted courtesy of The Construction Specifications Institute.

00300	BID FORMS

00400	SUPPLEMENTS TO BID FORMS
-410	Bid Security Forms
-420	Bidders Qualification Forms
-430	Subcontractor List
-440	Substitution List
-450	Equipment Suppliers List
-460	List of Alternates/Alternatives
-470	List of Estimated Quantities
-480	Noncollusion Affidavit

00500	AGREEMENT FORMS*

00600	BONDS AND CERTIFICATES*
-610	Performance Bonds
-620	Payment Bonds
-630	Warranty Bonds
-640	Maintenance Bonds
-650	Certificates of Insurance
-660	Certificates of Compliance

00700	GENERAL CONDITIONS*

00800	SUPPLEMENTARY CONDITIONS*
-810	Modifications to General Conditions
-820	Additional Articles
	Equal Employment Opportunity Requirement
	Insurance Requirements
	Non-Segregated Facilities Requirements
	Specific Project Requirements
	Statutory Requirements
	Wage Rate Requirements
-830	Wage Determination Schedule

00900	ADDENDA

SPECIFICATIONS— DIVISIONS 1-16

DIVISION 1 — GENERAL REQUIREMENTS

Section Number	Title
01010	SUMMARY OF WORK
	Work Covered by Contract Documents
	Contracts
	Work Under Other Contracts
	Future Work
	Work Sequence
	Contractor Use of Premises
	Occupancy Requirements
	Products Ordered in Advance
	Owner Furnished Products
01020	ALLOWANCES
-021	Cash Allowances
	Product Allowances
	Installation Allowances
	Inspection and Testing Allowances
	Contingency Allowances
-024	Quantity Allowances
01025	MEASUREMENT AND PAYMENT
	Schedule of Values
	Applications for Payment
	Unit Prices
01030	ALTERNATES/ ALTERNATIVES
01035	MODIFICATION PROCEDURES
	Change Orders
	Instructions
	Field Orders
	Directives
01040	COORDINATION
-041	Project Coordination
-042	Mechanical and Electrical Coordination
-043	Job Site Administration
-045	Cutting and Patching

01050	**FIELD ENGINEERING**		-425	Field Samples
			-430	Mock-Ups
01060	**REGULATORY**		-440	Contractor's Quality Control
	REQUIREMENTS		-445	Manufacturer's Field Services

01080   **IDENTIFICATION SYSTEMS**

01090   **REFERENCES**
-091   Reference Standards
-092   Abbreviations
-093   Symbols
-094   Definitions

01100   **SPECIAL PROJECT PROCEDURES**

  Airport Project Procedures
  Alteration Project Procedures
  Detention Project Procedures
  Environmental Protection
    Procedures
  Ground Transportation Project
    Procedures
  Hazardous Material Procedures
  Hospital Project Procedures
  Industrial Project Procedures
  Nuclear Project Procedures
  Preservation and Restoration
    Project Procedures
  Radiation Protection Procedures
  Security Project Procedures
  Shopping Mall Project Procedures

01200   **PROJECT MEETINGS**
-210   Preconstruction Conferences
-220   Progress Meetings
-245   Installation Meetings

01300   **SUBMITTALS**
-310   Progress Schedules
    Network Analysis Schedule
-320   Progress Reports
-330   Survey and Layout Data
-340   Shop Drawings, Product Data,
    and Samples
-360   Quality Control Submittals
    Design Data
    Test Reports
    Certificates
    Manufacturer's Instructions
-380   Construction Photographs

01400   **QUALITY CONTROL**
-410   Testing Laboratory Services
-420   Inspection Services

01500   **CONSTRUCTION FACILITIES AND TEMPORARY CONTROLS**

-505   Mobilization
-510   Temporary Utilities
    Temporary Electricity
    Temporary Lighting
    Temporary Heating, Cooling, and
      Ventilating
    Temporary Telephone
    Temporary Water
    Temporary Sanitary Facilities
    Temporary Fire Protection
-520   Temporary Construction
    Temporary Bridges
    Temporary Decking
    Temporary Overpasses
    Temporary Runarounds
-525   Construction Aids
    Construction Elevators, Hoists.
      and Cranes
    Scaffolding and Platforms
    Swing Staging
    Temporary Enclosures
    First Aid
-530   Barriers and Enclosures
    Barricades
    Fences
    Tree and Plant Protection
-540   Security
    Protection of Work and Property
-550   Access Roads and Parking
    Areas
    Access Roads
    Parking Areas
-560   Temporary Controls
    Construction Cleaning
    Dust Control
    Erosion and Sediment Control
    Noise Control
    Pest Control
    Pollution Control
    Rodent Control
    Surface Water Control
-570   Traffic Regulation
    Construction Parking Control
    Flagmen
    Flares and Lights
    Haul Routes
    Traffic Signals
-580   Project Identification and Signs
-590   Field Offices and Sheds

01600	**MATERIAL AND EQUIPMENT**
-610	Delivery, Storage, and Handling
	Packing
	Shipping
	Unloading and Acceptance
	Protection
-620	Installation Standards
-630	Product Options and Substitutions

01650	**FACILITY STARTUP/ COMMISSIONING**
-655	Starting of Systems
-660	Testing, Adjusting, and Balancing of Systems
-670	Systems Demonstrations

01700	**CONTRACT CLOSEOUT**
-710	Final Cleaning
-720	Project Record Documents
-730	Operation and Maintenance Data
	Operation Manuals
	Maintenance Instructions
-740	Warranties and Bonds
-750	Spare Parts and Maintenance Materials
-760	Warranty Inspections

01800	**MAINTENANCE**

DIVISION 2 - SITEWORK

Section Number	Title
02010	**SUBSURFACE INVESTIGATION**
-012	Standard Penetration Tests
	Borings
	Core Drilling
-016	Seismic Investigation
02050	**DEMOLITION**
-060	Building Demolition
-070	Selective Demolition
	Minor Demolition for Remodeling
	Selective Structural Demolition
-075	Concrete Removal
-080	Hazardous Material Abatement
02100	**SITE PREPARATION**
-110	Site Clearing
	Clearing and Grubbing
	Large Tract Tree Clearing
-115	Selective Clearing
	Sod Stripping
	Tree and Shrub Removal
	Tree Pruning
-120	Structure Moving
02140	**DEWATERING**
-042	Sand Drains
-044	Well Points
-046	French Drains
-048	Relief Wells
02150	**SHORING AND UNDERPINNING**
-152	Shores
-153	Needle Beams
-154	Grillage
-156	Underpinning
-158	Slabjacking
02160	**EXCAVATION SUPPORT SYSTEMS**
-162	Cribbing and Walers
-164	Soil and Rock Anchors
	Anchor Tieback Systems
-166	Ground Freezing
-167	Reinforced Earth
-168	Slurry Wall Construction

02170   **COFFERDAMS**
-172   Double Wall Cofferdams
-174   Cellular Cofferdams
-176   Piling with Intermediate Lagging
-178   Sheet Piling Cofferdams

02200   **EARTHWORK**
-210   Grading
      Rough Grading
      Finish Grading
-220   Excavating, Backfilling, and Compacting
      Borrow
      Elevator Jack Holes
      Embankment
      Excavating, Backfilling, and
        Compacting for Structures
      Excavating, Backfilling, and
        Compacting for Utilities
      Excavating, Backfilling, and
        Compacting for Pavement
      Rock Removal
-230   Base Courses
      Asphalt Base Course
      Caliche Base Course
      Granular Base Course
      Limerock Base Course
      Sand Clay Base Course
      Shell Base Course
      Soil Cement Base Course
      Subsoil Base Course
-240   Soil Stabilization
      Asphalt Soil Stabilization
      Cement Soil Stabilization
      Geotextile Soil Stabilization
      Lime Soil Stabilization
      Lime Slurry Soil Stabilization
      Pressure Grouting Soil
        Stabilization
-250   Vibro-Flotation
-270   Slope Protection and Erosion Control
      Gabions
      Membrane Systems
      Retaining Walls
      Riprap
      Sediment Control
      Silt Fences
      Slope Paving
      Stone Slope Protection
      Wire Mats
-280   Soil Treatment
      Rodent Control
      Termite Control
      Vegetation Control
-290   Earth Dams

02300   **TUNNELING**
-305   Tunnel Ventilation and Compression

-310   Tunnel Excavating
-320   Tunnel Lining
      Concrete Tunnel Lining
      Prefabricated Steel Tunnel Lining
-330   Tunnel Grouting
-340   Tunnel Support Systems
      Rock Bolting
      Steel Rings and Lagging

02350   **PILES AND CAISSONS**
-355   Pile Driving
      Pile Load Tests
      Pile Performance Specifications
-360   Driven Piles
      Composite Piles
      Concrete Displacement Piles
      Concrete Filled Steel Piles
      Precast Concrete Piles
      Prestressed Concrete Piles
      Rolled Steel Section Piles
      Sheet Piles
      Wood Piles
      Pressure Injected Footings
-370   Bored/Augered Piles
      Auger Cast Grout Piles
      Bored and Belled Concrete Piles
      Bored Friction Concrete Piles
      Cast-in-Place Concrete Piles -
        Uncased
      Drilled Concrete Piers
-380   Caissons
      Benoto Caissons
      Box Caissons
      Drilled Caissons
      Excavated Caissons
      Open Caissons
      Pneumatic Caissons
-390   Repair of Piles
      Extension and Repair of Concrete
        Piles
      Repair of Sheet Piles
      Repair of Wood Piles

02450   **RAILROAD WORK**
-452   Railroad Trackwork
-454   Railroad Service Facilities
      Fueling Depots
      Hi-Rail Access
      Yards
-456   Railroad Traffic Control

02480   **MARINE WORK**
-482   Dredging
-484   Seawalls and Bulkheads
-486   Groins and Jetties
-488   Docks and Facilities
      Marine Fenders
-490   Underwater Work

02500 PAVING AND SURFACING

-505 Granular Paving
   Crushed Stone Paving
   Cinder Surfacing

-510 Asphaltic Concrete Paving
   Asphaltic Concrete Base Course
   Asphaltic Concrete Surface
    Course
   Asphaltic Concrete Curb and
    Gutter
   Asphaltic Concrete Athletic Paving

-515 Unit Pavers
   Asphaltic Block Pavers
   Brick Pavers
   Concrete Pavers
   Stone Pavers

-520 Portland Cement Concrete
   Paving
   Concrete Curb and Gutter
   Integrally Colored Concrete
    Paving

-525 Prefabricated Curbs
   Granite Curbs
   Precast Concrete Curbs

-540 Synthetic Surfacing
   Synthetic Grass Surfacing
   Resilient Matting

-545 Bituminous Surface Treatment
   Single Bituminous Surface
    Treatment
   Double Bituminous Surface
    Treatment

-575 Pavement Repair
   Pavement Resurfacing
   Slurry Sealing

-580 Pavement Marking
   Tactile Warning Lines

02600 UTILITY PIPING MATERIALS

-605 Utility Structures
   Cleanouts
   Manholes and Covers
   Tunnels

-610 Pipe and Fittings
   Cast Iron Pipe
   Concrete Pipe
   Corrugated Metal Pipe
   Ductile Iron Pipe
   Mineral Fiber Reinforced Cement
    Pipe
   Plastic Pipe
   Pre-Insulated Pipe
   Steel Pipe
   Vitrified Clay Pipe

-640 Valves and Cocks
-645 Hydrants

02660 WATER DISTRIBUTION

-665 Water Systems
   Chilled Water Systems

   Cisterns
   Domestic Water Systems
   Fire Water Systems
   Heating Water Systems
   Thrust Restraints

-670 Water Wells
   Test Well Drilling
   Well Drilling and Casing

-675 Disinfection of Water Distribution
   Systems

02680 FUEL AND STEAM DISTRIBUTION

-685 Gas Distribution Systems
-690 Oil Distribution Systems
   Fuel Tanks
-695 Steam Distribution Systems

02700 SEWERAGE AND DRAINAGE

-710 Subdrainage Systems
   Disposal Wells
   Foundation Drainage Systems
   Retaining Wall Underdrains
   Tunnel Drainage Systems
   Underslab Drainage Systems

-720 Storm Sewerage
   Catch Basins, Grates, and Frames
   Culverts
   Curb Inlets
   Drainage Pipe
   French Drains
   Manhole Covers and Frames
   Precast Trench Drains
   Splash Blocks
   Surface Run-Off Collection

-730 Sanitary Sewerage
   Sewage Collection Lines
   Sewage Force Mains

-735 Combined Wastewater Systems
-740 Septic Systems
   Drainage Fields
   Grease Interceptors
   Leaching Cesspools
   Sand Filters
   Septic Tanks
   Siphon Tanks

02760 RESTORATION OF UNDERGROUND PIPE

-762 Inspection of Underground
   Pipelines
-764 Sealing Underground Pipelines
-766 Relining Underground Pipelines

02770 PONDS AND RESERVOIRS

-772 Ponds
   Cooling Water Ponds
   Fire Protection Reservoirs

	Stabilization Ponds
	Storm Water Holding Ponds
-774	Sewage Lagoons
-776	Pond and Reservoir Liners
-778	Pond and Reservoir Covers

02780	**POWER AND COMMUNICATIONS**
-785	Electric Power Transmission
	Overhead Electric Power Transmission
	Underground Electric Power Transmission
-790	Communication Transmission
	Fiber Optics Communications
	Microwave Communications
	Shortwave Communications
	Satellite Antennas

02800	**SITE IMPROVEMENTS**
-810	Irrigation Systems
-820	Fountains
-830	Fences and Gates
	Chain Link Fences and Gates
	Ornamental Metal Fences and Gates
	Tennis Court Windbreakers
	Wire Fences and Gates
	Wood Fences and Gates
-840	Walk, Road, and Parking Appurtenances
	Bicycle Racks
	Culvert Pipe Underpasses
	Guardrails
	Parking Barriers
	Parking Bumpers
	Signage
	Traffic Signals
-860	Playfield Equipment and Structures
	Playground Equipment
	Play Structures
	Recreational Facilities
-870	Site and Street Furnishings
	Prefabricated Planters
	Prefabricated Shelters
	Seating
	Tables
	Trash and Litter Receptors
	Tree Grates
-890	Footbridges

02900	**LANDSCAPING**
-910	Shrub and Tree Transplanting
-920	Soil Preparation
	Topsoil
-930	Lawns and Grasses
	Hydro-Mulching

	Plugging
	Seeding
	Sodding
	Sprigging
	Stolonizing
-950	Trees, Plants, and Ground Covers
	Ground Covers
	Plants and Bulbs
	Shrubs
	Trees
-970	Landscape Maintenance
	Fertilizing
	Liming
	Mowing
	Pruning
	Watering

DIVISION 3 - CONCRETE

Section Number	Title

03100 CONCRETE FORMWORK

-110 Structural Cast-in-Place Concrete Formwork
  Metal Pan Formwork
  Slip Formwork
-120 Architectural Cast-in-Place Concrete Formwork
-130 Permanent Forms
  Permanent Steel Forms
  Prefabricated Stair Forms

03200 CONCRETE REINFORCEMENT

-210 Reinforcing Steel
-220 Welded Wire Fabric
-230 Stressing Tendons
-240 Fibrous Reinforcing

03250 CONCRETE ACCESSORIES

  Anchors and Inserts
  Expansion and Contraction Joints
  Waterstops

03300 CAST-IN-PLACE CONCRETE

-310 Structural Concrete
  Heavyweight Structural Concrete
  Lightweight Structural Concrete
  Normalweight Structural Concrete
  Shrinkage Compensating Concrete
-330 Architectural Concrete
  Lightweight Architectural Concrete
  Normalweight Architectural Concrete
-340 Low Density Concrete
-345 Concrete Finishing
-350 Concrete Finishes
  Blasted Concrete Finishes
  Colored Concrete Finishes
  Exposed Aggregate Concrete Finishes
  Grooved Surface Concrete Finishes
  Heavy-Duty Concrete Floor Finishes
  Tooled Concrete Finishes
-360 Specially Placed Concrete
  Shotcrete
-365 Post-Tensioned Concrete

03370 CONCRETE CURING

03400 PRECAST CONCRETE

-410 Structural Precast Concrete - Plant Cast
  Precast Concrete Hollow Core Planks
  Precast Concrete Slabs
  Structural Precast Pretensioned Concrete - Plant Cast
-420 Structural Precast Post-Tensioned Concrete - Plant Cast
-430 Structural Precast Concrete - Site Cast
  Lift-Slab Concrete
  Precast Post-Tensioned Concrete - Site Cast
  Structural Precast Pretensioned Concrete - Site Cast
-450 Architectural Precast Concrete - Plant Cast
  Faced Architectural Precast Concrete-Site Cast
  Glass Fiber Reinforced Precast Concrete Site Cast
-460 Architectural Precast Concrete - Site Cast
-470 Tilt-Up Precast Concrete
-480 Precast Concrete Specialties

03500 CEMENTITOUS DECKS AND TOPPINGS

-510 Gypsum Concrete
  Gypsum Concrete Floor Underlayment
  Gypsum Concrete Roof Decks
-520 Insulating Concrete Decks
-530 Cementitious Wood Fiber Systems
  Cementitious Wood Fiber Planks
-540 Composite Concrete and Insulation Decks
-550 Concrete Toppings
  Cementitious Floor Underlayment

03600 GROUT

  Catalyzed Metallic Grout
  Epoxy Grout
  Nonmetallic Grout

03700 CONCRETE RESTORATION AND CLEANING

-710 Concrete Cleaning
-720 Concrete Resurfacing
-730 Concrete Rehabilitation

03800 MASS CONCRETE

DIVISION 4 - MASONRY

Section Number — Title

04100 MORTAR AND MASONRY GROUP

Cement and Lime Mortars
Chemical Resisting Mortars
Epoxy Mortars
High Bond Mortar
Masonry Grouts
Mortar Coloring Materials
Premixed Mortars

04150 MASONRY ACCESSORIES

Anchors and Tie Systems
Manufactured Control Joints
Joint Reinforcement
Weep Vents

04200 UNIT MASONRY

-210 Clay Unit Masonry
Brick Unit Masonry
Clay Tile Unit Masonry
Structural Clay Tile Unit Masonry
Terra Cotta Unit Masonry
-220 Concrete Unit Masonry
Exposed Aggregate Concrete Unit Masonry
Fluted Concrete Unit Masonry
Interlocking Concrete Unit Masonry
Molded Face Concrete Unit Masonry
Prefaced Concrete Unit Masonry
Preinsulated Concrete Unit Masonry
Sound Absorbing Concrete Unit Masonry
Split Face Concrete Unit Masonry
-230 Reinforced Unit Masonry
Reinforced Grouted Brick Masonry
Reinforced Grouted Concrete Unit Masonry
-235 Pre-assembled Masonry Panel Systems
-240 Non Reinforced Masonry Systems
Single Wythe Masonry Systems
Multiple Wythe Masonry Systems
Veneer Masonry Systems
Mortarless Concrete Unit Masonry
-270 Glass Unit Masonry
-280 Gypsum Unit Masonry
-290 Adobe Unit Masonry

04400 STONE
-410 Rough Stone
-420 Cut Stone
-440 Flagstone
-450 Stone Veneer
-455 Marble
-460 Limestone
-465 Granite
-470 Sandstone
-475 Slate

04500 MASONRY RESTORATION AND CLEANING
-510 Masonry Cleaning
-520 Masonry Restoration

04550 REFRACTORIES
-555 Flue Liners
-560 Combustion Chambers
-565 Firebrick
-570 Castable Refractories

04600 CORROSION RESISTANT MASONRY
-605 Chemical Resistant Brick
-610 Vitrified Clay Liner Plates

04700 SIMULATED MASONRY
-710 Simulated Stone
-720 Cast Stone

DIVISION 5 - METALS

Section Number	Title

05010 METAL MATERIALS
Aluminum
Brass
Bronze
Cast Iron
Copper
Ductile Iron
Lead
Stainless Steel
Steel
Zinc

05030 METAL COATINGS
Acrylic Coatings
Anodic Coatings
Enamel Coatings
Fluorocarbon Coatings
Galvanic Coatings
Metallic Coatings
Porcelain Enamel Coatings
Powdered Coatings
Urethane Coatings

05050 METAL FASTENING
Bolting
Brazing
Chemical Bonding
Riveting
Soldering
Special Fasteners
Welding

05100 STRUCTURAL METAL FRAMING
-120 Structural Steel
 Architecturally Exposed Structural Steel
 Prefabricated Fireproofed Steel Columns
 Tubular Steel
-140 Structural Aluminum
 Architecturally Exposed Structural Aluminum
-150 Steel Wire Rope
-160 Framing Systems
 Geodesic Structures
 Space Frames

05200 METAL JOISTS
-210 Steel Joists
 Longspan Steel Joists
 Deep Longspan Steel Joists
 Open Web Steel Joists
 Steel Joist Girders
-250 Aluminum Joists
-260 Composite Joist System

05300 METAL DECKING
-310 Steel Deck
 Steel Floor Deck
 Steel Roof Deck
-320 Raceway Deck Systems
-330 Aluminum Deck
 Aluminum Floor Deck
 Aluminum Roof Deck

05400 COLD FORMED METAL FRAMING
-410 Load-Bearing Metal Stud Systems
-420 Cold Formed Metal Joist Systems
-430 Slotted Channel Framing Systems
-450 Metal Support Systems
 Electrical Support Systems
 Mechanical Support Systems
 Medical Support Systems

05500 METAL FABRICATIONS
-510 Metal Stairs
-515 Ladders
-520 Handrails and Railings
 Pipe and Tube Railings
-530 Gratings
-535 Floor Plates
-540 Castings
-550 Stair Treads and Nosings

05580 SHEET METAL FABRICATIONS
-582 Sheet Metal Enclosures
-584 Heating/Cooling Unit Enclosures

05700 ORNAMENTAL METAL
-710 Ornamental Stairs
-715 Prefabricated Spiral Stairs
-720 Ornamental Handrails and Railings
-725 Ornamental Metal Castings
-730 Ornamental Sheet Metal

05800	**EXPANSION CONTROL**
-810	Expansion Joint Cover Assemblies
	Elastomeric Joint Cover Assemblies
	Metal Plate Cover Assemblies
	Strip Seal Floor Joint Covers
-820	Slide Bearings
-830	Bridge Expansion Joint Assemblies
	Bridge Bearings
	Bridge Sole Plates

05900	**HYDRAULIC STRUCTURES**
-910	Penstocks
-915	Bulkheads
-920	Trashracks
-925	Manifolds
-930	Bifurcations

DIVISION 6 - WOOD AND PLASTICS

Section Number	Title
06050	**FASTENERS AND ADHESIVES**
06100	**ROUGH CARPENTRY**
-105	Treated Wood Foundations
-110	Wood Framing
	Assembled Wood Components
-115	Sheathing
-120	Structural Panels
-125	Wood Decking
-128	Mineral Fiber Reinforced-Cement Panels
	Cementitious Reinforced Panels
06130	**HEAVY TIMBER CONSTRUCTION**
-132	Mill-Framed Structures
-133	Pole Construction
-135	Timber Trusses
-140	Timber Decking
-145	Timber Bridges and Trestles
06150	**WOOD AND METAL SYSTEMS**
	Wood Chord Metal Joists
06170	**PREFABRICATED STRUCTURAL WOOD**
-180	Glued-Laminated Construction
	Glued-Laminated Decking
	Glued-Laminated Structural Units
-190	Wood Trusses
	Prefabricated Architectural Wood Trusses
	Prefabricated Wood Trusses
-195	Prefabricated Wood Beams and Joists
	Plywood Web Joists
06200	**FINISH CARPENTRY**
-220	Millwork
-240	Laminates
	Plastic Laminates
	Wood Laminates
	Metallic Laminates
-250	Prefinished Wood Paneling
-255	Prefinished Hardboard Paneling
-260	Board Paneling

06300   **WOOD TREATMENT**
-310   Preservative Treatment
-320   Fire Retardant Treatment
-330   Insect Treatment

06400   **ARCHITECTURAL
            WOODWORK**
-410   Custom Casework
              *Plastic Laminate Faced Wood
                Cabinets
              Shop Finished Wood Cabinets
              Unfinished Wood Cabinets*
-420   Panelwork
              *Plastic Laminate Faced Paneling
              Stile and Rail Paneling
              Wood Veneer Faced Paneling*
-430   Stairwork and Handrails
-440   Wood Ornaments
-450   Standing and Running Trim
-460   Exterior Frames
-470   Screens, Blinds, and Shutters
-480   Custom Wood Turning

06500   **STRUCTURAL PLASTICS**

06600   **PLASTIC FABRICATIONS**
-610   Glass Fiber and Resin
              Fabrications
-620   Cast Plastic Fabrications
-630   Historic Plastic Reproductions

06650   **SOLID POLYMER
            FABRICATIONS**

DIVISION 7 - THERMAL AND MOISTURE PROTECTION

Section Number	Title
07100	**WATERPROOFING**
-110	Sheet Membrane Waterproofing
	Bituminous Sheet Membrane Waterproofing
	Elastomeric Sheet Membrane Waterproofing
	Modified Bituminous Sheet Membrane Waterproofing
	Thermoplastic Sheet Membrane Waterproofing
-120	Fluid Applied Waterproofing
-125	Sheet Metal Waterproofing
-130	Bentonite Waterproofing
-140	Metal Oxide Waterproofing
-145	Cementitious Waterproofing
07150	**DAMPPROOFING**
-160	Bituminous Dampproofing
-175	Cementitious Dampproofing
07180	**WATER REPELLENTS**
07190	**VAPOR RETARDERS**
07195	**AIR BARRIERS**
07200	**INSULATION**
-210	Building Insulation
	Batt Insulation
	Building Board Insulation
	Foamed-in-Place Insulation
	Loose Fill Insulation
	Sprayed Insulation
-220	Roof and Deck Insulation
	Asphaltic Perlite Concrete Deck
	Roof Board Insulation
07240	**EXTERIOR INSULATION AND FINISH SYSTEMS**
07250	**FIREPROOFING**
-252	Thermal Barriers for Plastics
-255	Cementitious Fireproofing
-260	Intumescent Mastic Fireproofing

-262	Magnesium Oxychloride Fireproofing
-265	Mineral Fiber Fireproofing

07270 FIRESTOPPING

Fibrous Fire Safing
Fire Penetration Sealants
Firestopping Mortars
Firestopping Pillows
Intumescent Firestopping Foams
Silicone Firestopping Foams
Mechanical Firestopping Devices
    for Plastic Pipe

07300 SHINGLES AND ROOFING TILES

-310 Shingles
    Asphalt Shingles
    Fiberglass Shingles
    Metal Shingles
    Mineral Fiber Cement Shingles
    Porcelain Enamel Shingles .
    Slate Shingles
    Wood Shingles
    Wood Shakes

-320 Roofing Tiles
    Clay Roofing Tiles
    Concrete Roofing Tiles
    Metal Roofing Tiles
    Mineral Fiber Cement Roofing
        Tiles
    Plastic Roofing Tiles

07400 MANUFACTURED ROOFING AND SIDING

-410 Manufactured Roof and Wall
    Panels
    Manufactured Roof Panels
    Manufactured Wall Panels
-420 Composite Panels
-440 Faced Panels
    Aggregate Coated Panels
    Porcelain Enameled Faced Panels
    Tile Faced Panels
-450 Glass Fiber Reinforced
    Cementitious Panels
-460 Siding
    Aluminum Siding
    Composition Siding
    Hardboard Siding
    Mineral Fiber Cement Siding
    Plastic Siding
    Plywood Siding
    Steel Siding
    Wood Siding

07480 EXTERIOR WALL ASSEMBLIES

07500 MEMBRANE ROOFING

-510 Built-Up Bituminous Roofing
    Built-Up Asphalt Roofing
    Built-Up Coal Tar Roofing
-515 Cold Applied Bituminous Roofing
    Cold Applied Mastic Roof
        Membrane
    Glass Fiber Reinforced Asphalt
        Emulsion
-520 Prepared Roll Roofing
-525 Modified Bituminous Sheet
    Roofing
-530 Single Ply Membrane Roofing
-540 Fluid Applied Roofing
-545 Coated Foamed Roofing
-550 Protected Membrane Roofing
-560 Roof Maintenance and Repairs
    Roof Moisture Survey
    Roofing Resaturants

07570 TRAFFIC COATINGS

-572 Pedestrian Traffic Coatings
-576 Vehicular Traffic Coatings

07600 FLASHING AND SHEET METAL

-610 Sheet Metal Roofing
-620 Sheet Metal Flashing and Trim
-630 Sheet Metal Roofing Specialties
-650 Flexible Flashing
    Laminated Sheet Flashing
    Plastic Sheet Flashing
    Rubber Sheet Flashing

07700 ROOF SPECIALTIES AND ACCESSORIES

-710 Manufactured Roof Specialties
    Copings
    Counterflashing Systems
    Gravel Stops and Fascias
    Relief Vents
    Reglets
    Roof Expansion Assemblies
-720 Roof Accessories
    Manufactured Curbs
    Roof Hatches
    Gravity Ventilators
    Penthouse Ventilators
    Ridge Vents
    Smoke Vents

07800 SKYLIGHTS

-810 Plastic Unit Skylights
    Domed Plastic Unit Skylights
    Pyramid Plastic Unit Skylights
    Vaulted Plastic Unit Skylights

-820   Metal Framed Skylights
Domed Metal Framed Skylights
Motorized Metal Framed Skylights
Ridge Metal Framed Skylights
Sloped Metal Framed Skylights
Vaulted Metal Framed Skylights

07900   **JOINT SEALERS**

-910   Joint Fillers and Gaskets
Compression Seals

DIVISION 8 - DOORS AND WINDOWS

Section Number	Title
08100	**METAL DOORS AND FRAMES**
-110	Steel Doors and Frames *Standard Steel Doors and Frames* *Custom Steel Doors and Frames*
-120	Aluminum Doors and Frames
-130	Stainless Steel Doors and Frames
-140	Bronze Doors and Frames
08200	**WOOD AND PLASTIC DOORS**
-210	Wood Doors *Flush Wood Doors* *Prefinished Wood Doors* *Plastic Laminate Faced Doors* *Metal Faced Wood Doors* *Stile and Rail Wood Doors*
-220	Plastic Doors
08250	**DOOR OPENING ASSEMBLIES**
-255	Packaged Steel Door Assemblies
-260	Packaged Wood Door Assemblies
-265	Packaged Plastic Door Assemblies
08300	**SPECIAL DOORS**
-305	Access Doors *Access Panels*
-310	Sliding Doors and Grilles *Sliding Metal Doors* *Sliding Wood Doors* *Sliding Glass Doors* *Sliding Grilles*
-315	Pressure Resistant Doors *Blast Resistant Doors* *Airtight Doors* *Watertight Doors*
-320	Security Doors
-325	Cold Storage Doors
-330	Coiling Doors and Grilles *Overhead Coiling Doors* *Overhead Coiling Grilles* *Side Coiling Doors* *Side Coiling Grilles* *Coiling Counter Doors* *Coiling Counter Grilles*
-350	Folding Doors and Grilles *Accordion Folding Doors*

	Panel Folding Doors
	Accordion Folding Grilles
-355	Chain Closures
-360	Sectional Overhead Doors
-365	Vertical Lift Doors
	Multileaf Vertical Lift Doors
	Telescoping Vertical Lift Doors
-370	Industrial Doors
-375	Hangar Doors
-380	Traffic Doors
	Flexible Traffic Doors
	Rigid Traffic Doors
	Flexible Strip Doors
-385	Sound Control Doors
-390	Storm Doors
-395	Screen Doors

08400 ENTRANCES AND STOREFRONTS

-410	Aluminum Entrances and Storefronts
-420	Steel Entrances and Storefronts
-430	Stainless Steel Entrances and Storefronts
-440	Bronze Entrances and Storefronts
-450	All-Glass Entrances
-460	Automatic Entrance Doors
-470	Revolving Entrance Doors
-480	Balanced Entrance Doors
-490	Sliding Storefronts

08500 METAL WINDOWS

-510	Steel Windows
-520	Aluminum Windows
-530	Stainless Steel Windows
-540	Bronze Windows

08600 WOOD AND PLASTIC WINDOWS

-610	Wood Windows
	Metal Clad Wood Windows
	Plastic Clad Wood Windows
-630	Plastic Windows

08650 SPECIAL WINDOWS

-655	Roof Windows
-660	Security Windows and Screens
	Security Windows
	Security Screens
-665	Pass Windows
-670	Storm Windows

08700 HARDWARE

-710	Door Hardware
	Hanging Hardware
	Latching Hardware
	Controlling Hardware
	Door Trim
	Weatherstripping and Seals
-740	Electro-Mechanical Hardware
	Electrical Locking Systems
	Electro-Magnetic Door Holders
-760	Window Hardware
	Automatic Window Equipment
	Window Operators
	Window Locks
	Window Lifts
-770	Door and Window Accessories

08800 GLAZING

-810	Glass
	Float Glass
	Rolled Glass
	Tempered Glass
	Laminated Glass
	Insulating Glass
	Coated Glass
	Mirrored Glass
	Wired Glass
	Decorative Glass
	Bent Glass
-840	Plastic Glazing
	Bullet Resistant Plastic Glazing
	Decorative Plastic Glazing
	Insulating Plastic Glazing
-850	Glazing Accessories

08900 GLAZED CURTAIN WALLS

-910	Glazed Steel Curtain Walls
-920	Glazed Aluminum Curtain Walls
-930	Glazed Stainless Steel Curtain Walls
-940	Glazed Bronze Curtain Walls
-950	Translucent Wall and Skylight Systems
-960	Sloped Glazing Systems
-970	Structural Glass Curtain Walls

DIVISION 9 - FINISHES

Section Number	Title
09100	**METAL SUPPORT SYSTEMS**
-110	Non-load Bearing Wall Framing Systems
-120	Ceiling Suspension Systems
-130	Acoustical Suspension Systems
09200	**LATH AND PLASTER**
-205	Furring and Lathing
	Gypsum Lath
	Metal Lath
	Veneer Plaster Base
	Plaster Accessories
-210	Gypsum Plaster
	Acoustical Plaster
	Fireproofing Plaster
-215	Veneer Plaster
-220	Portland Cement Plaster
-225	Adobe Finish
-230	Plaster Fabrications
09250	**GYPSUM BOARD**
-260	Gypsum Board Systems
-270	Gypsum Board Accessories
09300	**TILE**
-310	Ceramic Tile
	Ceramic Mosaics
	Conductive Tile
-320	Thin Brick Tile
-330	Quarry Tile
	Chemical Resistant Quarry Tile
-340	Paver Tile
-350	Glass Mosaics
-360	Plastic Tile
-370	Metal Tile
-380	Cut Natural Stone Tile
09400	**TERRAZZO**
-410	Portland Cement Terrazzo
-420	Precast Terrazzo
-430	Conductive Terrazzo
-440	Plastic Matrix Terrazzo
09450	**STONE FACING**
09500	**ACOUSTICAL TREATMENT**
-510	Acoustical Ceilings
	Acoustical Panel Ceilings

	Acoustical Tile Ceilings
	Acoustical Metal Pan Ceilings
-520	Acoustical Wall Treatment
-525	Acoustical Space Units
-530	Acoustical Insulation and Barriers
09540	**SPECIAL WALL SURFACES**
	Fiber Reinforced Plastic Coated Panels
	Reinforced Gypsum Units
	Aggregate Coatings
09545	**SPECIAL CEILING SURFACES**
	Linear Metal Ceilings
	Mirror Panel Ceilings
	Textured Metal Ceiling Panels
	Textured Gypsum Ceiling Panels
	Linear Wood Ceilings
	Suspended Decorative Grids
09550	**WOOD FLOORING**
-560	Wood Strip Flooring
-565	Wood Block Flooring
-570	Wood Parquet Flooring
	Acrylic Impregnated Wood Parquet Flooring
	Vinyl Bonded Wood Parquet Flooring
-580	Wood Composition Flooring
-590	Resilient Wood Flooring Systems
	Cushioned Wood Flooring
	Mastic Set Wood Flooring
	Spring Supported Wood Flooring
	Steel Channel Wood Flooring
	Steel Splined Wood Flooring
09600	**STONE FLOORING**
-610	Flagstone Flooring
-615	Marble Flooring
-620	Granite Flooring
-625	Slate Flooring
09630	**UNIT MASONRY FLOORING**
-635	Brick Flooring
	Chemical Resistant Brick Flooring
	Industrial Brick Flooring
-640	Pressed Concrete Unit Flooring
09650	**RESILIENT FLOORING**
-660	Resilient Tile Flooring
-665	Resilient Sheet Flooring
-670	Fluid-Applied Resilient Flooring
-675	Static Control Resilient Flooring
	Static Resistant Resilient Flooring

Conductive Resilient Flooring
Static Dissipative Resilient
Flooring
-678   Resilient Base and Accessories

09680   CARPET

-682   Carpet Cushion
-685   Sheet Carpet
-690   Carpet Tile
-695   Wall Carpet
-698   Indoor/Outdoor Carpet

09700   SPECIAL FLOORING

-705   Resinous Flooring
-710   Magnesium Oxychloride Flooring
-720   Epoxy-Marble Chip Flooring
-725   Seamless Quartz Flooring
-730   Elastomeric Liquid Flooring
         Conductive Elastomeric Liquid
         Flooring
-750   Mastic Fills
-755   Plastic Laminate Flooring
-760   Asphalt Plank Flooring

09780   FLOOR TREATMENT

-785   Metallic-Type Static
         Disseminating and Spark
         Resistant Finish
-790   Slip Resistant Finishes

09800   SPECIAL COATINGS

-810   Abrasion Resistant Coatings
-815   High Build Glazed Coatings
-820   Cementitious Coatings
-830   Elastomeric Coatings
-835   Textured Plastic Coatings
-840   Fire Resistant Paints
-845   Intumescent Paints
-850   Chemical Resistant Coatings
-860   Graffiti Resistant Coatings
-870   Coating Systems for Steel
         Exterior Coating System for Steel
         Storage Tanks
         Interior Coating System for Steel
         Storage Tanks
         Coating System for Steel Piping
-880   Protective Coatings for Concrete

09900   PAINTING

-910   Exterior Painting
-920   Interior Painting
-930   Transparent Finishes

09950   WALL COVERINGS

DIVISION 10 - SPECIALTIES

Section Number	Title
10100	**VISUAL DISPLAY BOARDS**
-110	Chalkboards
-115	Markerboards
-120	Tackboards
-130	Operable Board Units
-140	Display Track System
-145	Visual Aid Board Units
10150	**COMPARTMENTS AND CUBICLES**
-160	Metal Toilet Compartments
-165	Plastic Laminate Toilet Compartments
-170	Plastic Toilet Compartments
-175	Particleboard Toilet Compartments
-180	Stone Toilet Compartments
-185	Shower and Dressing Compartments
-190	Cubicles
	Cubicle Curtains
	Cubicle Track and Hardware
10200	**LOUVERS AND VENTS**
-210	Metal Wall Louvers
	Operable Metal Wall Louvers
	Stationary Metal Wall Louvers
	Motorized Metal Wall Louvers
-220	Louvered Equipment Enclosures
-225	Metal Door Louvers
-230	Metal Vents
	Metal Soffit Vents
	Metal Wall Vents
10240	**GRILLES AND SCREENS**
10250	**SERVICE WALL SYSTEMS**
10260	**WALL AND CORNER GUARDS**
	Corner Guards
	Bumper Guards
	Impact Resistant Wall Protection
10270	**ACCESS FLOORING**
-272	Rigid Grid Access Floor Systems
-274	Snap-on Stringer Access Floor Systems
-276	Stringerless Access Floor Systems

10290	**PEST CONTROL**		-456	Turnstiles
-292	Rodent Control		-458	Detection Specialties
-294	Insect Control			
-296	Bird Control		**10500**	**LOCKERS**

<table>
<tr><td>-505</td><td>Metal Lockers</td></tr>
<tr><td>-510</td><td>Wood Lockers</td></tr>
</table>

10300 FIREPLACES AND STOVES

-305 Manufactured Fireplaces
  Manufactured Fireplace Chimneys
  Manufactured Fireplace Forms
-310 Fireplace Specialties and
  Accessories
  Fireplace Dampers
  Fireplace Water Heaters
  Fireplace Screens and Doors
  Fireplace Inserts
-320 Stoves

10340 MANUFACTURED EXTERIOR
  SPECIALTIES

-342 Steeples
-344 Spires
-346 Cupolas
-348 Weathervanes

10350 FLAGPOLES

-352 Ground Set Flagpoles
-354 Wall Mounted Flagpoles
-356 Automatic Flagpoles
-358 Nautical Flagpoles

10400 IDENTIFYING DEVICES

-410 Directories
  Electronic Directories
-415 Bulletin Boards
-420 Plaques
-430 Exterior Signs
  Dimensional Letter Signs
  Illuminated Exterior Signs
  Non-illuminated Exterior Signs
  Post and Panel/Pylon Exterior
   Signs
  Electronic Message Signs
-440 Interior Signs
  Dimensional Letters
  Door Signs
  Engraved Signs
  Illuminated Interior Signs
  Non-Illuminated Interior Signs
  Electronic Message Signs

10450 PEDESTRIAN CONTROL
  DEVICES

-452 Portable Posts and Railings
-454 Rotary Gates

        Plastic Laminate Faced Lockers
-515 Coin-Operated Lockers
-518 Glass Lockers

10520 FIRE PROTECTION
  SPECIALTIES

-522 Fire Extinguishers, Cabinets, and
  Accessories
  Fire Extinguishers
  Fire Extinguisher Cabinets
-526 Fire Blankets and Cabinets
-528 Wheeled Fire Extinguisher Units

10530 PROTECTIVE COVERS

-532 Walkway Covers
-534 Car Shelters
-536 Awnings
-538 Canopies

10550 POSTAL SPECIALTIES

-551 Mail Chutes
-552 Mail Boxes
-554 Collection Boxes
-556 Central Mail Delivery Boxes

10600 PARTITIONS

-605 Wire Mesh Partitions
-610 Folding Gates
-615 Demountable Partitions
  Demountable Gypsum Board
   Partitions
  Demountable Metal Partitions
  Demountable Wood Partitions
-630 Portable Partitions, Screens, and
  Panels

10650 OPERABLE PARTITIONS

-652 Folding Panel Partitions
-655 Accordion Folding Partitions
-660 Sliding Partitions
-665 Coiling Partitions

10670 STORAGE SHELVING

-675 Metal Storage Shelving
-680 Storage and Shelving Systems

-683   Mobile Storage Systems
    Motorized Mobile Storage Systems
    Manual Mobile Storage Systems
-685   Wire Shelving
-688   Prefabricated Wood Storage Shelving

10700   **EXTERIOR PROTECTION DEVICES FOR OPENINGS**

-705   Exterior Sun Control Devices
-710   Exterior Shutters
-715   Storm Panels

10750   **TELEPHONE SPECIALTIES**

-755   Telephone Enclosures
-760   Telephone Directory Units
-765   Telephone Shelves

10800   **TOILET AND BATH ACCESSORIES**

-810   Toilet Accessories
    Commercial Toilet Accessories
    Detention Toilet Accessories
    Hospital Toilet Accessories
-820   Bath Accessories
    Residential Bath Accessories
    Shower and Tub Doors

10880   **SCALES**

10900   **WARDROBE AND CLOSET SPECIALTIES**

DIVISION 11 - EQUIPMENT

Section Number	Title
11010	**MAINTENANCE EQUIPMENT**
-012	Vacuum Cleaning Systems
-014	Window Washing Systems
-016	Floor and Wall Cleaning Equipment
-018	Housekeeping Carts
11020	**SECURITY AND VAULT EQUIPMENT**
-022	Vault Doors and Day Gates
-024	Security and Emergency Systems
-026	Safes
-028	Safe Deposit Boxes
11030	**TELLER AND SERVICE EQUIPMENT**
-032	Service and Teller Window Units
-034	Package Transfer Units
-036	Automatic Banking Systems
-038	Teller Equipment Systems
11040	**ECCLESIASTICAL EQUIPMENT**
-042	Baptisteries
-044	Chancel Fittings
11050	**LIBRARY EQUIPMENT**
-052	Book Theft Protection Equipment
-054	Library Stack Systems
-056	Study Carrels
-058	Book Depositories
	Automated Book Storage and Retrieval Systems
11060	**THEATER AND STAGE EQUIPMENT**
-062	Stage Curtains
-064	Rigging Systems and Controls
-066	Acoustical Shell Systems
-068	Folding and Portable Stages

11070 **INSTRUMENTAL EQUIPMENT**
-072   Organs
-074   Carillons
-076   Bells

11080 **REGISTRATION EQUIPMENT**

11090 **CHECKROOM EQUIPMENT**

11100 **MERCANTILE EQUIPMENT**
-102   Barber and Beauty Shop
        Equipment
-104   Cash Registers and Checking
        Equipment
-106   Display Cases
        Refrigerated Display Cases
-108   Food Processing Equipment
        Food Weighing and Wrapping
        Equipment

11110 **COMMERCIAL LAUNDRY AND**
        DRY CLEANING EQUIPMENT
-112   Washers and Extractors
-114   Dry Cleaning Equipment
-116   Drying and Conditioning
        Equipment
-118   Finishing Equipment
        Ironers and Accessories

11120 **VENDING EQUIPMENT**
-122   Money Changing Machines
-124   Vending Machines
        Beverage Vending Machines
        Candy Vending Machines
        Cigarette Vending Machines
        Food Vending Machines
        Stamp Vending Machines
        Sundry Vending Machines

11130 **AUDIO-VISUAL EQUIPMENT**
-132   Projection Screens
-134   Projectors
-136   Learning Laboratories

11140 **VEHICLE SERVICE**
        EQUIPMENT
-142   Vehicle Washing Equipment
-144   Fuel Dispensing Equipment
-146   Lubrication Equipment

11150 **PARKING CONTROL**
        EQUIPMENT
-152   Parking Gates
-154   Ticket Dispensers
-156   Key and Card Control Units
-158   Coin Machine Units

11160 **LOADING DOCK EQUIPMENT**
-161   Dock Levelers
-162   Dock Lifts
-163   Portable Ramps, Bridges, and
        Platforms
-164   Dock Seals and Shelters
-165   Dock Bumpers

11170 **SOLID WASTE HANDLING**
        EQUIPMENT
-171   Packaged Incinerators
-172   Waste Compactors
-173   Bins
-174   Pulping Machines and Systems
-175   Chutes and Collectors
-176   Pneumatic Waste Systems

11190 **DETENTION EQUIPMENT**

11200 **WATER SUPPLY AND**
        TREATMENT EQUIPMENT
-210   Pumps
        Axial Flow Pumps
        Centrifugal Pumps
        Deepwell Turbine Pumps
        Mixed Flow Pumps
        Vertical Turbine Pumps
-220   Mixers and Flocculators
-225   Clarifiers
-230   Water Aeration Equipment
-240   Chemical Feeding Equipment
        Coagulant Feed Equipment
-250   Water Softening Equipment
        Base-Exchange or Zeolite
        Equipment
        Lime-Soda Process Equipment
-260   Disinfectant Feed Equipment
        Chlorination Equipment
        pH Equipment
-270   Fluoridation Equipment

11280 **HYDRAULIC GATES AND**
        VALVES
-285   Hydraulic Gates
        Bulkhead Gates
        High Pressure Gates

Hinged Leaf Gates
Radial Gates
Slide Gates
Sluice Gates
Spillway Crest Gates
Vertical Lift Gates
-295 Hydraulic Valves
Butterfly Valves
Regulating Valves

**11300 FLUID WASTE TREATMENT
AND DISPOSAL EQUIPMENT**

-302 Oil/Water Separators
-304 Sewage Ejectors
-306 Packaged Pump Stations
-310 Sewage and Sludge Pumps
-320 Grit Collecting Equipment
-330 Screening and Grinding
Equipment
-335 Sedimentation Tank Equipment
-340 Scum Removal Equipment
-345 Chemical Equipment
-350 Sludge Handling and Treatment
Equipment
-360 Filter Press Equipment
-365 Trickling Filter Equipment
-370 Compressors
-375 Aeration Equipment
-380 Sludge Digestion Equipment
-385 Digester Mixing Equipment
-390 Package Sewage Treatment
Plants

11400 FOOD SERVICE EQUIPMENT

-405 Food Storage Equipment
-410 Food Preparation Equipment
-415 Food Delivery Carts and
Conveyors
-420 Food Cooking Equipment
-425 Hood and Ventilation Systems
Surface Fire Protection Systems
-430 Food Dispensing Equipment
-435 Ice Machines
-440 Cleaning and Disposal
Equipment
-445 Bar and Soda Fountain
Equipment

11450 RESIDENTIAL EQUIPMENT

-452 Residential Appliances
-454 Built-In Ironing Boards
-458 Disappearing Stairs

11460 UNIT KITCHENS

11470 DARKROOM EQUIPMENT

-472 Transfer Cabinets
-474 Darkroom Processing Equipment
-476 Revolving Darkroom Doors

**11480 ATHLETIC, RECREATIONAL,
AND THERAPEUTIC
EQUIPMENT**

-482 Scoreboards
-484 Backstops
-486 Gym Dividers
-488 Bowling Alleys
-490 Gymnasium Equipment
-492 Exercise Equipment
-494 Therapy Equipment
-496 Shooting Ranges

**11500 INDUSTRIAL AND PROCESS
EQUIPMENT**

11600 LABORATORY EQUIPMENT

11650 PLANETARIUM EQUIPMENT

11660 OBSERVATORY EQUIPMENT

11680 OFFICE EQUIPMENT

11700 MEDICAL EQUIPMENT

-710 Medical Sterilizing Equipment
-720 Examination and Treatment
Equipment
-730 Patient Care Equipment
-740 Dental Equipment
-750 Optical Equipment
-760 Operating Room Equipment
-770 Radiology Equipment

11780 MORTUARY EQUIPMENT

11850 NAVIGATION EQUIPMENT

11870 AGRICULTURAL EQUIPMENT

DIVISION l2 - FURNISHINGS

Section Number	Title

12050   **FABRICS**

12100   **ARTWORK**
-110   Murals
   Photo Murals
-120   Wall Decorations
   Paintings
   Prints
   Tapestries
   Wall Hangings
-140   Sculpture
   Carved Sculpture
   Cast Sculpture
   Constructed Sculpture
   Relief Artwork
-160   Ecclesiastical Artwork
-170   Stained Glass Work

12300   **MANUFACTURED CASEWORK**
-301   Metal Casework
-302   Wood Casework
-304   Plastic Laminate Faced Casework
-345   Laboratory Casework
   Laboratory Countertops, Sinks, and Accessories
-350   Medical Casework
   Dental Casework
   Hospital Casework
   Nurse Station Casework
   Optical Casework
   Veterinary Casework
-360   Educational Casework
   Library Casework
-370   Residential Casework
   Bath Casework
   Kitchen Casework
-380   Specialty Casework
   Bank Casework
   Display Casework
   Dormitory Casework
   Ecclesiastical Casework
   Hotel and Motel Casework
   Restaurant Casework

12500   **WINDOW TREATMENT**
-510   Blinds
   Horizontal Louver Blinds
   Vertical Louver Blinds

-515   Interior Shutters
-520   Shades
   Insulating Shades
   Lightproof Shades
   Translucent Shades
   Woven Wood Shades
-525   Solar Control Film
-530   Curtain Hardware
   Curtain Track
-540   Curtains
   Draperies
   Fabric Curtains
   Lightproof Curtains
   Vertical Louver Curtains
   Woven Wood Curtains

12600   **FURNITURE AND ACCESSORIES**
-605   Portable Screens
-610   Open Office Furniture
   Open Office Partitions
   Open Office Work Surfaces
   Open Office Storage Units
   Open Office Shelving
   Open Office Light Fixtures
-620   Furniture
   Classroom Furniture
   Dormitory Furniture
   Ecclesiastical Furniture
   Hotel and Motel Furniture
   Laboratory Furniture
   Library Furniture
   Lounge Furniture
   Medical Furniture
   Office Furniture
   Restaurant Furniture
   Residential Furniture
   Specialized Furniture
-640   Furniture Systems
   Integrated Work Units
-650   Furniture Accessories
   Ash Receptacles
   Clocks
   Desk Accessories
   Lamps
   Waste Receptacles

12670   **RUGS AND MATS**
-675   Rugs
-680   Foot Grilles
-690   Floor Mats and Frames
   Chair Pads
   Entrance Tiles
   Floor Mats
   Floor Runners
   Mat Frames

12700	**MULTIPLE SEATING**
-705	Portable Audience Seating
	Folding Chairs
	Interlocking Chairs
	Stacking Chairs
-710	Fixed Audience Seating
-730	Stadium and Arena Seating
-740	Booths and Tables
-750	Multiple Use Fixed Seating
-760	Telescoping Stands
	Telescoping Bleachers
	Telescoping Chair Platforms
-770	Pews and Benches
-775	Seat and Table Systems
	Pedestal Tablet Arm Chairs

12800	**INTERIOR PLANTS AND PLANTERS**
-810	Interior Plants
-815	Artificial Plants
-820	Interior Planters
-825	Interior Landscape Accessories
-830	Interior Plant Maintenance

DIVISION 13 - SPECIAL CONSTRUCTION

Section Number	Title
13010	**AIR SUPPORTED STRUCTURES**
13020	**INTEGRATED ASSEMBLIES**
-025	Integrated Ceilings
13030	**SPECIAL PURPOSE ROOMS**
-032	Athletic Rooms
-034	Sound Conditioned Rooms
-036	Clean Rooms
-038	Cold Storage Rooms
-040	Hyperbaric Rooms
-042	Insulated Rooms
-046	Shelters and Booths
-048	Planetariums
-050	Prefabricated Rooms
-052	Saunas
-054	Steam Baths
-056	Vaults
13080	**SOUND, VIBRATION, AND SEISMIC CONTROL**
13090	**RADIATION PROTECTION**
13100	**NUCLEAR REACTORS**
13120	**PRE-ENGINEERED STRUCTURES**
-121	Pre-Engineered Buildings
-122	Metal Building Systems
-123	Glazed Structures
	Greenhouses
	Solariums
	Swimming Pool Enclosures
-124	Portable and Mobile Buildings
-125	Grandstands and Bleachers
-130	Observatories
-132	Prefabricated Dome Structures
-135	Cable Supported Structures
-140	Fabric Structures
-142	Log Structures
-145	Modular Mezzanines

13150 AQUATIC FACILITIES

-152 Swimming Pools
Below Grade Swimming Pools
On Grade Swimming Pools
Elevated Swimming Pools
Recirculating Gutter Systems
Swimming Pool Accessories
Swimming Pool Cleaning Systems
-160 Aquariums
-165 Aquatic Park Facilities
Water Slides
Wave Pools
-170 Tubs and Pools
Hot Tubs
Whirlpool Tubs
Therapeutic Pools

13175 ICE RINKS

13180 SITE CONSTRUCTED INCINERATORS

-182 Sludge Incinerators
-184 Solid Waste Incinerators

13185 KENNELS AND ANIMAL SHELTERS

13200 LIQUID AND GAS STORAGE TANKS

-205 Ground Storage Tanks
-210 Elevated Storage Tanks
-215 Underground Storage Tanks
-217 Tank Lining Systems
-219 Tank Cleaning Procedures

13220 FILTER UNDERDRAINS AND MEDIA

-222 Filter Bottoms
-226 Filter Media
Anthracite Media
Charcoal Media
Diatomaceous Earth
Mixed Media
Sand Media

13230 DIGESTER COVERS AND APPURTENANCES

-232 Fixed Covers
-234 Floating Covers
-236 Gasholder Covers

13240 OXYGENATION SYSTEMS

-242 Oxygen Dissolution System
-246 Oxygen Generators
-248 Oxygen Storage Facility

13260 SLUDGE CONDITIONING SYSTEMS

13300 UTILITY CONTROL SYSTEMS

-310 Water Supply Plant Operating and Monitoring Systems
Display Panels
Metering Devices
Sensing and Communication Devices
-320 Wastewater Treatment Plant Operating and Monitoring Systems
Control Panels
Display Panels
Metering Devices
Sensing and Communication Devices
-330 Power Generating and Transmitting Control Systems
Control Panels
Display Panels
Meters
Relays

13400 INDUSTRIAL AND PROCESS CONTROL SYSTEMS

13500 RECORDING INSTRUMENTATION

-510 Stress Instrumentation
-515 Seismic Instrumentation
-520 Meteorological Instrumentation

13550 TRANSPORTATION CONTROL INSTRUMENTATION

-560 Airport Control Instrumentation
-570 Railroad Control Instrumentation
-580 Subway Control Instrumentation
-590 Transit Vehicle Control Instrumentation

13600 SOLAR ENERGY SYSTEMS

-610 Solar Flat Plate Collectors
Air Collectors
Liquid Collectors
-620 Solar Concentrating Collectors
-625 Solar Vacuum Tube Collectors

-630	Solar Collector Components
	Solar Absorber Plates and Tubing
	Solar Coatings and Surface Treatment
	Solar Collector Insulation
	Solar Glazing
	Solar Housing and Framing
	Solar Reflectors
-640	Packaged Solar Systems
-650	Photovoltaic Collectors

13700	**WIND ENERGY SYSTEMS**

13750	**COGENERATION SYSTEMS**

13800	**BUILDING AUTOMATION SYSTEMS**
-810	Energy Monitoring and Control Systems
-815	Environmental Control Systems
-820	Communications Systems
-825	Security Systems
-830	Clock Control Systems
-835	Elevator Monitoring and Control Systems
-840	Escalators and Moving Walks Monitoring and Control Systems
-845	Alarm and Detection Systems
-850	Door Control Systems

13900	**FIRE SUPPRESSION AND SUPERVISORY SYSTEMS**

13950	**SPECIAL SECURITY CONSTRUCTION**

DIVISION 14 - CONVEYING SYSTEMS

Section Number	Title
14100	**DUMBWAITERS**
-110	Manual Dumbwaiters
-120	Electric Dumbwaiters
-140	Hydraulic Dumbwaiters
14200	**ELEVATORS**
-210	Electric Traction Elevators
	Electric Traction Passenger Elevators
	Electric Traction Service Elevators
	Electric Traction Freight Elevators
-240	Hydraulic Elevators
	Hydraulic Passenger Elevators
	Hydraulic Service Elevators
	Hydraulic Freight Elevators
14300	**ESCALATORS AND MOVING WALKS**
-310	Escalators
-320	Moving Walks
14400	**LIFTS**
-410	People Lifts
	Counterbalanced People Lifts
	Endless Belt People Lifts
-420	Wheelchair Lifts
	Inclined Wheelchair Lifts
	Vertical Wheelchair Lifts
-430	Platform Lifts
	Orchestra Lifts
	Stage Lifts
-440	Sidewalk Lifts
-450	Vehicle Lifts
14500	**MATERIAL HANDLING SYSTEMS**
-510	Automatic Transport Systems
	Guided Vehicle Systems
	Track Vehicle Systems
-530	Postal Conveying Systems
-540	Baggage Conveying and Dispensing Systems

-550	Conveyors	

-550 Conveyors
    Belt Conveyors
    Bucket Conveyors
    Container Conveyors
    Hopper and Track Conveyors
    Monorail Conveyors
    Oscillating Conveyors
    Pneumatic Conveyors
    Roller Conveyors
    Scoop Conveyors
    Screw Conveyors
    Selective Vertical Conveyors
-560 Chutes
    Dry Bulk Material Chutes
    Escape Chutes
    Laundry and Linen Chutes
    Package Chutes
-570 Feeder Equipment
    Apron Feeders
    Reciprocating Plate Feeders
    Rotary Airlock Feeders
    Rotary Flow Feeders
    Vibratory Feeders
-580 Pneumatic Tube Systems

14600 HOISTS AND CRANES
-605 Crane Rails
-610 Fixed Hoists
    Electric Fixed Hoists
    Manual Fixed Hoists
    Air Powered Fixed Hoists
-620 Trolley Hoists
    Electric Trolley Hoists
    Manual Trolley Hoists
    Air Powered Trolley Hoists
-630 Bridge Cranes
    Top Running Overhead Cranes
    Underslung Overhead Cranes
-640 Gantry Cranes
-650 Jib Cranes
-670 Tower Cranes
-680 Mobile Cranes
-690 Derricks

14700 TURNTABLES

14800 SCAFFOLDING
-810 Suspended Scaffolding
    Beam Scaffolding
    Carriage Scaffolding
    Hook Scaffolding
-820 Rope Climbers
    Manual Rope Climbers
    Powered Rope Climbers
-830 Telescoping Platforms
    Electric and Battery Telescoping Platforms
    Pneumatic Telescoping Platforms

14900 TRANSPORTATION SYSTEMS
-910 People Mover Systems
-920 Monorail Systems
-930 Funicular Systems
-940 Aerial Tramway Systems
-950 Aircraft Passenger Loading Systems

DIVISION 15 -
MECHANICAL

Section Number	Title

15050 BASIC MECHANICAL MATERIALS AND METHODS

-060 Pipes and Pipe Fittings
   Aluminum and Aluminum Alloy Pipe and Fittings
   Concrete Pipe and Fittings
   Copper and Copper Alloy Pipe and Fittings
   Ferrous Pipe and Fittings
   Fiber Pipe and Fittings
   Glass Pipe and Fittings
   Hoses and Fittings
   Plastic Pipe and Fittings
   Pre-Insulated Pipe and Fittings
-100 Valves
   Manual Control Valves
   Self Actuated Valves
-120 Piping Specialties
-130 Gages
-140 Supports and Anchors
-150 Meters
-160 Pumps
-170 Motors
-175 Tanks
-190 Mechanical Identification
-240 Mechanical Sound, Vibration, and Seismic Control

15250 MECHANICAL INSULATION

-260 Piping Insulation
-280 Equipment Insulation
-290 Ductwork Insulation

15300 FIRE PROTECTION

-310 Fire Protection Piping
-320 Fire Pumps
-330 Wet Pipe Sprinkler Systems
-335 Dry Pipe Sprinkler Systems
-340 Pre-Action Sprinkler Systems
-345 Combination Dry Pipe and Pre-Action Sprinkler Systems
-350 Deluge Sprinkler Systems
-355 Foam Extinguishing Systems
-360 Carbon Dioxide Extinguishing Systems
-365 Halogen Agent Extinguishing Systems

-370 Dry Chemical Extinguishing Systems
-375 Standpipe and Hose Systems

15400 PLUMBING

-410 Plumbing Piping
-430 Plumbing Specialties
-440 Plumbing Fixtures
-450 Plumbing Equipment
   Domestic Water Heat Exchangers
   Drinking Water Cooling Systems
   Pumps
   Storage Tanks
   Water Conditioners
   Water Filtration Devices
   Water Heaters
-475 Pool and Fountain Equipment
-480 Special Systems
   Compressed Air Systems
   Deionized Water Systems
   Distilled Water Systems
   Fuel Oil Systems
   Gasoline Dispensing Systems
   Helium Gas Systems
   Liquified Petroleum Gas Systems
   Lubricating Oil Systems
   Natural Gas Systems
   Nitrous Oxide Gas Systems
   Oxygen Gas Systems
   Reverse Osmosis Systems
   Vacuum Systems

15500 HEATING, VENTILATING, AND AIR CONDITIONING

-510 Hydronic Piping
-515 Hydronic Specialties
-520 Steam and Steam Condensate Piping
-525 Steam and Steam Condensate Specialties
-530 Refrigerant Piping
-535 Refrigerant Specialties
-540 HVAC Pumps
-545 Chemical Water Treatment

15550 HEAT GENERATION

-555 Boilers
-570 Boiler Accessories
-575 Breechings, Chimneys, and Stacks
-580 Feedwater Equipment
-590 Fuel Handling Systems
-610 Furnaces

-620   Fuel Fired Heaters
Duct Furnaces
Gas Fired Unit Heaters
Oil Fired Unit Heaters
Radiant Heaters

15650   REFRIGERATION

-655   Refrigeration Compressors
-670   Condensing Units
-680   Water Chillers
Absorption Water Chillers
Centrifugal Water Chillers
Reciprocating Water Chillers
Rotary Water Chillers
-710   Cooling Towers
Mechanical Draft Cooling Towers
Natural Draft Cooling Towers
-730   Liquid Coolers
-740   Condensers

15750   HEAT TRANSFER

-755   Heat Exchangers
-760   Energy Storage Tanks
-770   Heat Pumps
Air Source Heat Pumps
Rooftop Heat Pumps
Water Source Heat Pumps
-780   Packaged Air Conditioning Units
Computer Room Air Conditioning Units
Packaged Rooftop Air Conditioning Units
Packaged Terminal Air Conditioning Units
Unit Air Conditioners
-790   Air Coils
-810   Humidifiers
-820   Dehumidifiers
-830   Terminal Heat Transfer Units
Convectors
Fan Coil Units
Finned Tube Radiation
Induction Units
Unit Heaters
Unit Ventilators
-845   Energy Recovery Units

15850   AIR HANDLING

-855   Air Handling Units with Coils
-860   Centrifugal Fans
-865   Axial Fans
-870   Power Ventilators
-875   Air Curtain Units

15880   AIR DISTRIBUTION

-885   Air Cleaning Devices
Dust Collectors
Filters
-890   Ductwork
Metal Ductwork
Nonmetal Ductwork
Flexible Ductwork
Ductwork Hangars and Supports
-910   Ductwork Accessories
Dampers
Duct Access Panels and Test Holes
Duct Connection Systems
Flexible Duct Connections
Turning Vanes and Extractors
-920   Sound Attenuators
-930   Air Terminal Units
Constant Volume
Variable Volume
-940   Air Outlets and Inlets
Diffusers
Intake and Relief Ventilators
Louvers
Registers and Grilles

15950   CONTROLS

-955   Building Systems Control
-960   Energy Management and Conservation Systems
-970   Control Systems
Electric Control Systems
Electronic Control Systems
Pneumatic Control Systems
Self-Powered Control Systems
-980   Instrumentation
-985   Sequence of Operation

15990   TESTING, ADJUSTING, AND BALANCING

-991   Mechanical Equipment Testing, Adjusting, and Balancing
-992   Piping Systems Testing, Adjusting, and Balancing
-993   Air Systems Testing, Adjusting, and Balancing
-994   Demonstration of Mechanical Equipment
-995   Mechanical System Startup/ Commissioning

DIVISION I6 - ELECTRICAL

Section Number	Title
16050	**BASIC ELECTRICAL MATERIALS AND METHODS**
-110.	Raceways
	Cable Trays
	Conduits
	Surface Raceways
	Indoor Service Poles
	Underfloor Ducts
	Underground Ducts and Manholes
-120	Wires and Cables
	Fiber Optic Cable
	Low Voltage Wire
	600 Volt or Less Wire and Cable
	Medium Voltage Cable
	Undercarpet Cable Systems
-130	Boxes
	Floor Boxes
	Outlet Boxes
	Pull and Junction Boxes
-140	Wiring Devices
	Low Voltage Switching
-150	Manufactured Wiring Systems
-160	Cabinets and Enclosures
-190	Supporting Devices
-195	Electrical Identification
16200	**POWER GENERATION - BUILT-UP SYSTEMS**
-210	Generators
	Hydroelectric Generators
	Nuclear Electric Generators
	Solar Electric Generators
	Steam Electric Generators
-250	Generator Controls
	Instrumentation
	Starting Equipment
-290	Generator Grounding
16300	**MEDIUM VOLTAGE DISTRIBUTION**
-310	Medium Voltage Substations
-320	Medium Voltage Transformers
-330	Medium Voltage Power Factor Correction
-340	Medium Voltage Insulators and Lightning Arrestors
-345	Medium Voltage Switchboards
-350	Medium Voltage Circuit Breakers
-355	Medium Voltage Reclosers
-360	Medium Voltage Interrupter Switches
-365	Medium Voltage Fuses
-370	Medium Voltage Overhead Power Distribution
-375	Medium Voltage Underground Power Distribution
-380	Medium Voltage Converters
	Medium Voltage Frequency Changers
	Medium Voltage Rectifiers
-390	Medium Voltage Primary Grounding
16400	**SERVICE AND DISTRIBUTION**
-410	Power Factor Correction
-415	Voltage Regulators
-420	Service Entrance
-425	Switchboards
-430	Metering
-435	Converters
-440	Disconnect Switches
-445	Peak Load Controllers
-450	Secondary Grounding
-460	Transformers
-465	Bus Duct
-470	Panelboards
	Branch Circuit Panelboards
	Distribution Panelboards
-475	Overcurrent Protective Devices
	Circuit Breakers
	Fuses
-480	Motor Control
-485	Contactors
-490	Switches
	Transfer Switches
	Isolation Switches
16500	**LIGHTING**
-501	Lamps
-502	Luminaire Accessories
	Ballasts
	Lenses
	Lighting Maintenance Equipment
	Light Louvers
	Posts and Standards
-510	Interior Luminaires
	Fluorescent Luminaires
	High Intensity Discharge Luminaires
	Incandescent Luminaires
	Luminous Ceilings
-520	Exterior Luminaires
	Aviation Lighting
	Flood Lighting
	Navigation Lighting
	Roadway Lighting

Signal Lighting
Site Lighting
Sports Lighting
-535 Emergency Lighting
-545 Underwater Lighting
-580 Theatrical Lighting

16600 SPECIAL SYSTEMS

-610 Uninterruptible Power Supply Systems
-620 Packaged Engine Generator Systems
-630 Battery Power Systems
Central Battery Systems
Packaged Battery Systems
-640 Cathodic Protection
-650 Electromagnetic Shielding Systems
-670 Lightning Protection Systems
-680 Unit Power Conditioners

16700 COMMUNICATIONS

-720 Alarm and Detection Systems
Fire Alarm Systems
Smoke Detection Systems
Gas Detection Systems
Intrusion Detection Systems
Security Access Systems
-730 Clock and Program Systems
-740 Voice and Data Systems
Telephone Systems
Paging Systems
Call Systems
Data Systems
Local Area Network Systems
Door Answering Systems
Microwave and Radio Systems
Central Dictation Systems
Intercommunication Systems
-770 Public Address and Music Systems
-780 Television Systems
Master Antenna Systems
Video Telecommunication Systems
Video Surveillance Systems
Broadcast Video Systems
-785 Satellite Earth Station Systems
-790 Microwave Systems

16850 ELECTRIC RESISTANCE HEATING

-855 Electric Heating Cables and Mats
-880 Electric Radiant Heaters

16900 CONTROLS

-910 Electrical Systems Control
-915 Lighting Control Systems
Dimming Systems
-920 Environmental Systems Control
-930 Building Systems Control
-940 Instrumentating

INDEX

INDEX

465